Robots to Re-construction

The Roadmap to Robotized Asbestos Removal

Burkhard Corves
Tobias Haschke
and Mathias Hüsing
(Editors)

now
the essence of knowledge

Published, sold and distributed by:
now Publishers Inc.
PO Box 1024
Hanover, MA 02339
United States
Tel. +1-781-985-4510
www.nowpublishers.com
sales@nowpublishers.com

Outside North America:
now Publishers Inc.
PO Box 179
2600 AD Delft
The Netherlands
Tel. +31-6-51115274

ISBN: 978-1-68083-714-8
E-ISBN: 978-1-68083-715-5
DOI: 10.1561/9781680837155

Suggested citation: Burkhard Corves, Tobias Haschke and Mathias Hüsing (ed.). (2021). ***Robots to Re-Construction – The Roadmap to Robotized Asbestos Removal.*** Boston–Delft: Now Publishers

Table of Contents

Acknowledgements

The research leading to these results has received funding from the European Union's Horizon 2020 research and innovation programmes under grant agreement No. 687593.

DOI: 10.1561/9781680837155.ch1

Chapter 1

Introduction and Motivation

By Burkhard Corves, Tobias Haschke and Mathias Hüsing

The contents of this book have been developed within the project "Robots to Re-Construction", which was funded by the European Union under grant agreement No. 687593. The intention of the project was to automate the removal of asbestos from homes using mobile robots. The aim was to eliminate the risks to human health when removing asbestos manually and to make the removal process more economical. In order to achieve this goal, a European consortium of seven partners was formed to combine the necessary competences in one team. In addition to the main goal of autonomous asbestos removal, there was also the objective to combine radar technologies with laser technologies and to give and implement new approaches to mobile robotics.

The project was led by the RWTH Aachen University (Germany), which also took over the central control, semantic mapping and task planning for the mobile robot fleets. The construction company Bouygues Construction (France) contributed its expertise in process experience and took on the necessary economic feasibility study with a focus on the market and subsequent implementation. The robotic systems, which are tailored to the removal process, were largely developed by the companies Danieli Telerobot Labs (Italy), Robotnik (Spain) and indurad (Germany). Danieli Telerobot Labs implemented the required robot arms, end effectors and the necessary process peripherals. Robotnik designed the mobile robots as an agile carrier platform and developed localisation and path planning algorithms for operation. indurad's robust radar sensors enabled the robots to reliably detect the environment even in the demanding environment of construction sites and to control the processes in a stable way. Furthermore, indurad implemented the graphical user interface for the system. The integration of all components into a complete system and the performance of real scenario tests were carried out by Eurecat (Spain). The research institute Sigma Clermont (France) supported the consortium in the design of the control system during the removal process, the stability analysis and the basic design of the mechanical system components.

In total, the consortium developed two different prototypes to get closer to the automated solution of asbestos removal. The first prototype (in the context of this book always called V1) was developed in particular for the fundamental acquisition of knowledge for use on construction sites in order to collect data on the removal process and the associated process steps. The second prototype (in this book always abbreviated as V2) was then developed and optimised on the basis of the knowledge gained. This book presents both development cycles and shows the different levels of knowledge and decisions made during the development process. Due to the versatile requirements of the process, numerous methods in the context of modelling and algorithms have also been developed, which are also illustrated in this book, in addition to the described development of the mechatronic components. Very successfully the radar technology could be implemented in mobile robotics.

The content of this book is also structured according to these diverse contents. First, the Asbestos Removal Process is presented, and then the robotic systems developed for it are introduced. Based on this, the algorithms are explained which help the robots to perform the desired process. The chapters on the test results and the subsequent summary as well as the outlook discussion round off this publication.

DOI: 10.1561/9781680837155.ch2

Chapter 2

Asbestos Removal Process

By Florian Dufour, Francesco Becchi, Tobias Haschke and Tim Detert

2.1 Market

2.1.1 Asbestos Contamination of active sites of BYCN around 2016

The following study is made on projects at a study state. It represent all the projects impacted by asbestos, since 2016 (with few other before). The study include all recent projects realized by Bouygues Bâtiment Nord Est and therefore cover the following market area:

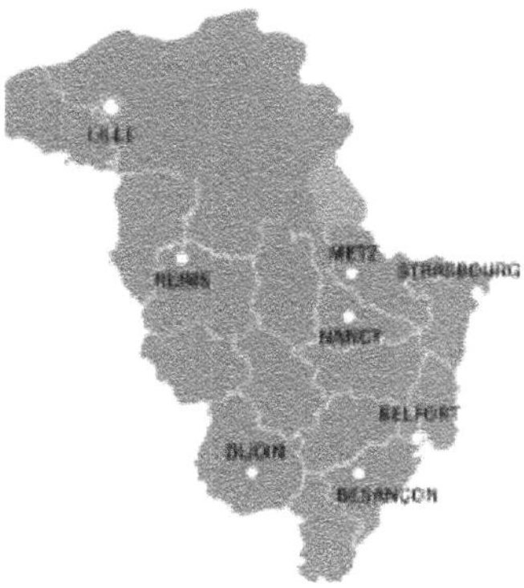

It will soon be extended to the other Bouygues Bâtiment subsidiaries to cover the biggest area possible. The projects studies and their completeness are lists in the table below:

Name of the project	Date of the data	Asbestos study	Quantites	Plans
Abbé Lemire	2016	Yes	Yes	Yes
CH Le Quesnoy	2014 – 2016	None	Yes	Yes
Colmar Luxembourg	2016	None	Yes	No
Dijon Ilot Corse	2017	Yes	Yes	No
Dole				
EDF Partner Nogent	2017	None	Yes	No
EHPAD Vertus	2015	None	Yes	Yes
ICF 240	2017	Yes	Yes	Yes
Koenigshoffen	2016	None	Yes	No
La Madeleine 96	2016	None	Yes	Yes
Lille Jacquard	2017	Yes	Yes	Yes
Lille Kolb	2016	Yes	Yes	Yes
Lunéville	2016	Yes	None	No
Maternité Thionville	2016	None	Yes	Yes
Mons Anémone	2016	None	Yes	No
Reims Jardelles	2016	Yes	None	No
Roubaix 339	2016 – 2017	Yes	Yes	Yes
Rouvroy 279	2017	No asbestos	No asbestos	Yes
Villeneuve d'Ascq Le Château	2015	Yes	Yes	No
Wattignies	2017	Yes	None	Yes

2.1.2 Building Information of Contaminated Sites

The lines shown in this document are the recapitulative ones. In the Excel where was made the study, it is possible to open those lines to see the detail below, floor by floor.

General information	
Name	Number of flats
Abbé Lemire	"Rooms"
CH Le Quesnoy	Hospital
Colmar Luxembourg	
Dijon Ilot Corse	
Dole	
EDF Partner Nogent	
EHPAD Vertus	Hospital
ICF 240	239
Koenigshoffen	
La Madeleine 96	96
Lille Jacquard	107
Lille Kolb	76
Lunéville	
Maternité Thionville	Hospital
Mons Anémone	
Reims Jardelles	
Roubaix 339	339
Rouvroy 279	279
Villeneuve d'Ascq Le Château	
Wattignies	674

2.1.2.1 General Information

For this study, we took the same project we studied before in the asbestos contamination study. For some, the plans were not available, which explain the red lines, for whom we don't have information. In this study, contaminated building can be apartment building, hospitals or flats used for social insertion. When it is an apartment building, the number of flats has been written, for the two others, the structure of the buildings are not details, as it is a special typology.

2.1.2.2 Asbestos Information

In this part of the table, we summarized the diagnostic stat of each project. This information is not always available, so there are only a few lines filled. The first column is about the amount of flat that has been diagnosed. The second the number of flat that are contaminated. The third one is the percentage of flat diagnosed compared to the total amount of flat. The last one is the percentage of flats contaminated compared to the amount of flat diagnosed.

General informat	Asbestos information			
Name	Number of flats diagnosed	Number of flats contaminated	% diagnosed/N° tot	% contaminated /N° diagnosed
Abbé Lemire	No information			
CH Le Quesnoy	No information			
Colmar Luxembourg				
Dijon Ilot Corse				
Dole				
EDF Partner Nogent				
EHPAD Vertus	No information			
ICF 240	183	71	76,6%	38,8%
Koenigshoffen				
La Madeleine 96	No information			
Lille Jacquard	97	No information		
Lille Kolb	76	76	100,0%	100,0%
Lunéville				
Maternité Thionville	No information			
Mons Anémone				
Reims Jardelles				
Roubaix 339	278	278	82,0%	100,0%
Rouvroy 279	No asbestos			
Villeneuve d'Ascq Le Château				
Wattignies	No information			

2.1.2.3 Technical Building Analysis

This part summarize the analysis of floor plans and their specificities.

General informat	Technical specificities									
Name	General informatio	Staircase and elevator			Flat geometry				Other specificities
	Number of level	Round staircase ?	Number of staircases	Elevator	Smallest door size (cm)	Smaller corridor width (m)	Corridor turns inside (Yes/No)	special wall angle in a flat ?	Step inside the flat (Yes/No)
Abbé Lemire	R+8	Yes (1 to 2)	1 to 5	Yes (2)	70	1,2	Yes	77°	No
CH Le Quesnoy	R+3	No	3	Yes (2)	90	1,25	Door 90 to door 90	77°	No
Colmar Luxembourg				No					
Dijon Ilot Corse				No					
Dole				No					
EDF Partner Nogent				No					
EHPAD Vertus	R+1	No	3 to 4	Yes (1)	No dimensions but seems quite small.		Yes but large	No	No
ICF 240	R+6	No	17 and 4 for R+6	Yes (4)	70	0,9	Yes	No	No
Koenigshoffen				No					
La Madeleine 96	R+8	Yes (only)	4	Yes (4)	70	0,95	Yes 0,95 to 0,95	No	No
Lille Jacquard	R+12	Yes (1)	1 to 2	Yes	No dimensions		Yes but not small	No	No
Lille Kolb	R+7	Yes (only)	2 to 3	Yes (2)					
Lunéville				No					
Maternité Thionville	R+2	No	1	Yes (2)	No dimensions		Yes	No	No
Mons Anémone				No					
Reims Jardelles				No					
Roubaix 339	R+10	Yes (only)	7	Yes (7)	70 (60 sometimes)	100	Yes	Yes see nota	No
Rouvroy 279	R+1	No	5 to 3	No	73	80	Yes	No	No
Villeneuve d'Ascq Le Château				No					
Wattignies	7, R+3/+4/+7	Different sites - see details below							

General informat	
Name	General informatio Number of level
Abbé Lemire	R+8
CH Le Quesnoy	R+3
Colmar Luxembourg	
Dijon Ilot Corse	
Dole	
EDF Partner Nogent	
EHPAD Vertus	R+1
ICF 240	R+6
Koenigshoffen	
La Madeleine 96	R+8
Lille Jacquard	R+12
Lille Kolb	R+7
Lunéville	
Maternité Thionville	R+2
Mons Anémone	
Reims Jardelles	
Roubaix 339	R+10
Rouvroy 279	R+1
Villeneuve d'Ascq Le Château	
Wattignies	?, R+3/+4/+7

2.1.2.4 Level Information

This part shows the number of floors for each building. R+ "x" means that there is a ground floor + "x" floors. For example, R + 3 means a ground floor + 3 floors up to it.

General informat			
Name	Staircase and elevator		
	Round staircase ?	Number of staircases	Elevator
Abbé Lemire	Yes (1 to 2)	1 to 5	Yes (2)
CH Le Quesnoy	No	3	Yes (2)
Colmar Luxembourg			No
Dijon Ilot Corse			No
Dole			No
EDF Partner Nogent			No
EHPAD Vertus	No	3 to 4	Yes (1)
ICF 240	No	17 and 4 for R+6	Yes (4)
Koenigshoffen			No
La Madeleine 96	Yes (only)	4	Yes (4)
Lille Jacquard	Yes (1)	1 to 2	Yes
Lille Kolb	Yes (only)	2 to 3	Yes (2)
Lunéville			No
Maternité Thionville	No	1	Yes (2)
Mons Anémone			No
Reims Jardelles			No
Roubaix 339	Yes (only)	7	Yes (7)
Rouvroy 279	No	5 to 3	No
Villeneuve d'Ascq Le Château			No
Wattignies			

2.1.2.5 Staircase and Elevator

In the column "round staircase" we had a look of the presence or not of circular staircases and if yes, how many. The number of staircases is the total amount of staircases, including round ones. The last column shows the presence of elevator or not in the building, and if yes the number of elevators.

2.1.2.6 Flat Geometries and specificities

General informat	Technical specificities				
	Flat geometry				Other specificities
Name	Smallest door size (cm)	Smalller corridor width (m)	Corridor turns inside (Yes/No)	special wall angle in a flat ?	Step inside the flat (Yes/No)
Abbé Lemire	70	1,2	Yes	77°	No
CH Le Quesnoy	90	1,25	Door 90 to door 90	77°	No
Colmar Luxembourg					
Dijon Ilot Corse					
Dole					
EDF Partner Nogent					
EHPAD Vertus	No dimensions but seems quite small,		Yes but large	No	No
ICF 240	70	0,9	Yes	No	No
Koenigshoffen					
La Madeleine 96	70	0,95	Yes 0,95 to 0,95	No	No
Lille Jacquard	No dimensions		Yes but not small	No	No
Lille Kolb					
Lunéville					
Maternité Thionville	No dimensions		Yes	No	No
Mons Anémone					
Reims Jardelles					
Roubaix 339	70 (60 sometimes)	100	Yes	Yes see nota	No
Rouvroy 279	73	80	Yes	No	No
Villeneuve d'Ascq Le Château					
Wattignies	Different sites - see details below				

This part is about the special geometrical details in the buildings. The two first columns are about the smallest door and corridor width, in centimetres. Those values are mostly linked, the smaller is the corridor, and the smaller the door is. The third column recapitulate if there are corridor turns or not in the flats, and if yes, what are the dimensions of those corridor, or door if it is from one door to another. The "special wall angle in the flat" column shows the special angles that can be in a flat. This can have an influence on the behavior of the robot during the grinding. For the Roubaix 339 project, the shape of the building is special and the floor plans are recapitulated in the nota tab of the Excel. The last column is about the presence or not of steps inside the flat.

2.1.2.7 Corridor Turns

For the corridor turns topic, the different cases are explained bellow with the linked plans:

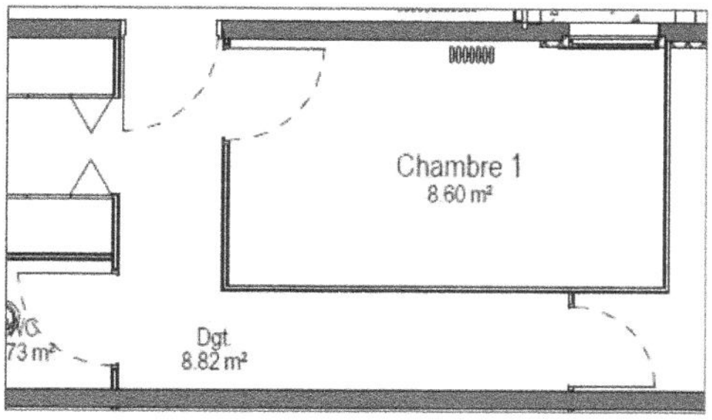

2.1.2.7.1 Lille Jacquard

In this case, there is no measure, but the shape seems to be acceptable for the intervention of the robot.

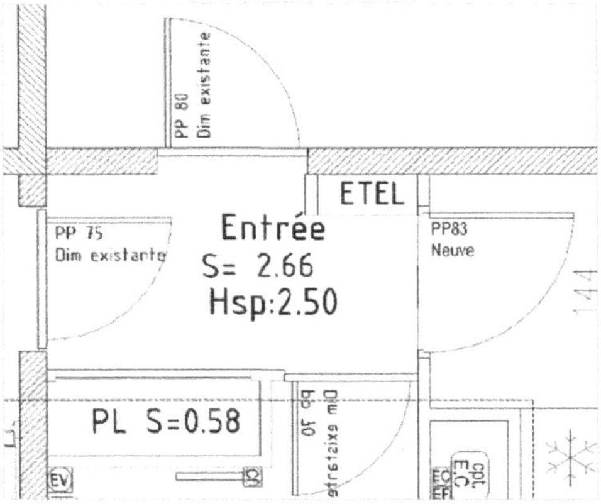

2.1.2.7.2 Roubaix 336

In this case, the dimensions of doors and corridor should allow the robot to go everywhere, without problems.

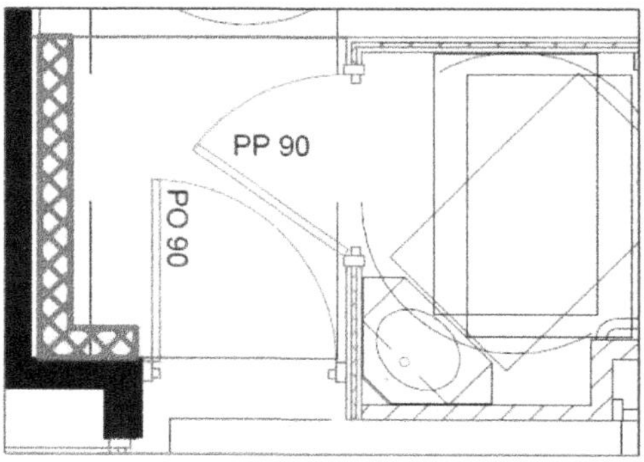

2.1.2.7.3 CH Le Quesnoy

In this case, it is a hospital. All the rooms are made accessible for disabled people, which means the robot can go everywhere, like it is shown on the picture on the left.

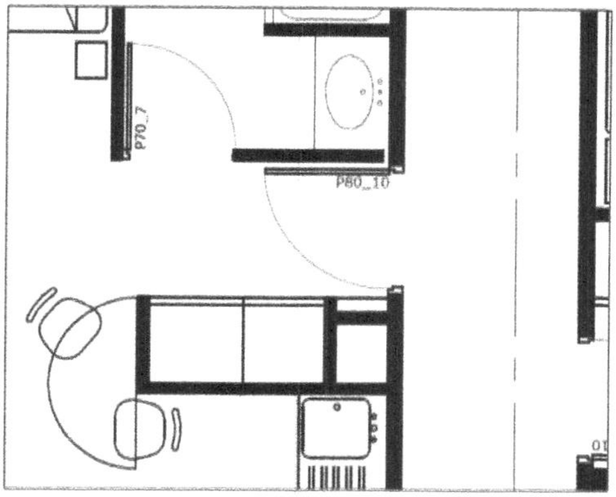

2.1.2.7.4 Abbé Lemire

In this case, the corridor is 1, 2 m wide, and the flat door is 80 cm large, which cause no problem for the robot intervention. For the 70 cm large door, as there is the place in the room on the left to turn, going in the bathroom should not be a problem.

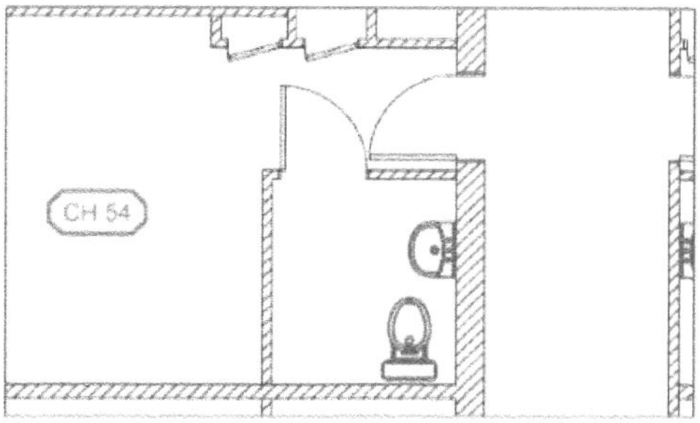

2.1.2.7.5 EHPAD Vertus

Ibn this case, there is no dimensions, but we can see that the corridor is wide, so the robot can easily go inside the room. As for the Abbe Lemire project, the place in the room on the left should allow the robot to go inside the bathroom normally.

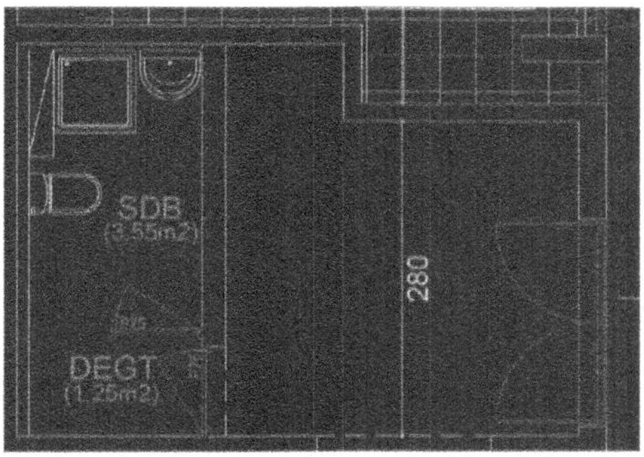

2.1.2.7.6 ICF 240

This case could be a problem for the robot. Both doors are 75 cm wide, and the corridor does not seem to be very big, so it could be quite tricky situation for the robot.

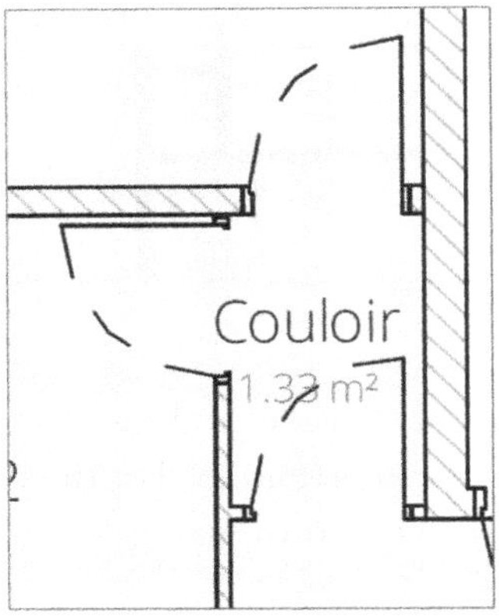

2.1.2.7.7 La Madeleine

This case could be tricky for the robot too. Indeed, the corridor is 90 cm wide and the door is 70 cm wide.

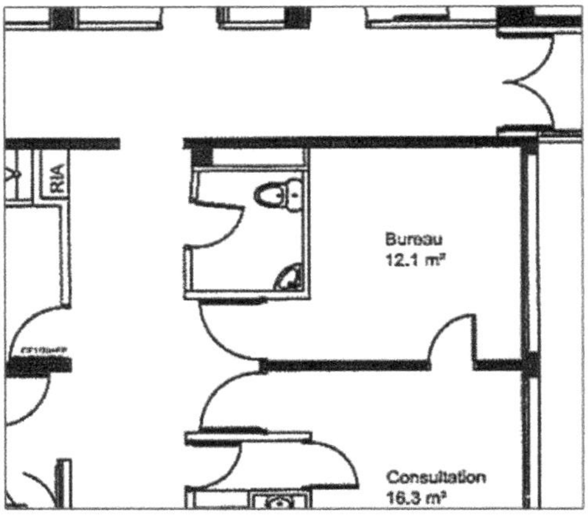

2.1.2.7.8 Maternity of Thionville

In this case, we do not have any measures, but the width of both corridors seems to be acceptable for the robot to go through the narrow part in the middle.

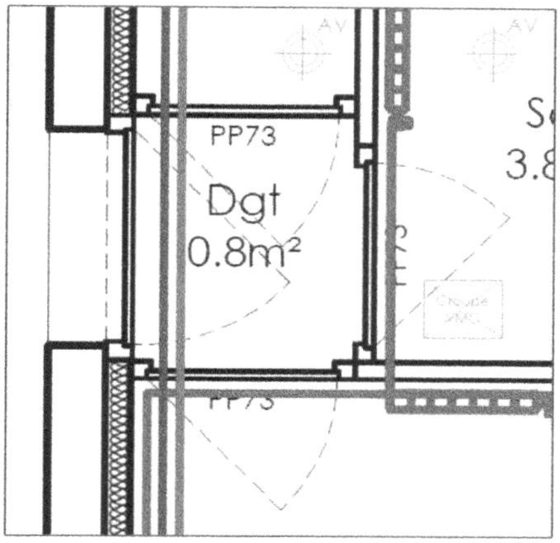

2.1.2.7.9 Rouvroy 279

In this case, the robot would have some problem to go from one door to another. A possibility would be to use the left door, which is a door opening on the outside, to turn. But it is a quite complicated situation for the robot.

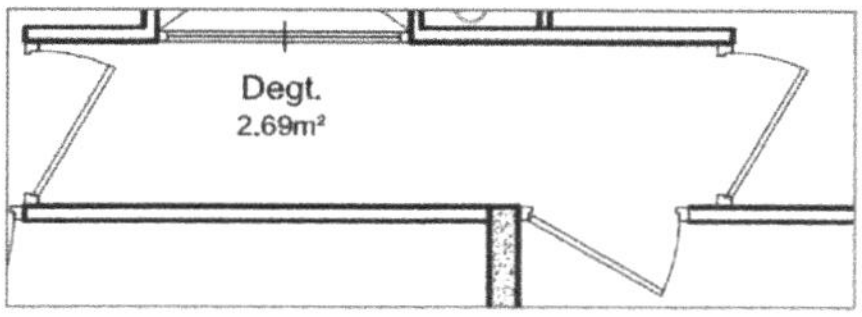

2.1.2.7.10 Wattignies

For this case, we do not have any measures. So if the doors are 70 cm wide, it would be problematic for the robot to turn. If the doors are bigger, then this configuration would not be a problem.

2.1.2.8 Special Angles

There are 3 projects that are concerned by special angles. Here are the details for each:

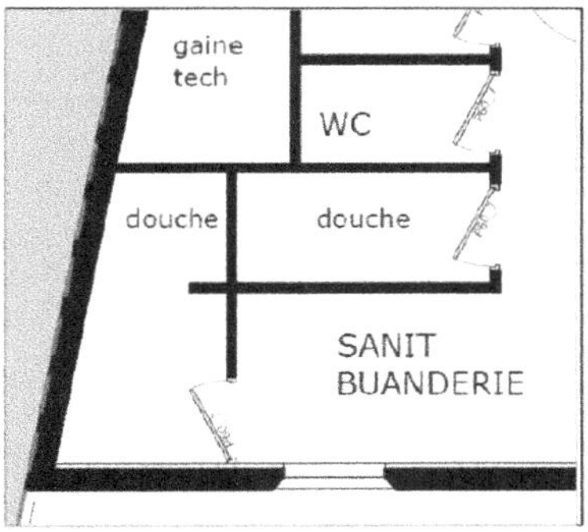

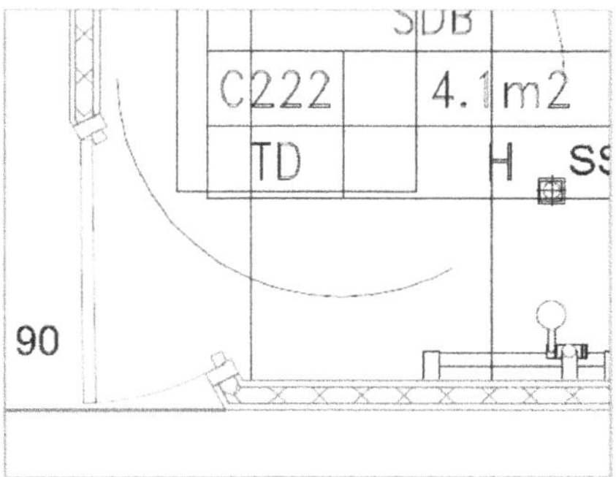

2.1.2.8.1 Abbé Lemire and CH Le Quesnoy

Those two projects have a specific 77° angle between two walls. We will have to consider if this angle could be grinded by the robot, or if it means that we willl need a human íntervention for it.

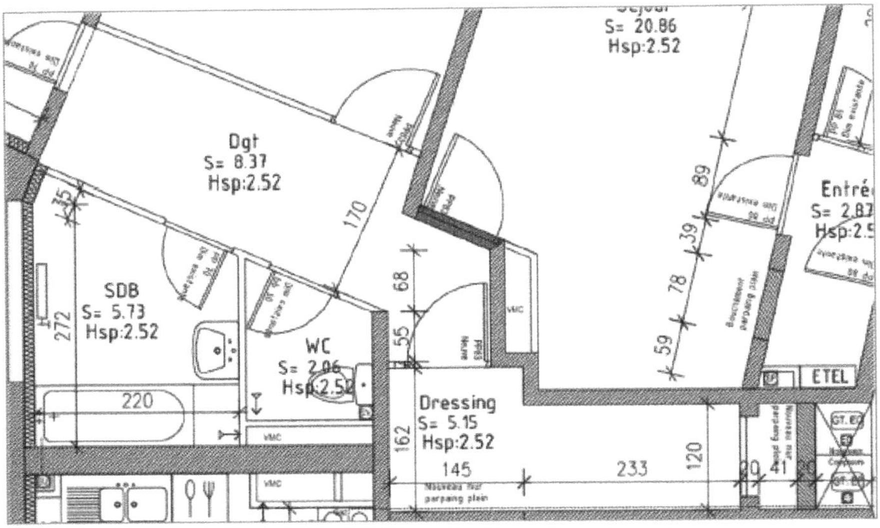

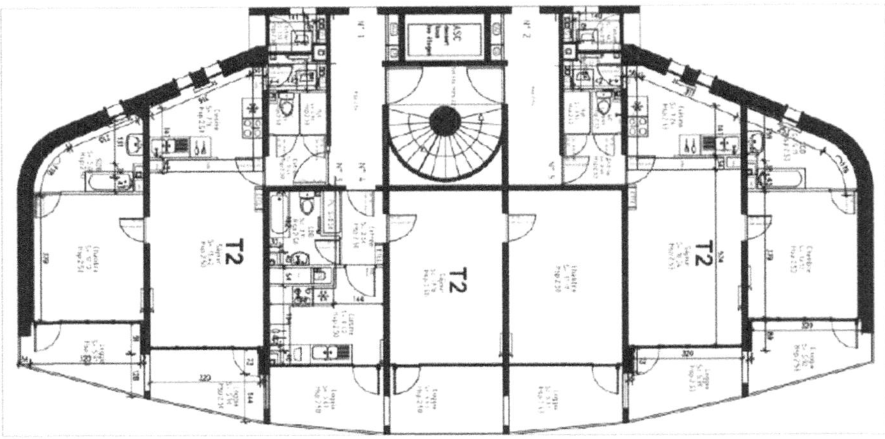

2.1.2.8.2 Roubaix 339

The Roubaix project is containing special shapes at the extremities of the building. According to a simple robotic solution, these special cases are excluded.

2.1.3 Conclusions of Analysis

2.1.3.1 Staircase and Elevator

In some buildings, there are circular staircases. In those cases, bringing the robot inside flat for removing asbestos could be a problem. In the projects studied, when there are some round staircase, there are some elevators, which avoid problems. The panel of projects analyzed in this study is wild, but limited, as it is only based on north east of France projects. That means that we could encounter buildings with round staircases, but no elevator. We would then have to possibilities: the project is out of the scope of intervention of the robot, or we would have to find another solution to bring the robot onto site. Considering the height of most apartment building studied, the only case where there is no elevator is on the Rouvroy projects. However, seeing the size and the organization of the building (small houses with very small rooms on the first floor), the intervention with the robot would not be very relevant. For the rest, we can see that we can mostly count on the elevator to bring the robot where it needs to go. For the other building, that are not for apartments, they are made to receive public, with means there will be elevators in most cases. All this means that we can count on elevators to bring the robot in the intervention zones, in most cases.

2.1.3.2 Flat Geometry

For the inside sizes of flats, the hypothesis that we took at the beginning of the project are validated by the study. Indeed, the smallest door size is 70 cm. The only exception is some doors that measures 60 cm, but it is from showers, which means they could be dismantled before the robot intervention. Anyway, this case occurred only once, which means it is in most case valid for the robot. For the corridor width, the only exception is Rouvroy, where the corridor is 80 cm large, but it is a project that would not be a target for the robot (see comment above). For the rest, the smallest width is 90 cm, which is good for the robot.

2.2 Environment and Requirements

Deliverable Di1.1

Technical report with protocols of the trials and specifications of the process parameters Concrete surfaces processing

Authors: Francesco Becchi, Tim Detert

2.2.1 Deliverable in the Context of the Work Program

The scope of this internal deliverable is to resume the results of the preliminary trials performed for asbestos removal on concrete. The main target of this preliminary test activity was to identify the tool behavior and relevant process parameters for the design considerations of the robotic arm, the tool integration and control of the robotic arm. Surface processing trials on a representative mock-up using a standard industrial robot with different tool and test configurations were performed. The report is based on the intermediate presentations already distributed between the parterns.

This same deliverable should be updated once available test results on tiles removal.

2.2.2 Methods and Work Carried Out

To perform the discussed tests two different setup were prepared.

The first mock-up was used for concrete wall processing and is in the following described.

A second similar mock-up for tiles removal was prepared but not used within the scope of the discussed tests.

2.2.2.1 Plaster Mock-up

The internal surface of a “C” shaped1200x600h1000 mm concrete structure was used as mock-up for vertical and horizontal (downward and upward) surfaces.

The rough concrete surface was painted in bright red and then covered by a 4 mm layer of plaster finally painted in blue.

2.2.2.2 Tool Used

According to D1.1 indications a hand held Blastrac BGV180 grinder was used for the tests.Tools main data are below reported for reference.

Figure 2.1. Plaster mock-up.

Technical Specifications of the BGV-180AV Hand Held Grinder

Name	Description
Working width	∅ 180mm
Motor	2.6kW / 230V - 50Hz & 60Hz / 16A - Single phase 2.4kW / 110V – 50Hz & 60Hz / 32A – Single phase
Working speed	6600 rpm
Weight	5.4kg
Dimensions (L.W.H)	495mm / 185mm / 325mm
Drive system / speed	Manual / N/A
Working direction	Forward (push)
Application	Concrete / Stone / Asphalt / Elastic materials

Options / Accessories for the BGV-180AV Hand Held Grinder

Name	Description
Dust collector	BDC-1114 / BDC-1216 / BDC-1220
∅180 diamond cup wheel	BG707107 – Designed for soft floors / materials. BG707108 – Designed for medium floors / materials. BG707106 – Designed for hard floors / materials.
∅180 PCD cup wheel	BG707198 – Designed to remove elastic materials
∅180 turbo cup wheel	E01401 – Medium bond with a special segment pattern, for a more even grinding surface
Upper handle	E06005
Flexible shroud	E04748
Suction hose ∅38	E5423235
Suction hose ∅51	004583
Personal safety features	On request / See manual

Figure 2.2. Tool specifications.

The tool was adapted to be installed on a robotic setup removing the rear handle that integrates a safety starter.

2.2.2.3 Robot Arm Used

Tests were performed using a KUKA KR60 HA that was available in Telerobot labs. The payload and range of the robot (60 kg max payload) were more than adequate for the task. No specific torque or sensor control was added to the robotic system. The tests are run with a standard position control implemented to the Kuka control.

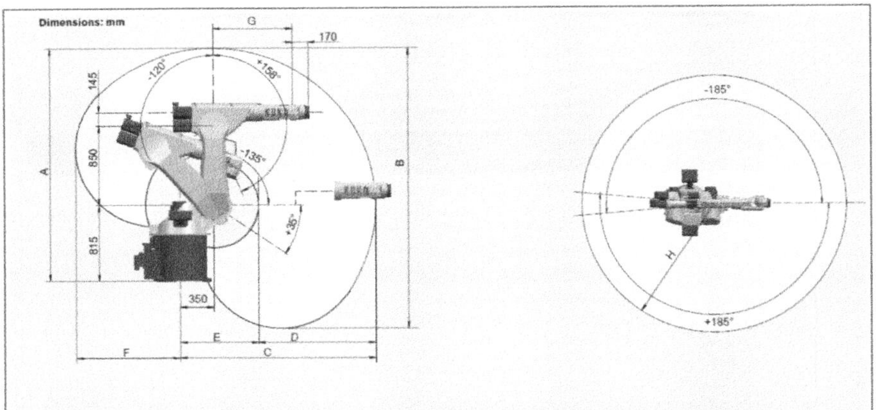

Work envelope	Dimensions[1]							Volume
	A	B	C	D	E	F	G	
KR 60-3	2,498 mm	3,003 mm	2,033 mm	1,218 mm	815 mm	1,084 mm	820 mm	27.2 m³

Figure 2.3. Robot specifications.

2.2.3 Concrete Trials

This section describes the trials performed on plaster mock-up.

The general feeling based on some test performed with the tool in hand give the idea that the tool lay on the grinding disc naturally sliding smoothly on the surface.

Goal of the trials is to try to reproduce this same behaviour using the robot arm.

2.2.3.1 Compliant Tool with Spring Support

In the first test the tool was visually aligned with the surface to check the attitude of the tool itself to slide on the wall.In this sequence of test performed the tool was connected to the robot with an intermediate spring support used to leave a small compliance between the tool and the robot.A force torque sensor was also added to monitor interaction forces.

This first test setup shows that it is hard if not impossible to ensure the correct tool alignment by visual feedback "looking at the tool". The small misalignment is amplified by the tool traces that quickly cut slots in the surface where the cutting disc locks.The following pictures ?show the result of the first test performed. Visible cuts are evident and deep.

Figure 2.4. Tool with spring support.

2.2.3.2 Frist Test, First Try

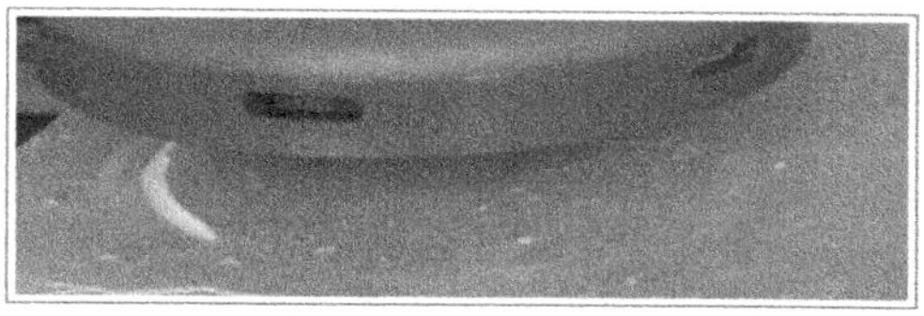

Figure 2.5. First try.

First try: Tool is not flat on the surface, the alignment is improved and the test repeated

2.2.3.3 Frist Test, Second Try

Figure 2.6. Second try.

Second try: Tool orientation was modified but still not perfectly flat on the surface.

Remark: The force reading is noisy. Working with fixed wrist position and spring compensation the force drops rapidly as the tool removes the concrete.

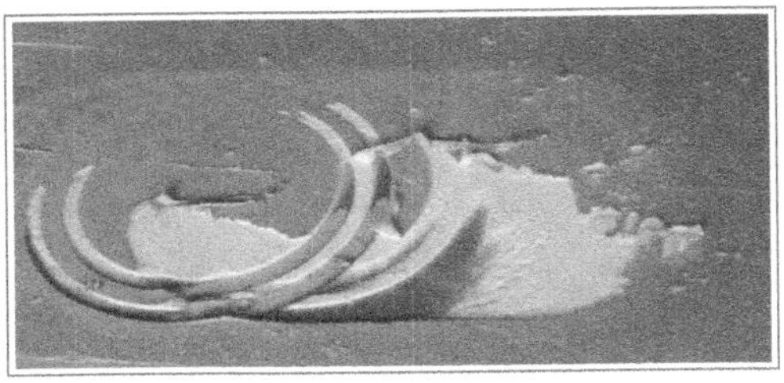

Figure 2.7. Third try.

2.2.3.4 Frist Test, Third Try

Third try: With the same orientation of previous test the tool is fed along one direction (speed is not always constant and removed depth changes- red and white zones).

2.2.3.5 Frist Test, Fourth Try

Figure 2.8. Fourth try.

Fourth try: The disc approach the wall with few degrees of inclination. The spring cannot compensate and the tool goes down deeper (concrete layer under red coating is visible).

2.2.3.6 Cardan Supported Tool

In the second test a cardan support was added between the tool and the robot with the aim of a simpler alignment of the tool.

Also in this case the result are not in line with the expectations with uneven surface processing.

Adding a cardan support to the tool to allow a nearly free and passive alignment of the tool, results in a tendency of the disc to "seat" in its cut: Asingle hinge floating

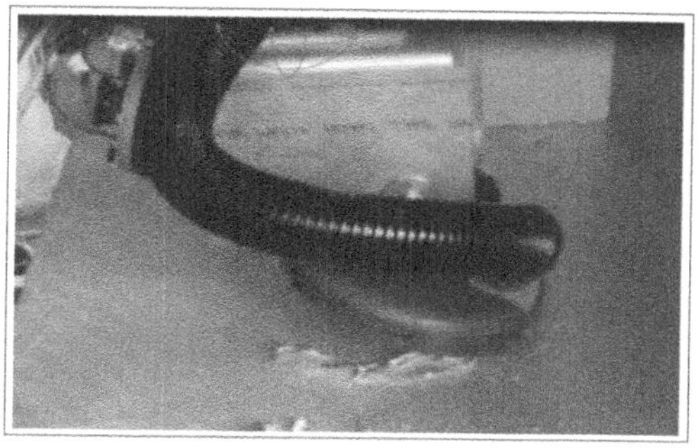

Figure 2.9. Second test tool configuration.

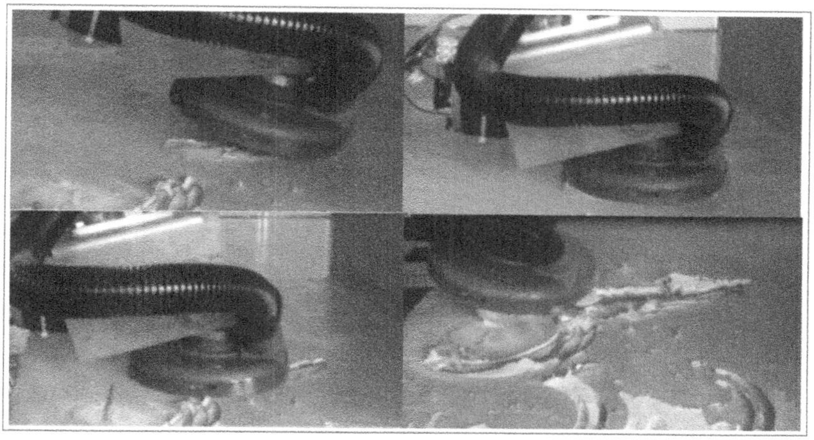

Figure 2.10. Second test.

Figure 2.11. Second test results.

tool doesn't seem to be a good solution.Both the suction cup and the disc itself lock pretty easily.The human compliance in tool handling follows a more structured trajectory adapting to the tool attitude. This does not seems to be easy to implement in a robot control.

2.2.3.7 Cardan Supported Tool

In the last test performed an external tool guide was added.Four idle wheels on the four corners of the disc limit the max tool depth and help the tool alignment in any direction.

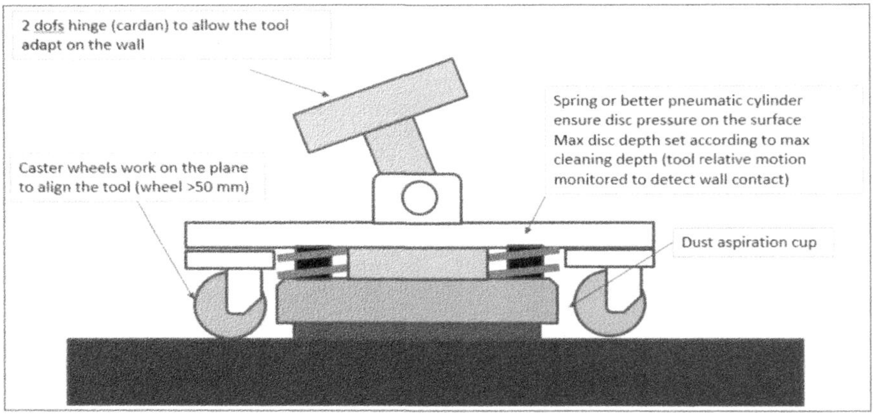

Figure 2.12. Third test conceptual.

Figure 2.13. Third test configuration.

Grinding depth was tuned by adding spacers between the wheel and support plate.The full system can be optimized and a softer suspension must be used.Overall this solution looks a first good and simple solution.

The result in this case are positive with several successful trials on the different orientation.

The main drawback that should be solved in the development of the tool support is the elimination of dust accumulation in the caster wheels used.

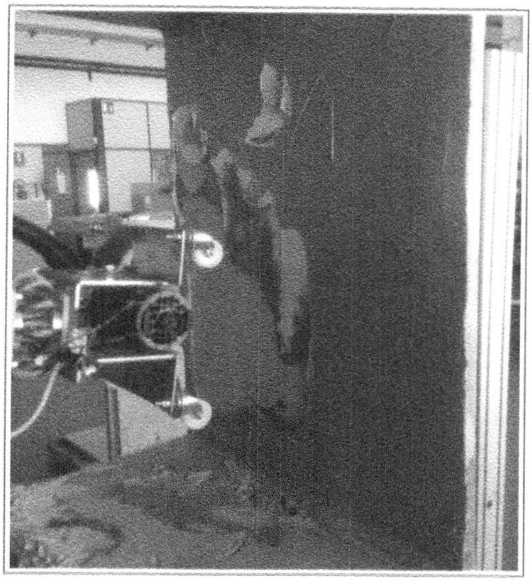

Figure 2.14. third test results.

2.2.3.8 Diamond Disc Test on Tiles

Two tests were performed using the Blastrac diamond disc tool already used on concrete on a tiles mock-up following Hilti suggestion.

Tiles removal was tested both on "floor" and on vertical wall to check:

- tool ability to remove tiles down to the concrete;
- stability of the tool and process efficiency;
- force needed and removal speed.

The following pictures show the two test performed and the result on the mock-up

The main conclusions of the tests performed are:

- the diamond disc **can remove** tiles completely;
- the efficiency of the tool is **significantly lower** that with concrete (see video);
- The force needed is significantly **higher** (estimated at least 20 kg);
- Tiles surface is **slippery** and tool stability is low (any tool force perturbance make the tool slide -see video);
- Tiles are cut down to **dust** particle that are aspirated by dust extractor with good efficiency.

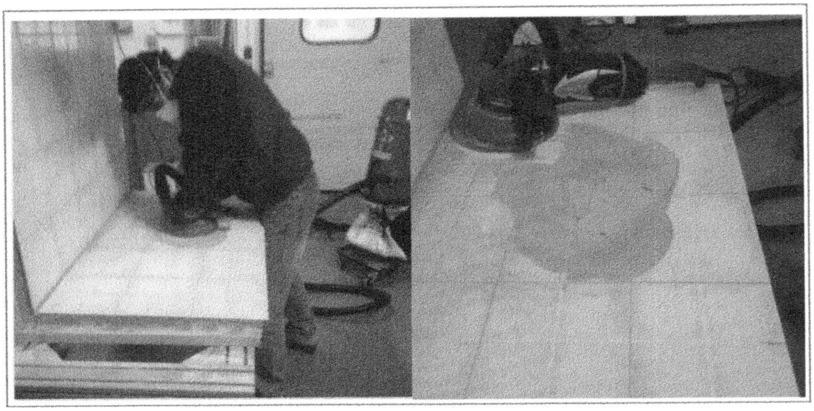

Figure 2.15. Test on the Floor.

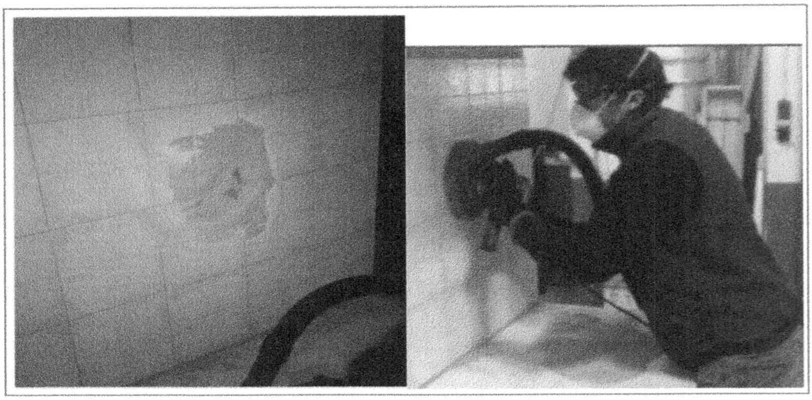

Figure 2.16. Test on the Wall.

2.2.4 Outlook and Future Work to be Carried Out

As anticipate the reported work covers only test performed with different tool adaptation strategies on plaster/concrete wall. Tests on active attitude control and tiles removal are planned to complete the deliverable.

Future investigations clearly need to aim at a well-developed and advance control of the robot arm. Different strategies of passive alignment, active force control using force-torque sensors, mechanical or control compliance and so on need to be considered to be able to implement a working process.

2.2.5 Summary and Conclusion

Different tool/robot configuration were tested with mechanical/passive tool attitude compensation and simple robot arm position control.

The only solution that gave promising result requires some external caster wheels for a mechanical guidance of the tool.The first implementation of this solution already outlined the need of a dust proof caster wheel design.

The active attitude control of the tool with Force torque sensor closed loop control was not tested in this stage and will be validated in the following tasks.

Deliverable D1.2
Testing rehabilitation site available for regular tests

Authors: Florian Dufour, Tim Detert

2.2.6 Deliverable in the Context of the Work Program

The operational process for asbestos removal will be developed, tested and validated on a testing rehabilitation site. This facility, provided by Bouygues, will represent an exemplary mid-sized room, with typical surfaces and materials. It will allow to perform tests on a site looking alike future real rehabilitation site, but without any hazardous contamination. The development and validation of the operational process will be done in several iterations, with regular workshops. All the partners will participate to those testing iterations (from task 3.5 and 6.5). The complexity of the tests will increase gradually. It starts with the tele-operated execution of a single asbestos-removal-task and ends with processing an autonomous task planning and process control (Milestone 8).The increasing complexity of the process will be addressed by step-by-step improvements of all components, the control and an advanced scenario at the testing site.

2.2.7 Methods and Work Carried Out

To be able to provide a testing site, suitable for all developments, it was first necessary to define the technical requirements and more generally everything that was needed for such testing site, for example the room dimensions, the power supply needs and internet access. The current specifications and requirements are given in the annex, even though they might be adjusted during further planning. After capitalizing all the information about the requirements, we came to the conclusion that a big technical area would be the most suitable, considering the needs defined.

From this statement, we contacted the technical site managers within Bouygues Company and evaluated the feasibility of using one of the several material bases within Bouygues Company. The main one is situated in Chilly-Mazarin next to

Paris. The problem of this base is that it is the main one in France, what means that it is actively used, by all kind of projects. Consequently, we couldn't have a big enough place available for our tests, especially for a period of 18 months. Another consideration was erecting a testing site on wasteland. However, using such an area would mean providing the whole necessary infrastructure: a closed area for the tests, the offices, security and so on, for efficiency reasons this approach was discarded.We finally found a place inside the Bouygues Travaux Public facility, in Bourg lès Valence. It is a technical storage and maintenance area for Bouygues TP machines. Thus, all the technical supplies needed for our project are available at this site. We decided to visit the site, with different partners, to have different points of view for the evaluation.The following people were present during the visit:

- Tim DETERT, RWTH
- Sami CHARAF EDDINE, RWTH
- Quentin TREBOT, SIGMA
- Bruno LINEATTE, BYCN
- Florian DUFOUR, BYCN

After a day of discussions between partners and local Bouygues persons, we concluded than the place will need only few developments in order to be suitable for our tests.

We completed our requirements with regard to the last modification, especially following the visit.

Consequentially, we presented an overview of the site and discussed the place and linked facilities with all the partners.

2.2.8 Testing Rehabilitation Site Available for Regular Tests

The testing site is situated in Bourg lès Valence, 115 km in the south of Lyon, in France.It is situated on a civil work equipment facility, from Bouygues Travaux Publics.

The site will be organized as shown in Figure 2.17. We will build a hangar, next to the current Bouygues offices, providing all necessary infrastructures. In the hangar, all technical supplies needed to perform the tests will be available. We will bring precast wall and prepare the surfaces to be able to simulate a typical room for the tests.The first draft of the organization of the hangar is the following one:

We plan to separate the hangar in two parts, one for the operational part and one for the robotic one. Those two areas will be separated for security reasons. Three concrete precast walls will be erected to perform the tests. We will simulate special typologies like corridors, turns or small rooms by lighter structures. There will be

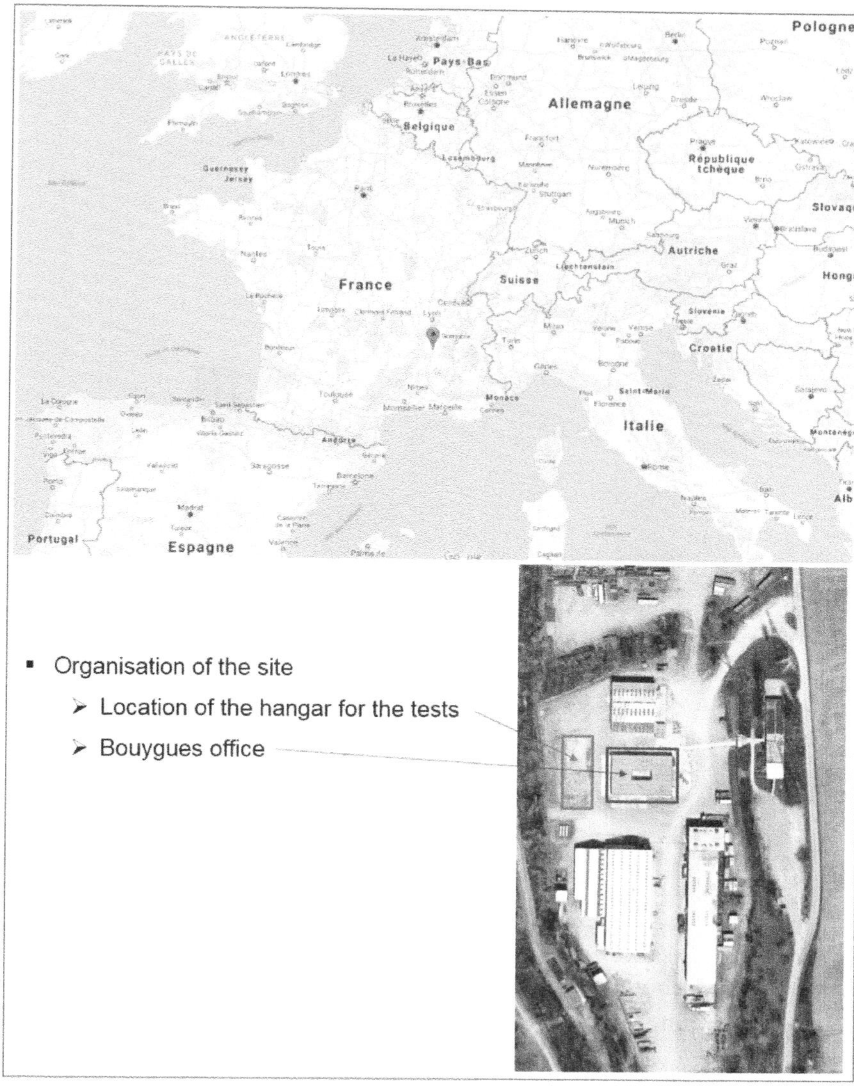

Figure 2.17. Placement of the site.

a slab below and above those three walls to allow stability and testable surfaces as ceiling and floor. The following scheme is a 3D example of the concrete box:

Linked to this concrete testing box, we will need an extra intervention for the preparation of the support, the replacement of the removed testing materials (plaster, tiles, …) and so on. These works will be performed either by Bouygues professionals or subcontractors. Those tasks will be anticipated and planned in advance to the test, for the representative materials to be as close to reality as possible considering dry time, final hardness and so on.

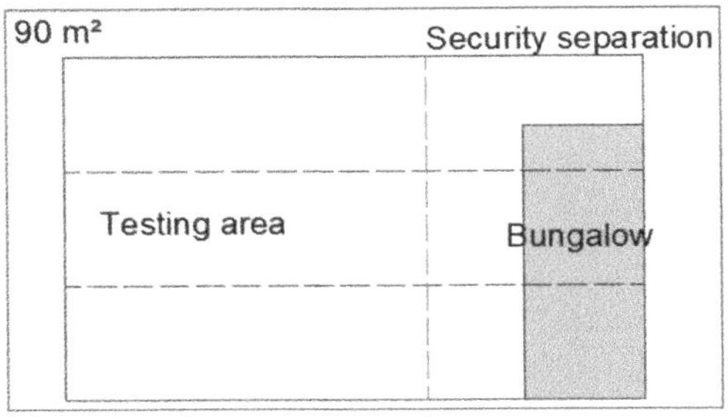

Figure 2.18. Organization of the hangar.

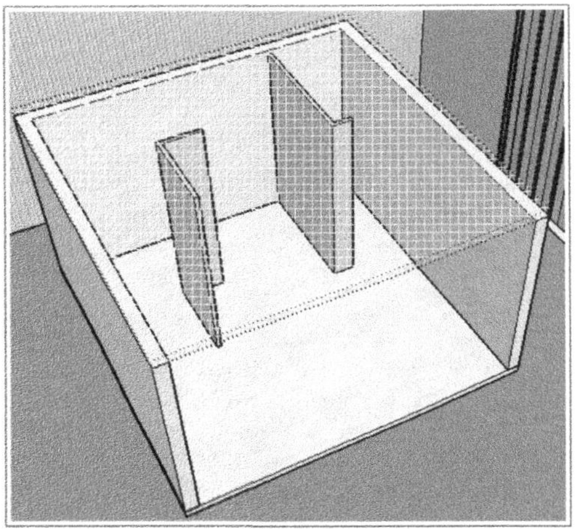

Figure 2.19. 3D example of the concrete box.

2.2.9 Outlook and Future Work to be Carried Out

We plan to start to build the hangar in February/March 2017. We need to prepare the area for the hangar, and then build it. In the same time, we will choose and order the concrete walls, and then prepare them, to be able to begin the tests in May 2017 according to the developed testing time plan.

In parallel to the construction of the testing hangar, we will define testing scenario, serving as a base to compare and improve the process. Consequently, we will define the robotic workspace and the way we will perform the tests. Linked to the tests, the safety measures will be determined to protect workers and the environment against hazardous dust and separate them safely from the robotic workspace.

When all those stages are completed, we will perform the tests, raise the complexity of the robotic process, and develop it. There will be some workshops to confirm the direction taken and follow the evolution of the robot.

2.2.10 Summary and Conclusion

In this deliverable report, we described the requirements for the testing site and possible sites, which were investigated. We have detailed the plans for the testing site, which will lead to the development of the robotic system.

We benchmarked the different places within Bouygues facilities and defined precisely how we could perform the tests, to be sure to have a suitable place. The chosen testing site will now be erected and made available in the next months, to be able to start performing first tests, according to the experimental plan.

2.3 Process of Automated Asbestos Removal

Authors still have to provide the content

Deliverable 1.3
Document with benchmarking directions, specifying detailed performance objectives and metrics

Authors: Florian Dufour, Tim Detert

2.3.1 Deliverable in the Context of the Work Program

To judge the economic viability and to enhance the number of buildings cleaned, it must be guaranteed that the robotic process and system can achieve the best performances in terms of quality and in terms of time saving.

Performance objectives and metrics were developed for this deliverable to benchmark the operational process, the system, and its integration to an intervention on an asbestos contaminated site (as specified in task 1.1) under real life conditions, as referred to in task 1.5.

The task 1.3 is performed parallel to the development of the operational process (task 1.4) to be able to identify important factors for the automated process.

2.3.2 Methods and Work Carried Out

Before being able to define any objectives, we had to define a typical process for the robot, as a basis for measurements and improvements.

The first step was to analyze and document the process, from the arrival of the transport of the robot on site to the end of the intervention, with the robot back to the transport. In parallel to this, we analyzed and documented a typical day for a human asbestos removal worker, as a comparison base.

Based on this, we will perform a precise time analysis of the overall process and its sub-tasks, as main indicator for the performances assessment. Additional to this time analysis, we will define the performance objectives needed to ensure a safe process in accordance with regulations. Particular attention will be paid to the dust levels monitoring and the way to manage wastes.

Finally, the objectives for the degree of autonomy for the process control and the usability of the user interface will be defined.

2.3.3 Benchmarking Directions, Specifying Detailed Performance Objectives and metrics

2.3.3.1 Robotic Process

As a first step for the performance objectives and metrics, we analyzed and documented a general process, taking in account all the stages of the project, from the arrival of the robot on site, passing by its routing to the working area, the removal action, till the way back to the transport means.

This process is divided in sub-tasks, to define on which ones we can focus and which we can't impact. The documentation of the process was done as a maximum detailed one, with different cases and way we can deal with the tasks.

We start by the most basic level, as following:

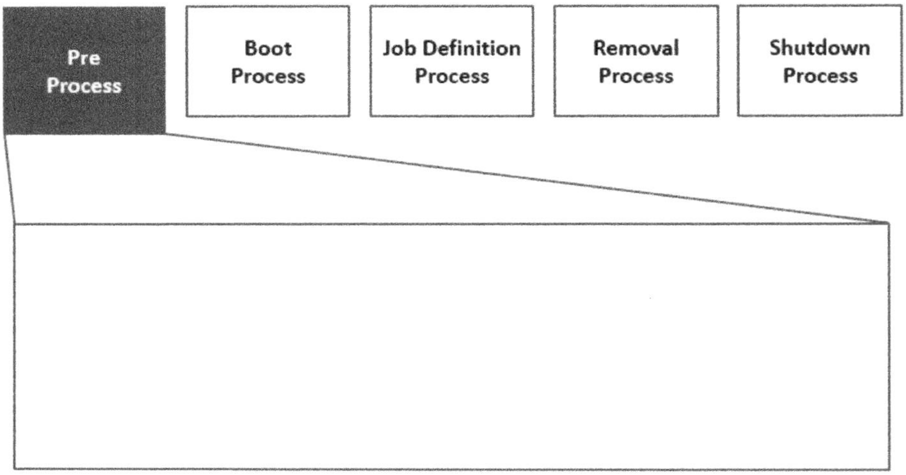

Then, we got deeper and deeper in the process, to finish at a detailed level like following:

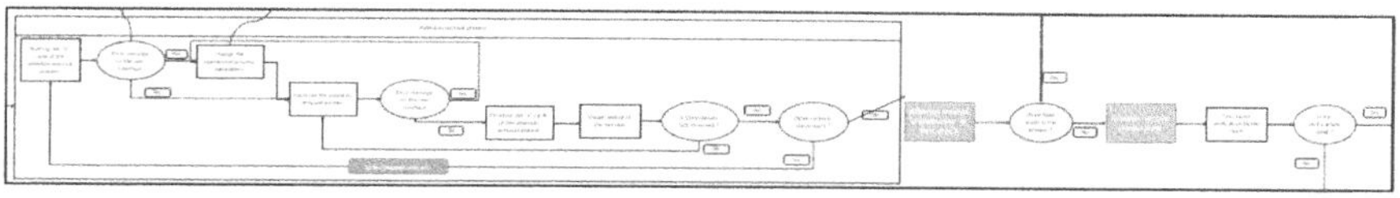

2.3.3.2 Overall Timing Estimation

After documentation of the process, we summarized the different main actions in a simpler process:

Thanks to this, we were able to estimate what would impact, in terms of timing, the process, and to compare it with a human. For example, we know that a worker needs 15 to 20 min to dress before getting into the secluded area, oppositely to the robot, which would be able to go inside faster. However, a worker can go faster than the robot to the right floor, so we will have to be as efficient as possible to bring the robot there, to lose as less time as possible.

In parallel, we will create a table that will allow us to compare the robot with a human worker more precisely during the different stages. First, we will perform identical tests for the main process tasks in the Bourg les Valence testing facility with both, human workers and the robotic solution. The idea is to perform a typical process, grinding 3 walls for example, in the exactly same conditions. This will take into account the time to get inside the seclusion, the time to process and the time to clean and get out. Thanks to this, we will be able to compare, action by action, the difference between the robot and a human, and so, on what topics we should improve the robot.

When the robot will be more developed and that we will be able to proceed tests on site, we will do again the comparison test, to be as close as possible to the final application context.

2.3.3.3 Tasks Automation

After the documentation of the process and the time estimation of each task, we had a look to the different tasks that might be automated, from the most obvious one, like removing the contaminated material, to more complex ones, like moving from a secluded area to another.

Afterward, we prioritized them, from the most important to least. This step is currently under investigation and will include the user feedback form asbestos removal experts. The purpose is to be able to concentrate and to automate what is the most important, then go deeper and deeper into the automation. This is the preliminary prioritized table at the moment:

Task	Priority	Remarks
Removing asbestos	1.1	Assisted teleoperation for grinding process at the wall/ceiling - autonomous distance /tool-force control Positioning of the platform relative to the wall could be tele-operated
Autonomy for the grinding process at the surface	1.2	Autonomous grinding process at the wall/ceiling with autonomous feed and distance /tool-force control
Autonomous platform positioning	1.3	Autonomous maneuvering of the mobile platform to progress the working area at one wall or ceiling and move relative to the specified area
Moving inside a secluded room autonomously	1.4	Autonomous maneuvering of the mobile platform to change to a different working area
Moving inside the seclusion	2	Being able to move from one room to another inside a secluded area
Managing wastes	3	Handling, dropping and storing the waste containers. The alternative is the tele-operated dropping of the waste containers.
Moving in between secluded area	4	Decontamination subject linked, the robot leaves the contaminated site temporarily
Cleaning the confinement	5	Vacuum cleaning protection sheets and non-protected surfaces
Moving everywhere	6	Crossing obstacles like stairs, steps, gaps. This capability can be applied in the above mentioned autonomous tasks. For now they would be interrupted and need manual teleoperation in tricky situations.

Task	Priority	Remarks
Booting	7	Getting ready to process the materials. This includes the whole set-up and ramp up of the robot and process and would include full self-diagnostics and fully autonomous mapping and avigation based on existing task plan.
Dust measuring while processing	8	Existing but not officially used in the legislation

2.3.3.4 Benchmarking Tables for the Robot

As a benchmark of our solution to asbestos removal, we will produce at least 2 different tables.

One to compare our solution with other possible solutions for every main task of the robot. The purpose of this is to be able to enlighten and measure the efficiency of possible improvement for our robot, and the following ones, for any other fields. We will compare the solutions through the main topics that are the timeline and the cost, but we will also take into account the special constraints linked to the task (dust production, geometry, decontamination…).

On another hand, we will develop a table to be able to measure the success of our project. We will list the main goals the robot has to reach, on every tasks, and rate it, with a comparison with a human when possible, to be able to scale the success of the overall project. The main goals will come from the initial description of the tasks, and from the different topics that appeared during the development.

Those two tables will allow us to measure the success, and to compare the different solutions that might be used to improve the robot.

2.3.4 Outlook and Future Work to be Carried Out

The overall process and different steps of the robot intervention were analyzed and defined on different levels. We got deeper and deeper, from an overview process to a more detailed one.

We started the time analysis of the process, and defined performance objectives and metrics based on it. We specified the degree of autonomy we are aiming to, in order to process the asbestos removal as efficiently as we can.

We will now develop the robot, and build the tables that will allow us to benchmark our robot, following the main thematic described before.

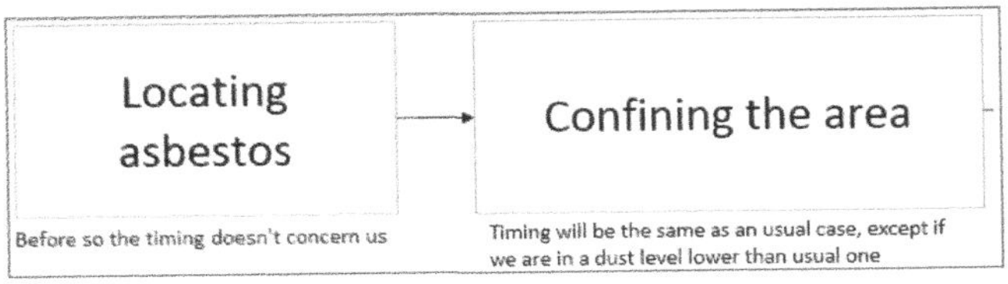

Figure 2.20. Manual Steps.

2.3.5 Summary and Conclusion

The benchmarking of the system will be based on the overall process and not only consider the technical implementations to make sure all user needs and the full process are considered.

In this deliverable, we have analysed and described the overall process including the lower level details to benchmark the full process according to it.

Afterward, we made a first estimation of the time linked to the actions, and planed tests to make it more precisely. Then, we defined the level of automation of each task, and their priority.

We have now everything to test our robot, benchmark it through table we will create and develop, and measure all part of the process.

Thanks to those different points, we will be able to develop a robot very close to the definitive use case.

Task 1.3 Specification of performance objectives and metrics for the process validation

The specification of performance objectives and metrics of the task 1.3 is closely connected with the operational process of task 1.4. Indeed, each part of the process that is similar to a human intervention has to be more efficient and/or cheaper, and the ones specific to the robotic intervention has to be as fast and limited as possible.

We took the overall process developed previously, and especially its main tasks. When the robot will be fully operational and on site, we will be able to make a chrono analysis of each subtasks, but for the current state of the project, a main tasks analysis will be sufficient. Some of the tasks of the project are identical as a classic asbestos removal intervention. Therefore, we will not define any objectives on them. One thing that could change is if the robot allows us to have less residual dust in the air than a normal intervention. This would then allow the removing company to lower the seclusion and so the cost of the intervention.

Both steps are typical and well-known steps from the current asbestos removal process and won′t be changed by this project. “Locating asbestos” is done during the initiation of the construction process to identify the areas and materials where asbestos is used. The tests behind are fully established and will create the output like:

Figure 2.21. Secluded room.

"Asbestos is used within the plaster on walls and the plastic tiles in the kitchens". Out of this the operators are informed which surfaces need to be cleaned and which ones can be secluded as they are not processed.

After this first step the construction site is secluded based on the gained information of asbestos contamination. When e.g. the floor does not need to be processed, it is going to be secluded to keep the process as easy as possible. In this case the time consuming step of cleaning the floor after the removal tasks would be unnecessary when the simple plastic sheet can be removed straight away.

After these manual processes the first adapted step is required for dealing with the automated asbestos removal task. As the robotic system need a global reference model of the environment, the data mining step of semantic mapping is performed on this step. It can either be performed with an industrial indoor laser scanner that creates a high definition point cloud of the entire flat or the automated solution developed by the consortium can be used. Both ways are possible as the input data for the semantic mapping algorithms is a simple point cloud data type.

Basically the captured point cloud is loaded to the algorithm and started to be processed. Therefore the first step is a seed growing to determine points in the point cloud that most likely belong to the same element like a wall, ceiling or special items like heaters. After this seed growing the found sub-point clouds are surrounded by so called bounding boxes to have a more simple 3d object to handle. Afterwards each bounding box is processed and reduced to a simple polygon to only keep the essence of information that is necessary for the process. The result is a light-weight descriptions of surfaces like walls, ceilings and floors and also special elements like doors and windows. Just as a last step the already found semantic elements are bound together to rooms and they are connected to flats. Out of this a full semantic map for the further process is created. Just in addition, this process of semantic mapping can take place independently towards the work with the removal robot so that time optimization for handling multiple construction sites as an operator may come in handy.

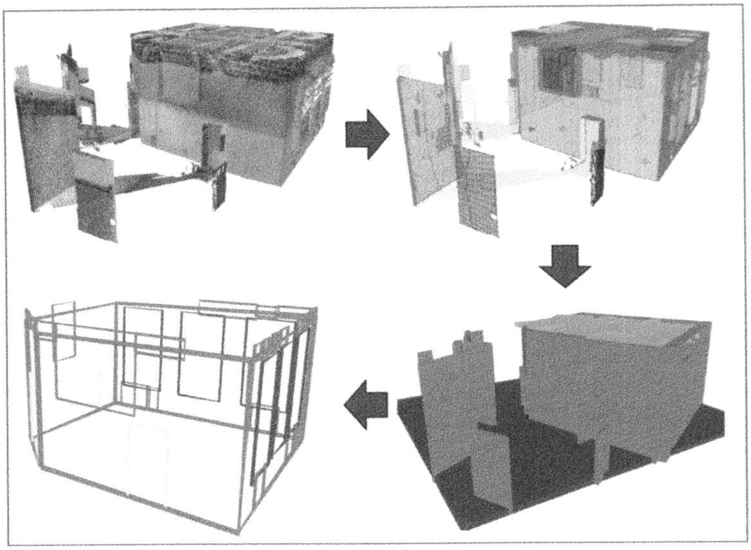

Figure 2.22. Semantic Mapping – Main Steps.

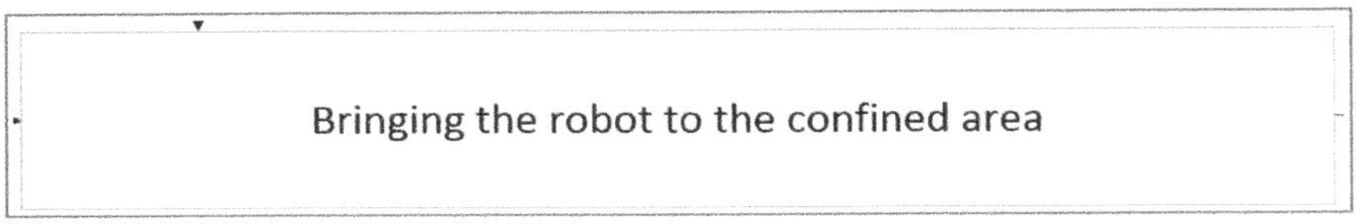

Figure 2.23. Transfer into confined area.

After the semantic mapping the real removal task can start. Therefore the removal robots need to get on the construction site.

The similar human task is very fast; workers go to the site, get ready and start the process, all in approximately 30min. The time for the robot to do the same is hard to define, as each building is different from another. However, the easier the access is (no step, elevator, large confined area…) the faster the robot will be and the more competitive the solution will be. When the robot is in the confined area, the booting process can start.

When the robotic unit will be operational, it will be easy to time it. It is a task that can be compared with the briefing of the human working team. Anyway, this will be a short task, and therefore will not have a big impact on the overall time of intervention. Mainly it is the setup of the safety infrastructure, the removal robot and the CPCS workspace.

To have a mobile solution for having an entrance control of the processed area the consortium established a robust safety relay for controlling the entrance of human

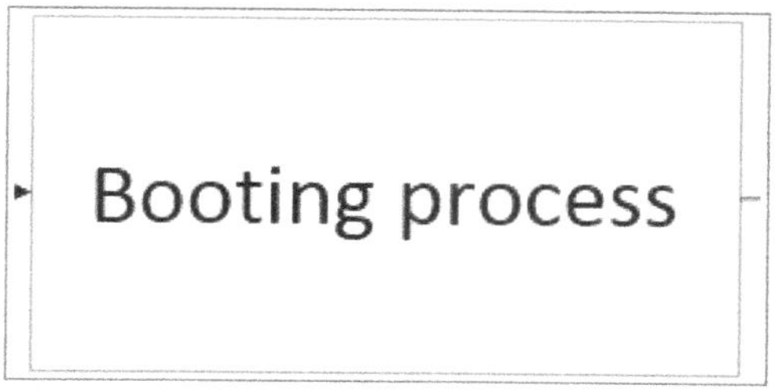

Figure 2.24. Booting.

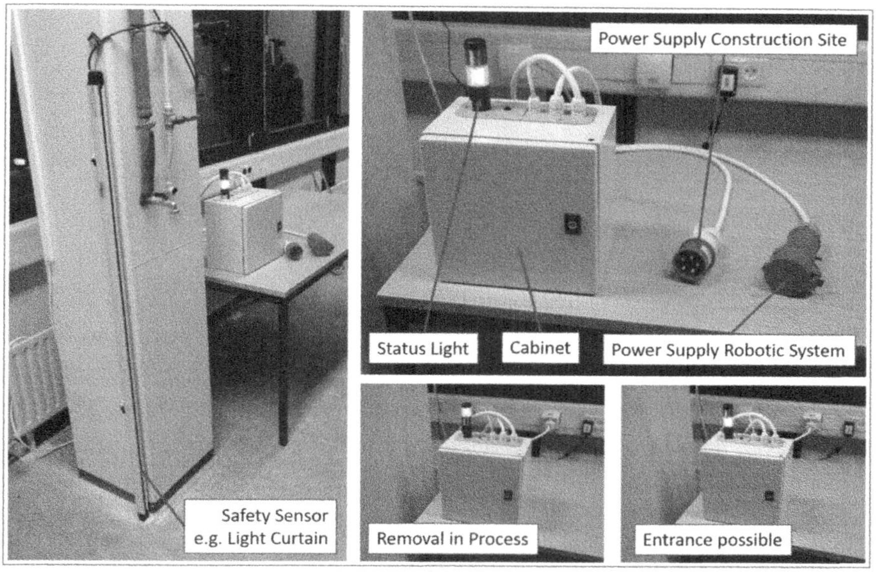

Figure 2.25. Safety installation for removal areas.

workers into the automated area. Therefore the safety relay is connecting the power supply of the construction site and the power supply of the robotic system. The relay in between can be triggered by any typicall safety device. In the following figure a typical light curtain is used but it also could be any safety door with a status relay for closed and open conditions. This safety setup is e.g. positioned in front of the currently closed removal area that can be a single flat or an entire corridor. A light is demonstrating the accessibility of the removal area where red is the sign for removal tasks being in process and green is free to enter as the robot is not working and parked safely.

Beside the setup of the safety environment and the access control for human workers the CPCS Workstation need to be setup as well. Therefore the consortium is using a typical server rack to store the central machines for the CPCS.

Figure 2.26. Mobile CPCS Setup.

The workstation enables the external control view onto the system and will have the user interface for task control and monitoring around. The mobile server rack is having its own cooling system and all the equipment (PCs, Screens, Mouse and Keyboard) can be placed inside the metal container during moving from site to site. From here the created semantic map is loaded and can be specified for the removal. The task planning is run based on this user information and will create the global removal strategy plan to be performed.

The booting process also contains the setup of the robotic tandem for operation. This includes the connection to the power supply and the central aspiration system. Furthermore, the system needs to be calibrated within the new site. On the one hand this is the calibration of the arm joints to identify them as homed and on the other hand it is the starting position of the mobile bases within the semantic map. Everything is designed to be done through the user interface.

After the full booting process the removal task can start.

This task is the core of the robot intervention. The faster or the more efficient the robot is, the more suitable the solution will be. There are two approaches to define the objectives for this task. If the grinding process itself is slower than a human one, the robot will have to work longer to be competitive. Indeed, workers can only work by two hours slots, with a daily total of 6 hours. The robot on the opposite, is not limited, and can work 8 hours per day, or even 24/7.

If the grinding process is faster than human one, then, the system is even more relevant. It can work a normal day schedule and be more efficient, and could even

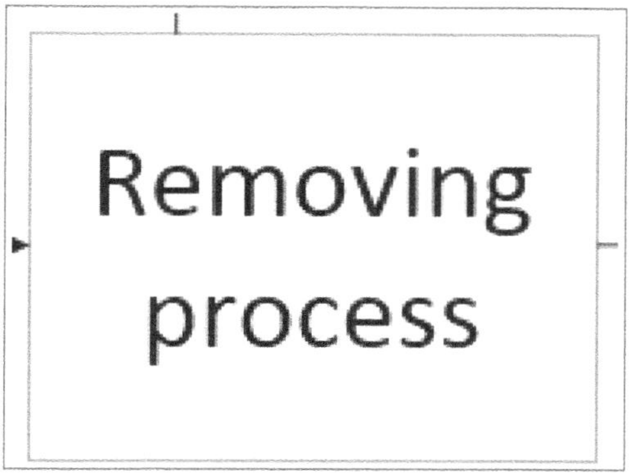

Figure 2.27. Robotized asbestos removal.

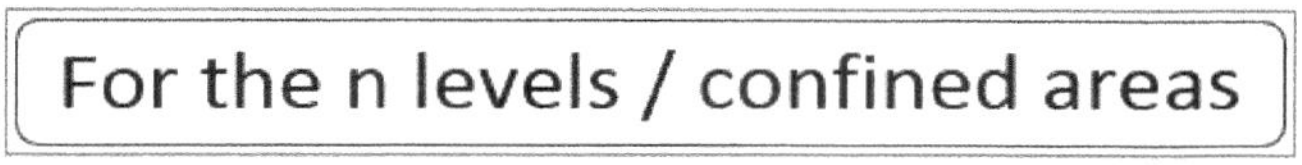

Figure 2.28. Optimized confined areas.

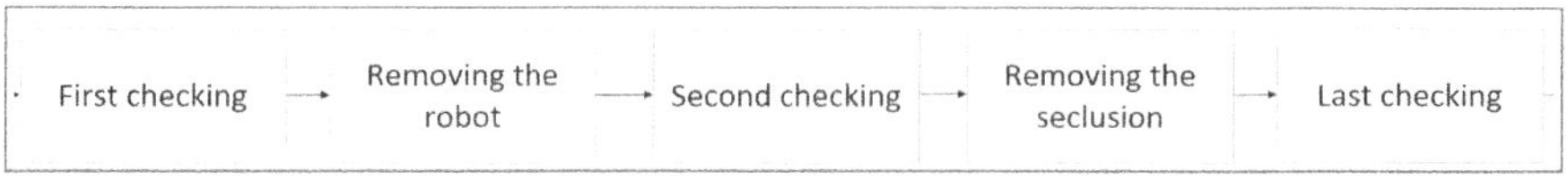

Figure 2.29. Manual after processes.

be push to 24/7 to achieve extremely fast removal. The case we are in will be defined when the robot will be operational and will allow us to target the good approach to the overall removal of asbestos.

As soon as the removal is complete, the robotic unit will have to go to the next secluded area. This task is similar to the "bringing the robot on site one" and has the same constrains. To limit as much as possible the impact of the task, an overall work will have to be carried out for each building, to optimise as much as possible the secluded areas, to have to move the robot the less possible.

The rest of the process is about checking the work done and removing the robot if the removal was properly performed and is done manually by experts.

All those tasks are mostly comparable to human ones. The only special one is the removing of the robot, and it can be done outside of the timed intervention, as the process itself will be finished.

For the health care and safety target of the process, the approach is to have a dust level at least equal to a classic intervention. Indeed, most of the price of a normal removal of asbestos is the seclusion and the tools that go with. If the robotic intervention need a stronger seclusion, the time and cost competitiveness will be even harder to reach. However, if the robotic intervention allows us to reduce the

Task	Priority	Autonomy	Remarks
Asbestos removal			
Processing flat surfaces	1	Full	Within the security distance to objects and edges
Removal feedback and deepth control	1	Full	
Feedback options for removal process success	1	Manual	
Feedback options for removal process success	3	Semi	Deep learning/automatic image processing
Processing around known objects (doors/windows	2	Semi	teleoperation of cartesian movement and autonomous process control (edges processing tool)
Processing around known objects (doors/windows	3	Full	Edges processing tool and detection
Processing around unknown objects	3	Semi	
Movement			
Moving inside the room	1	Full	Being able to move from one room to another inside a secluded area
Moving from one room to another	3	Full	
Moving from one room to another	1	Semi	
Task definition			
Task definition	3	Full	Definition with AR tags
Task definition	1	Semi	Defined in user interface with macro
Other thematics			
Managing wastes	4		
Moving in between secluded area	4		Decontamination subject linked
Cleaning the confinement	4		Vaccum cleaning protection sheets and non protected surfaces
Moving everywhere	4		Crossing obstacles like stairs
Booting	4		Getting ready to process the materials
Dust measuring while processing	4		Existing but not officially used in the legislation

Figure 2.30. Targets of automation.

seclusion needed, it will be a strong asset, from a safety point of view as well as competitiveness one.

It is obvious that the more autonomous the robot is, the better and the more competitive it will be.

DOI: 10.1561/9781680837155.ch3

Chapter 3

Robotic Systems for Removing Asbestos

By Francesco Becchi, Miguel Moreno, Matthias Rabel, Jean-Christophe Fauroux, Daniel Martin, Tobias Haschke, Mathias Hüsing, Tim Detert, Oriol Orra, Néstor García, Ramón Carrasco Jimenez, Masrur Doostdar, Lucas Rohe, Robin Bernau, Martin Gritzan and Siddharth Maraje

3.1 First Robotic Prototype (V1)

3.1.1 Mobile Base

Deliverable i2.1

Di2.1 Specifications and design concepts for mechatronic stabilizers

Authors:
FAUROUX Jean-Christophe, SIGMA Clermont
MARAJE Siddharth, SIGMA Clermont

3.1.2 Deliverable in the Context of the Work Program

The deliverable D2.2 takes input from *Task 1.2 – Initial trials of operational process, Task 2.1 – Preliminary simulations of the robotic units and Task 3.2 – Definition of technical requirements for the customization.*

For this deliverable D2.2, static and dynamic stability requirements that will not be fulfilled by the standard base platform are investigated and a concept for mechatronic stabilizers will be developed. The static requirements refer to the tilting limits of the platform due to gravity in the different arm and tool states. As the speed and acceleration of the platform will be low, the dynamic requirements will expectably not be critical regarding tilting and overturning limits. However, the forces to be exerted by the robot pressing the tool against the walls at all possible heights affect the robot stability. The standard way to cope with stability problems for civil engineering machines is to use landing gears. In the context of construction machinery, landing gears are a kind of stabilizer legs that are deployed to increase the platform stability. In this case, those devices must be modified for indoor use and taking advantage of contact points not exclusively located on the ground (walls, ceiling). The designed system should compensate the tooling forces against the walls during the removal operations and also the dynamic forces during locomotion / arm repositioning evaluated in task 1.2 and 3.2. The work requires evaluating a global stability criterion during asbestos removal and specifying mechatronic stabilizers (D 2.2):

- Choice for a suitable stability margin (static, dynamic) and the associated sensors (e.g. measurements on geometry, force pushing)
- Dynamic modelling for efficiency assessment based on the existing simulations from task 2.1 and further co-simulations:
  - 3D virtual model in statics and dynamics
  - Multi-body modelling with calculation of the stability margin during the task
  - Associated control methods
  - Specification and research of concepts for mechatronic stabilizers with a low number of DoF, stabilizing the robotic units against the solid environment (e.g. pushing against the ground or ceiling or attaching to a wall).

A tall (comparable to the kinematic dimensions from tentative dimensioning in section 1.4.4) robotic arm will be considered for preliminary evaluation. A virtual environment will be reconstructed from a digital map of the testing area and additional attributes from task 1.1 to represent the spatial dimensions of the scenario. The simulations are executed parallel to the initial trials of the operational process in task 1.2, to allow modifications and enhancements of the simulation at an early stage.

This deliverable has outputs to tasks 2.2, 2.3.

3.1.3 Methods and Work Carried Out

An extensive study on stability margins can be found in [Garcia *et al.*, 2002] and [Chebab, 2013]. As our robotic unit V1 should be slow and with no particular long dimension with respect to the others, a quasi-static stability margin criterion described for the first time by [McGhee and Frank, 1968] was used. For future work at higher speeds, the ZMP (Zero Moment point) criterion [Sardain and Bessonnet, 2004] could be used as well as non dimensional criteria such as the LLT (Lateral Load Transfer) indicator [Bouton *et al.*, 2007], that can be easily extended to the front direction.

The expression of the LLT can be found in equation (Equation 3.1), where F_{n1} and F_{n2} are the normal forces applied on the left and right sides of the vehicle. When the LLT equals to 1, the robot is at its stability limit, although a value of 0.8 or 0.9 should never be exceeded for safety. When the LLT exceeds 1, the two wheels on one side of the robot lift off and the robot can rollover. This non-dimensional margin is very convenient to use, provided the normal forces are measured by suitable force sensors in the wheel hubs or calculated by the multibody model.

$$LLT = \left(\frac{F_{n2} - F_{n1}}{F_{n2} + F_{n1}} \right) \tag{3.1}$$

The visit organized by Bouygues in a flat that had been cleaned (Clermont-Ferrand, 01/06/2016) confirmed the fact that pushing against the ceiling seems feasible in many cases, at least in apartments and less in offices. For the flats located on top of a building, pushing under main beams is recommended, although beam detection might be tricky. Pushing against thin walls (particularly gypsum plates) is not recommended, so a map of the walls that can serve as a support should be provided to the robot before cleaning. The door frames often provide a strong structure on which a stabilizer could push, provided the paint is kept intact.

3.1.4 Specifications and Design Concepts for Mechatronic Stabilizers

3.1.4.1 Central Stabilizer Pushing on Ceiling

This concept of central stabilizer (Figure 3.1) is extremely interesting as is applies a vertical force to all the wheels, thus transforming the unilateral contact between the wheel and the ground into a bi-lateral contact, much safer. The dispatching of the forces on all the wheels is difficult to model and can change instantaneously with a quasi-null displacement, depending on the degree of overconstraint of the wheels (suspension mechanism, deformable chassis, etc.).

A proposed implementation of the central stabilizer is a vertical jack, actuated by screw, capable to extend to the altitude of 3 m, and equipped in its end by a rubber bushing supporting a pushing plate. The rubber bushing is similar to the fixture that fixes the mast of a windsurf rig to the windsurf board. It is kinematically equivalent to a spherical joint but provides three translational and three rotational stiffnesses: the vertical translation decreases the pushing stiffness of the device, avoiding to

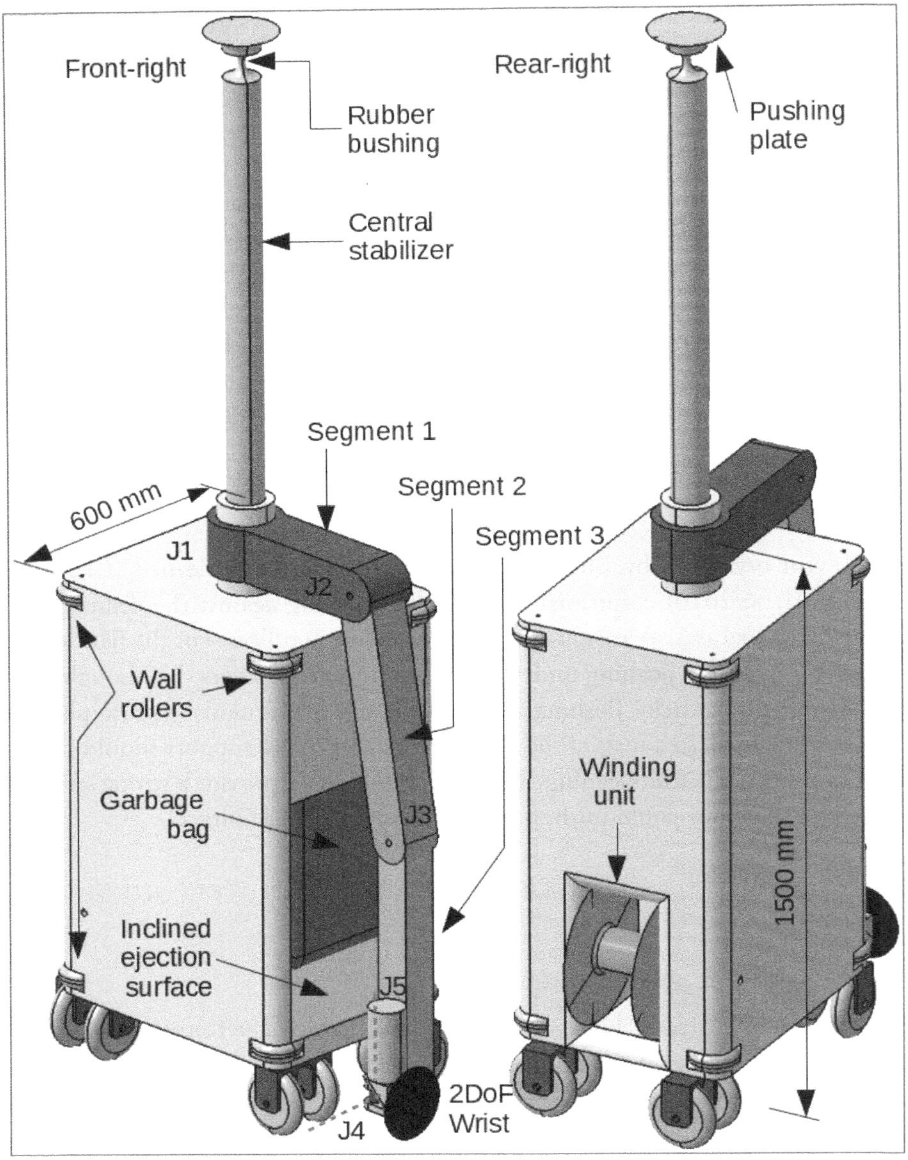

Figure 3.1. CAD preliminary implementation of the robotic unit with a central stabilizer, a winding unit and a garbage bag.

mark the ceiling too much; the rotational stiffnesses maintain the pushing plate towards the top whereas authorizing small rotations if the ceiling is irregular or not parallel to the ground. The pushing plate should be cushioned or made in a sufficiently soft material, with smooth edges, in order that a non-centered pushing force does not make a decentered mark on the ceiling.

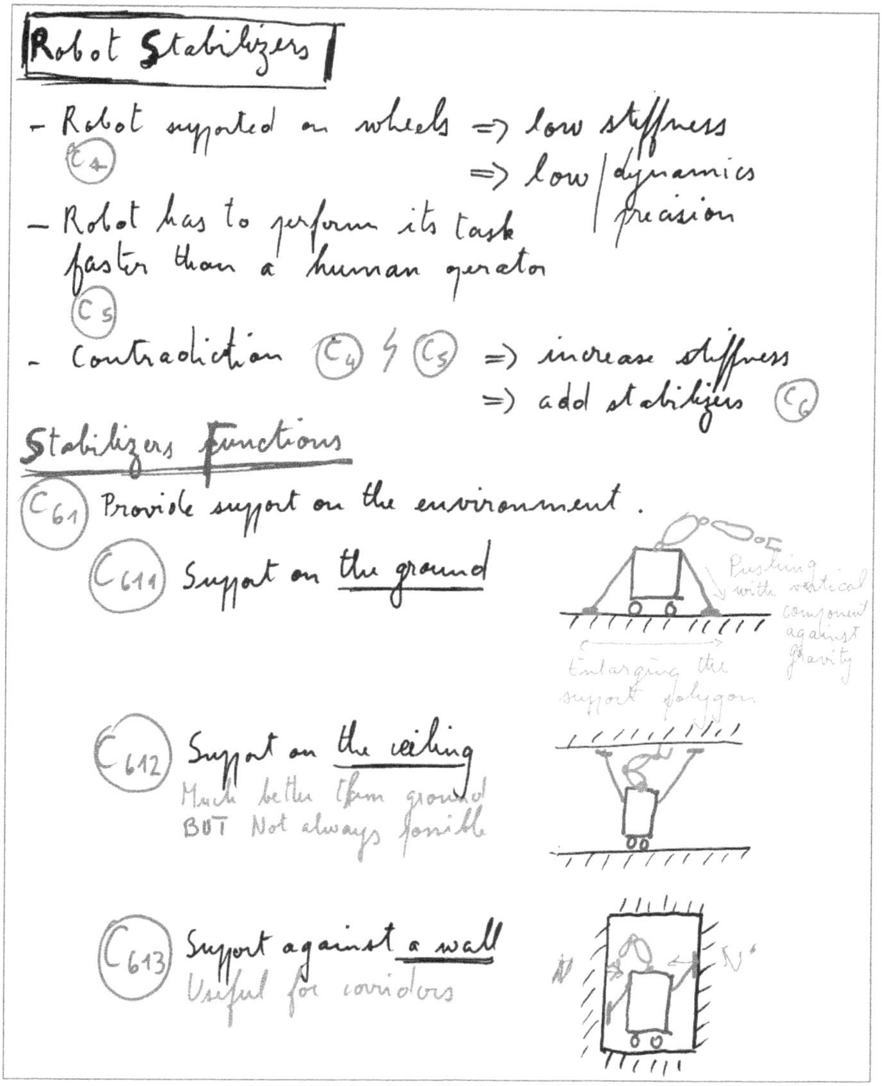

Figure 3.2. Brainstorming about stabilizer functions (Aachen meeting, 17/02/2016).

Obviously, pushing on the ceiling may also be forbidden in certain cases (weak ceiling, fake ceiling...) so this proposal for prototype V1 must be completed by other solutions for prototype V2.

3.1.4.2 More Advanced Stabilizers Concepts

In order to design suitable stabilizers, all the uses case had to be defined and specified. A brainstorming session in Aachen (Figure 3.2) permitted to classify the stabilizer functions.

3.1.4.3 Stabilizers Concepts for Cooperative Tasks

The considered cooperative tasks CT_i at the moment are the following:

- CT0. Several robots working separately in the same room with cable management
- CT1. Cancelling dynamic effect by opposite motions (arms or rotating tools)
- CT2. Manipulating large tools with dual / multiple arms
- CT3. Extending support polygon ? better stability
- CT4. Collaborative obstacle crossing
- CT5. Collaborative cable management
- CT6. Collaborative garbage manipulation
- CT7. Collaborative tool changing
- CT8. Multi-arm tasks: scratching / tearing / observing

The green scenarios are priority in this project. The other scenarios will be developed if enough time remains.

A proposition for CT1 and CT3 can be seen in Figure 3.3, where the stabilizers serve both for connecting two mono-robots to each other, thus forming a poly-robot, and to enlarge the supporting polygon of this poly-robot, which considerably enhances its stability without requiring any assumption on a strong ceiling above. Combined with the concept of dynamic balancing of antagonist arm motions, the resulting poly-robot may succeed in a considerable decrease of its tangential forces on the ground. The distance between the mono-robots should be calculated in order to avoid collision of the two respective arm workspaces without letting an uncleaned band between them. Similarly, the robots can be combined in a group of three for covering corners efficiently and increasing the support polygon in two directions at the same time.

An integrated proposition of a top module regrouping two stabilizing legs can be seen in Figure 3.4. The feet can be locked in suitable locking areas of a neighboring robotic unit for strong attachment between units. The legs can also be controlled for obstacle crossing, a requirement that is not covered at the moment by prototype V1. The axi-symmetric setting of the top module allows connections of the legs in any direction, either with another robot or with the environment.

3.1.5 Implementation of Testing Stabilizers for V1 Prototype

For the testing of first prototype, simple manual stabilizers will be implemented to have a back-up against unpredicted instability scenarios. In this report, initial ideas for implementing manual stabilizers are proposed.

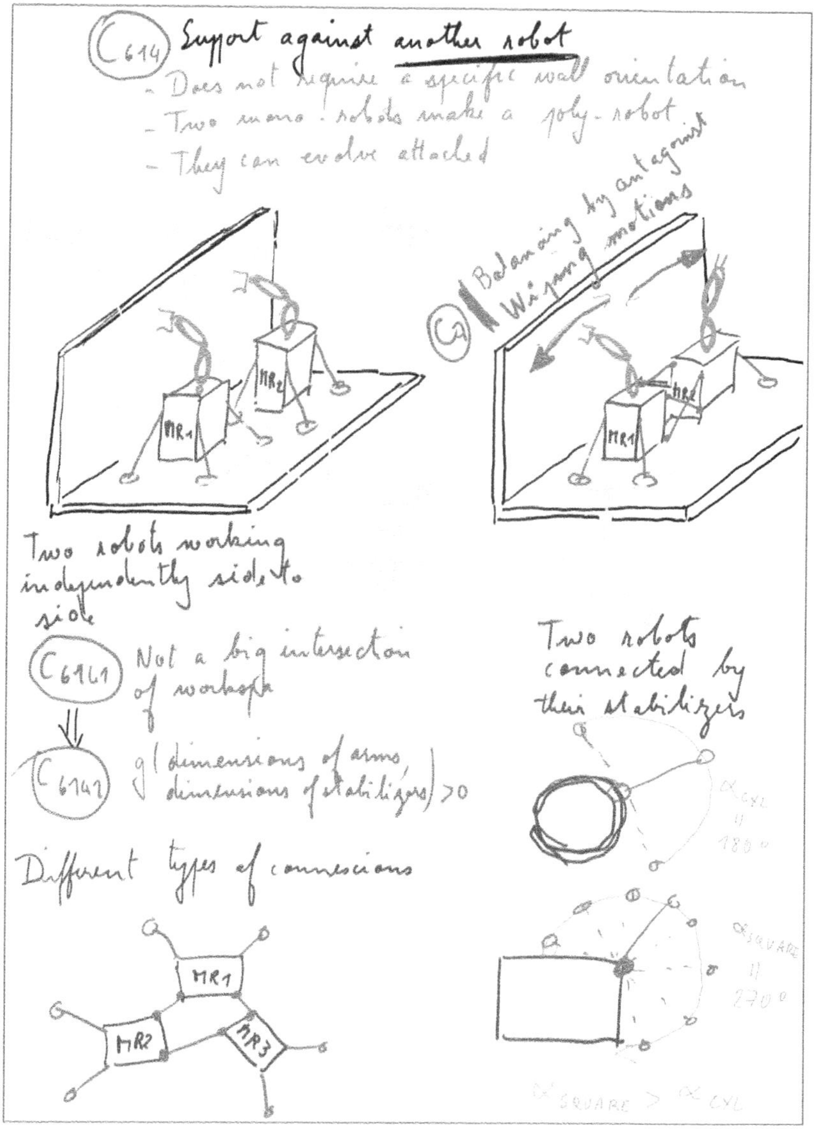

Figure 3.3. Concept of using stabilizers for cooperative tasks.

Requirements on Stabilizers

1. The first and the most important requirement is to increase the support polygon and hence the stability of the robotic unit.
2. Stabilizers should be capable of bearing the load of the robotic unit, as well as the forces/ torques disturbing the stability of the robotic unit.
3. Stabilizers pushing against the ceiling such that the mobile platform bears the pushing load are not allowed.

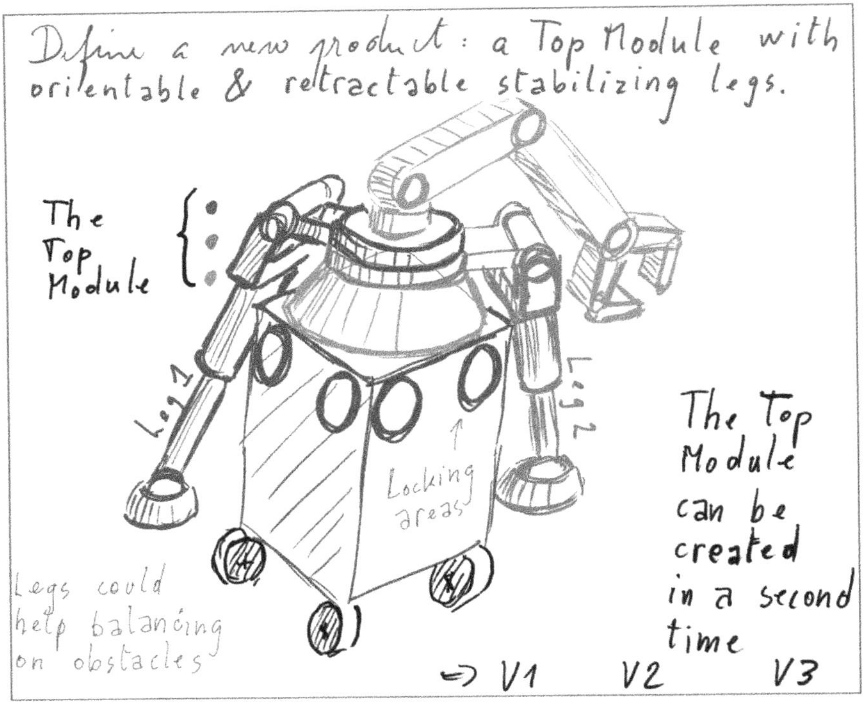

Figure 3.4. Concept of a top module regrouping the attachment points for two stabilizing legs and their power supply and control box. This module can be considered as an additional equipment.

4. After integrating with the robotic unit and in the retracted position, constraint on the overall size of the robotic unit should be respected.
5. Stabilizers should be as light as possible so as to follow the constraint on the overall weight of the robotic unit.
6. Stabilizers should not interfere with the operating robotic arm.
7. Stabilizers should not interfere into the positioning of the mobile platform w.r.t the wall.
8. Stabilizers should allow cleaning in the narrow corridors.
9. Stabilizers should be suitable for performing collaboration scenarios.

Keeping the above requirements in mind, few stabilizer architectures are proposed below.

Few key points on which the design of the stabilizers will be based are:

1. Position (attachment point) of the stabilizers on the robotic platform
2. Kinematic architecture of the stabilizer
3. Determining the resting points of stabilizer legs on the ground and ceiling

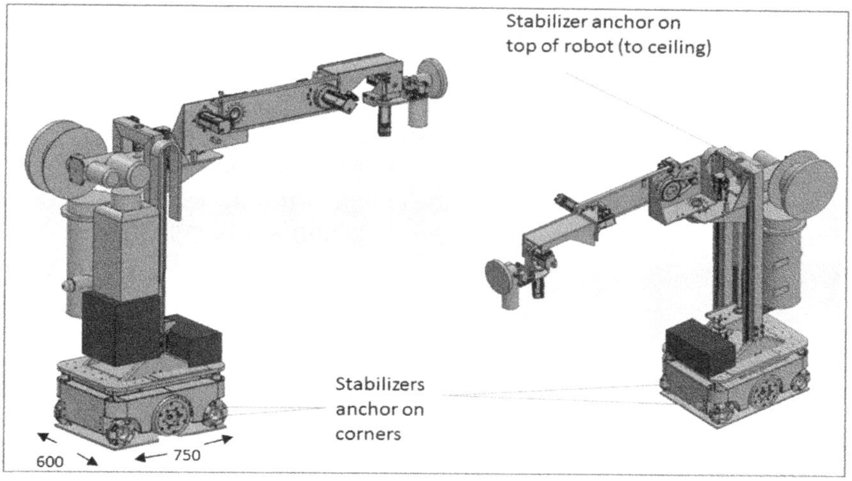

Figure 3.5. Attachments points for stabilizers.

3.1.5.1 Attachment Points of the Stabilizers on the Mobile Platform

In the current mock-up of the robotic unit, the attachment points are identified as above. There are in total 5 attachment points; one at each corner of the robotic unit plus one on the top plate of the vertical slider. However, as pushing on the mobile platform is not recommended, implementation of the vertical slider is doubtful. This allows us to put 4 stabilizers, one at each corner of the mobile platform.

3.1.5.2 Initial Proposition of Stabilizers

In this section, few ideas regarding the kinematic architecture of the stabilizer are proposed. The ideas are mostly concerning the implementation of the manual stabilizers.

Proposition 1: Out trigger Stabilizers

Figure 3.6. Example of outtrigger stabilizers.

- The kinematic architecture of the stabilizer consists of two prismatic joints. This simple architecture will serve as good option for the testing of first prototype.
- There is of course, a limit on the achievable extension of the horizontal prismatic joint which determines the increase in the support polygon.
- According to the dimensions of the mobile platform, we have a rectangular support polygon of the size approximately **600 mm × 750 mm**.

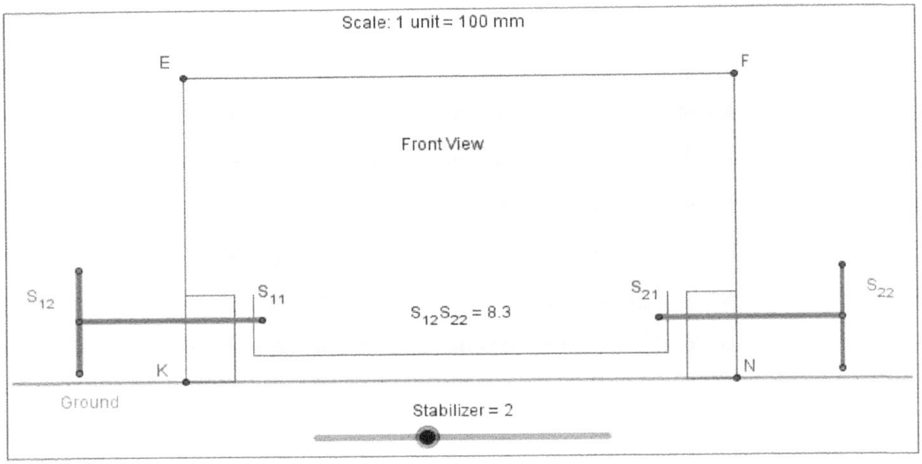

Figure 3.7. Extended pose of the stabilizers.

From Figure 3.8, it is clear that, for avoiding collision, the maximum retraction length on stabilizer is around 200 mm. This puts a constraint on the maximum length of extension (unless we design a telescopic structure). Distance $S_{12}S_{22}$ indicates the total distance between the two points on the stabilizers which is 830 mm approx. Thus, the width of the new support polygon is increased by 230 mm.

Proposition 2: Rotating Stabilizers
The kinematic architecture of this type of stabilizer is Revolute-Prismatic. The stabilizers are attached to the mobile platform using revolute joint; for e.g the stabilizer link $S_{11'}$–$S_{12'}$. At point $S_{12'}$, there is a prismatic joint perpendicular to the link to ground the stabilizer. The maximum length of the stabilizer links can be 600 mm. Links $S_{11'}$–$S_{12'}$ and $S_{32'}$-$S_{12'}$ define new support polygon.

Figure 3.10 shows the folded position of the stabilizer, thus satisfying the constraint on the overall size of the robotic unit. For avoiding interference in the folded

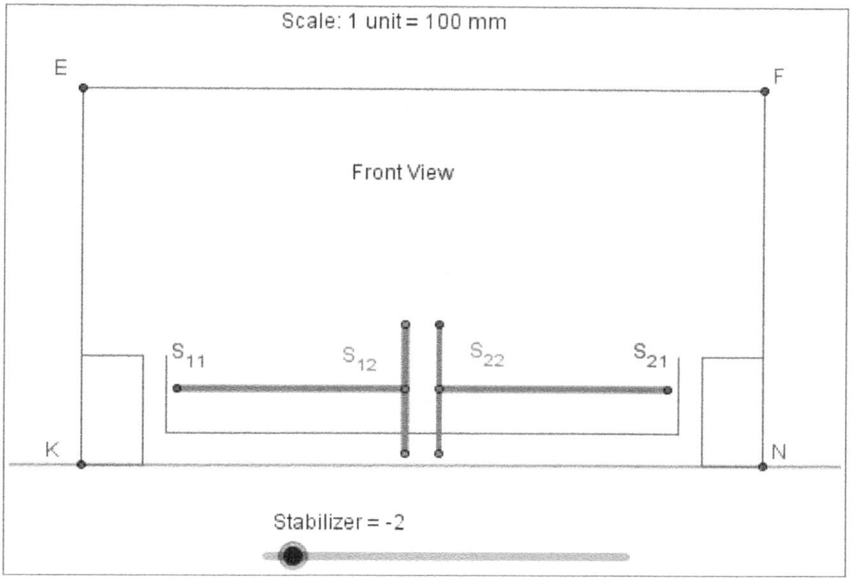

Figure 3.8. Retracted pose of the stabilizers.

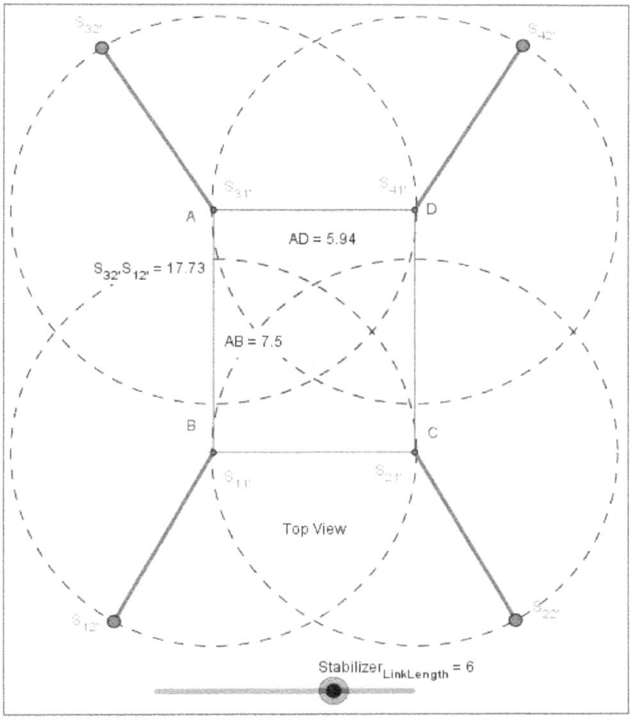

Figure 3.9. Stabilizers extended position.

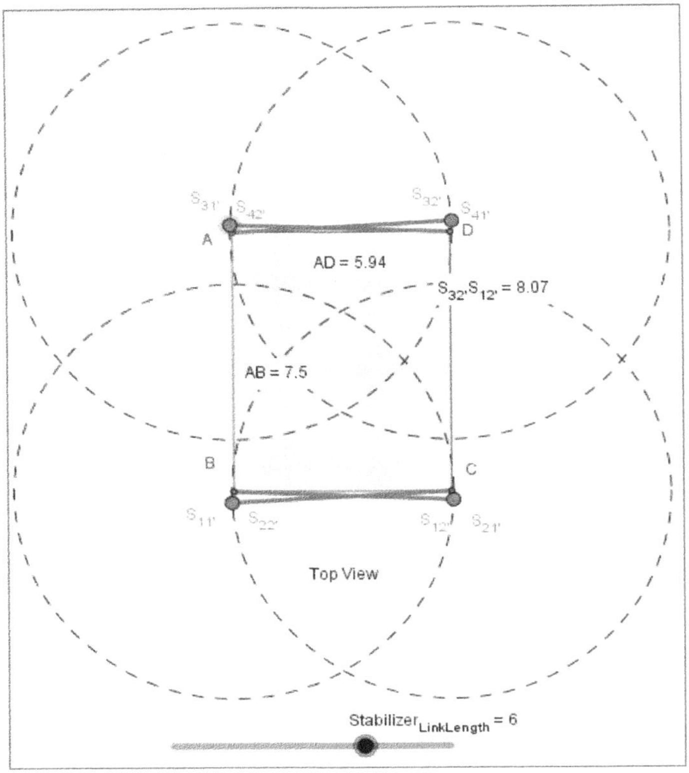

Figure 3.10. Stabilizers retracted position.

position i.e. to enable folding of stabilizers, the two front/rear stabilizers can be fixed at different altitudes on the two corners of the mobile platform. The long link lengths will allow more increase in the support polygon.

Proposition 3: 2-R Stabilizers

Here, the kinematics of the stabilizers is consisted of 2-R architecture as shown in Figure 3.11.

By considering the factor of safety as 1.3, the width of the extended support polygon will be 850 mm approx. i. e addition of 130 mm on both sides. The link lengths of the stabilizers are determined based on this requirement on reachability as 160 mm each.

3.1.5.3 Antagonist Pushing

The central vertical stabilizer installed attached to the top plate of the vertical slider will exert extra efforts on the mobile platform while pushing on the ceiling. This

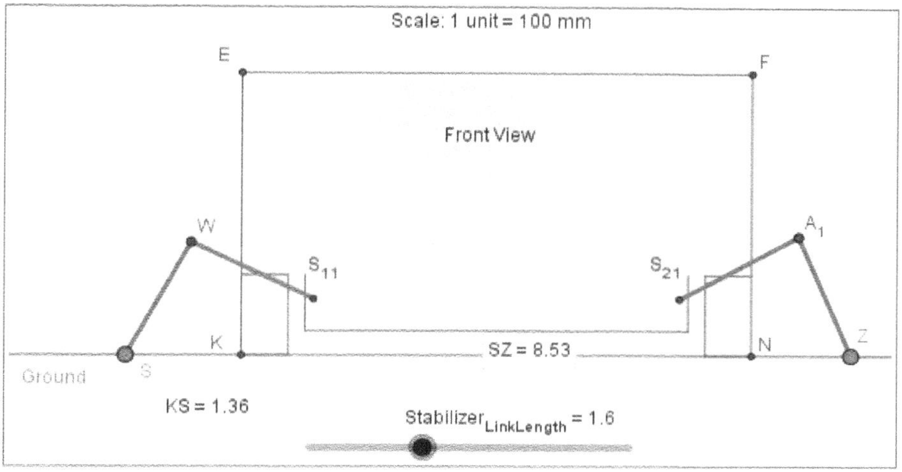

Figure 3.11. Stabilizers working position.

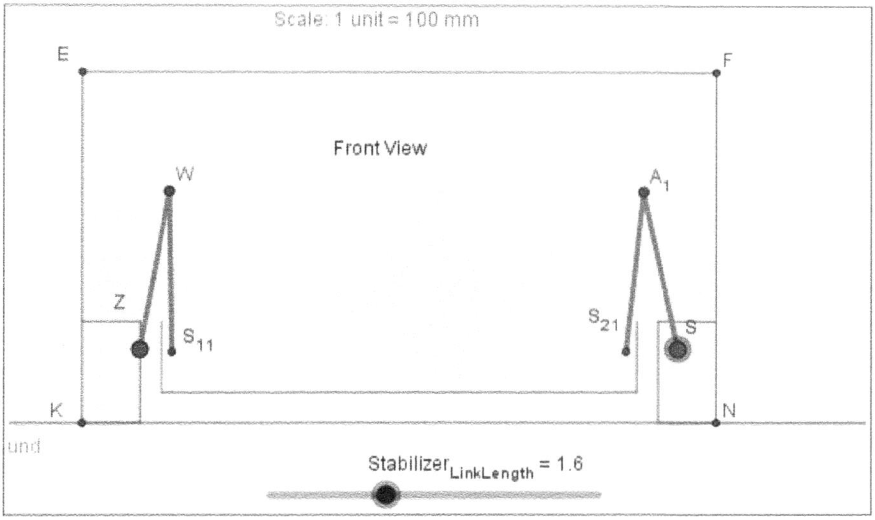

Figure 3.12. Stabilizers retracted position.

may damage the mobile platform as it is not designed for this extra effort. Thus to achieve the pushing against ceiling without putting extra load on the mobile platform, an idea of antagonist pushing mechanism in proposed. The two prismatic joints will act opposite to each other pushing on the ceiling and the ground simultaneously. Figures 3.13 and 3.14 show the pairs of antagonistic pushing links (Links: WE_1-WD_1 and A_1G_1-A_1F_1).

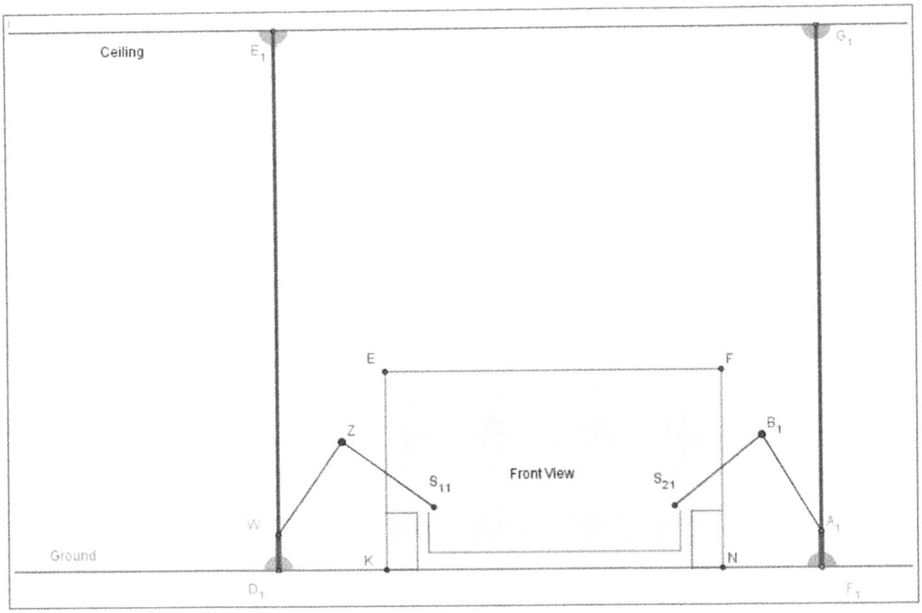

Figure 3.13. Antagonist pushing in 2-R architecture.

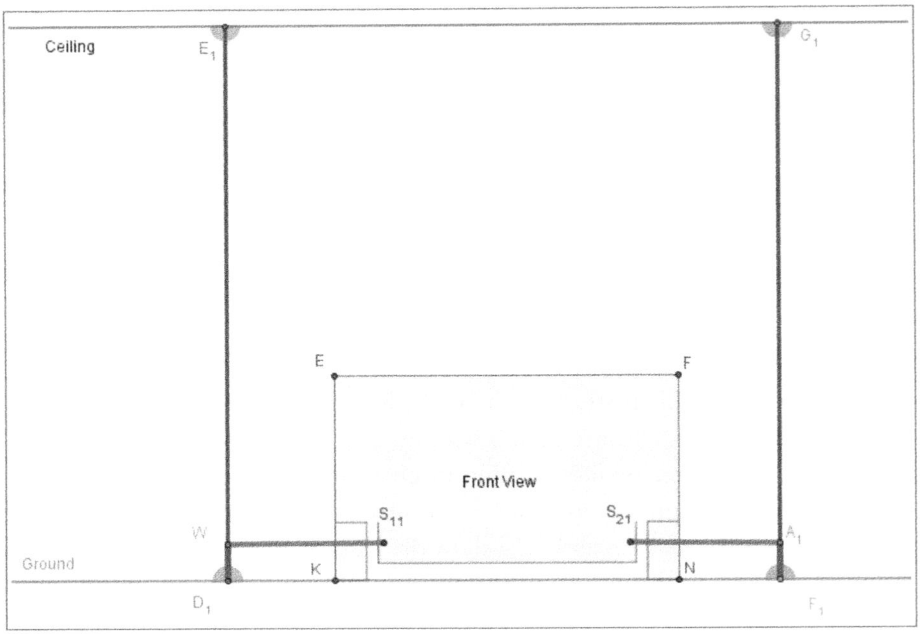

Figure 3.14. Antagonist pushing in outtrigger and rotating stabilizer architecture.

CAD Model of the Robotic unit equipped with stabilizers

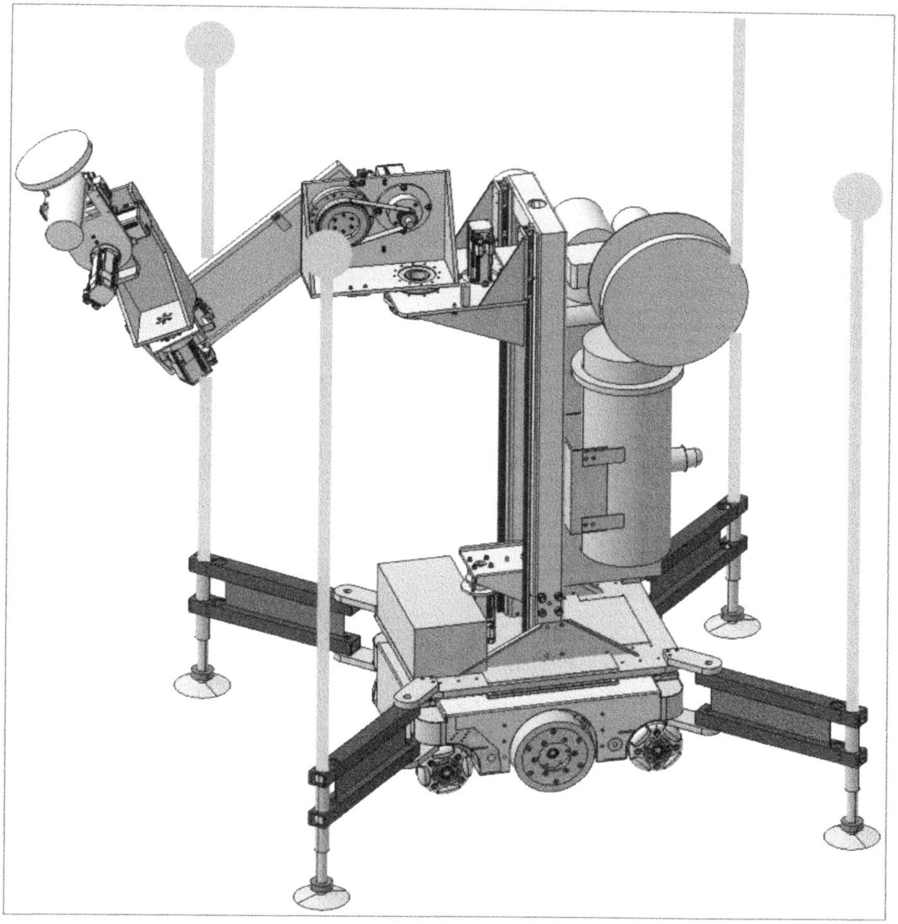

Figure 3.15. Antagonist pushing in rotating stabilizer architecture.

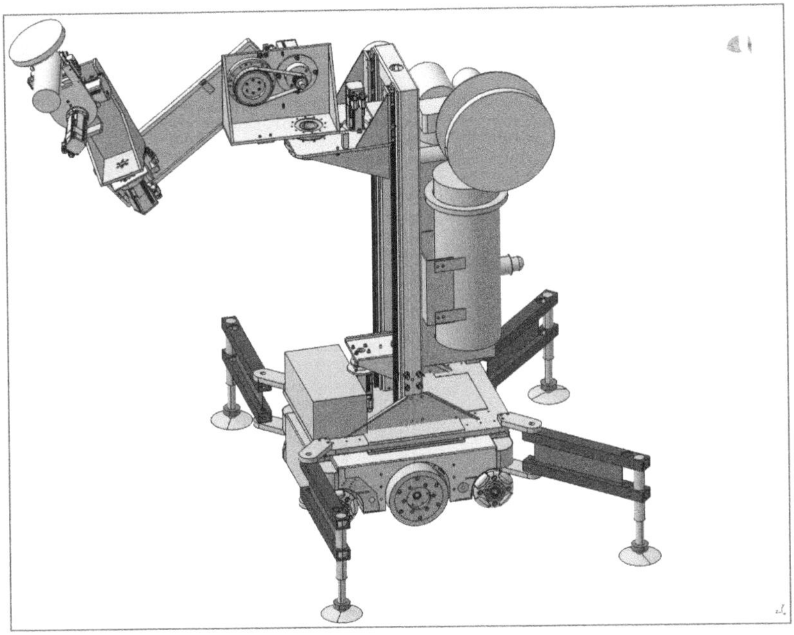

Figure 3.16. Flat alignment of front stabilizers for getting closer to the wall.

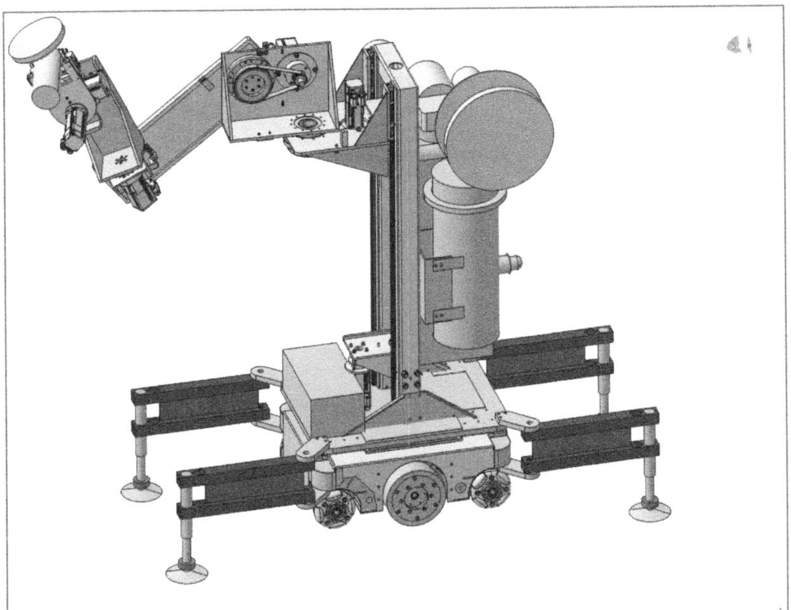

Figure 3.17. Alignment of stabilizers in narrow corridors for frontward cleaning.

3.1.5.4 Resting Points for Stabilizers

Resting points will define the contact between each stabilizer leg and ground. Resting points are nothing but contacts points between stabilizer legs and ground. They will define a new support polygon larger in dimensions than the one defined by wheel contact points of the robotic unit. This will give us the increased stability. Thus, the dimensional synthesis of the stabilizers is closely related to the assessment of stability as well as productivity of the robotic unit without stabilizers. Stability margin required in the worst case instability scenarios will lead us to the dimensional synthesis of the stabilizers. Similarly, stability margin expected in other unstable cases will determine the resting points for stabilizers. This will be determined through dynamic simulation of the cleaning process.

3.1.5.5 Collision Between Stabilizer legs and Robotic Arm

Stabilizers are deployed in the workspace of the robotic arm and thus there are occasions where robotic arm can collide with the antagonistic pushing links. This is realized using Geogebra models presented in Figure 3.18. This figure shows a top

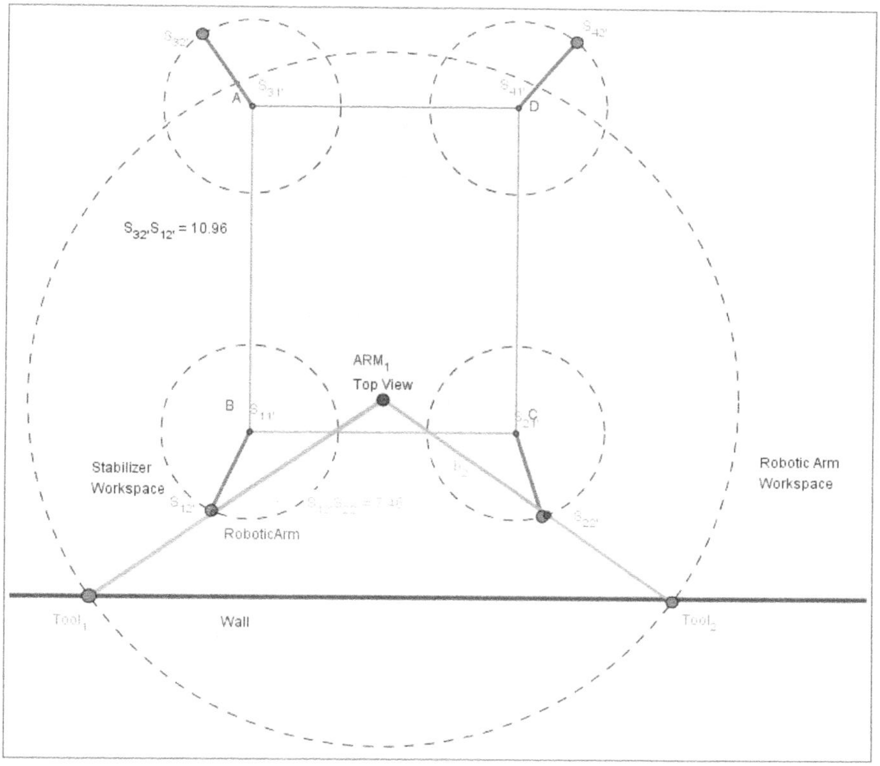

Figure 3.18. Collision between robotic arm and stabilizer.

view of the scenario where robotic arm is colliding with the antagonistic pair of the stabilizers. Thus, while determining the resting points for stabilizers, the scenarios of collision must also be considered.

3.1.5.6 Outlook and Future Work to be Carried Out

All these concepts require to be prepared and integrated into the CAD mock-up of prototype V2. They will be further developed during the next year, as this second prototype is only expected for month #30. They will strongly benefit from the experience and feedback to be expected from prototype V1, particularly for characterizing the dynamics, contacts and external forces of the real robot. The stabilizers will also serve to implement several collaborative scenarii listed in §3.3 within Task 2.4.

3.1.5.7 Summary and Conclusion

This deliverable is still under progress because all the dynamic data have to be confirmed by experimental tests. Several models for quasi-static stability have been constructed and provide encouraging results: overall, all our robotic units are stable within their complete workspace for slow motions, without any tool effort, for light terminal segments of the arm and without stabilizers. A proposition of a central stabilizer pushing against the ceiling has been made, which provides a good stabilization and transforms the unilateral contacts of the wheels on the ground into much safer bi-lateral contacts, because the robot cannot lift its wheels any more. However, because of limitations on the load bearing capacity of the platform, the real implementation of this system will be made into prototype V2. The proposition for V1 can be seen in figure 16, with four stabilizers providing antagonistic pushing against ground-ceiling.

Future work in task T2.2 will cover the structural and dimensional synthesis of advanced stabilizing legs serving both for stabilization and multi-robot interconnection for five collaborative scenarii. This is compatible with the expected planning for prototype V2 (30 months after the project beginning).

References

[Geogebra, 2016] Geogebra, the graphing calculator for functions, geometry, algebra, calculus, statistics and 3D maths. https://www.geogebra.org. Retrieved online in 2016.

[Bouton *et al.*, 2007] N. Bouton, R. Lenain, B. Thuilot, and J. Fauroux, "A rollover indicator based on the prediction of the load transfer in presence of sliding: application to an all terrain vehicle," in Robotics and Automation, 2007 IEEE International Conference on. IEEE, 2007, pp. 1158–1163.

[Chebab, 2013] Zine Elabidine Chebab, "Modelling and optimization of a climbing process", Master 2 report, Pierre and Marie Curie University, UPMC, Paris VI. Supervisor: J.C. Fauroux. 60p. 2013. on. IEEE, 2006, pp. 3405–3410.

[Garcia *et al.*, 2002] E. Garcia, J. Estremera, and P. G. De Santos, "A comparative study of stability margins for walking machines," Robotica, vol. 20, no. 6, pp. 595–606, 2002.

[McGhee and Frank, 1968] R. B. McGhee and A. Frank, "On the stability properties of quadruped creeping gaits," Mathematical Biosciences, vol. 3, pp. 331–351, 1968.

[Prause *et al.*, 2015] Isabel Prause, Jean-Christophe Fauroux, Mathias Hüsing and Burkhard Corves. "Using Geometry Sketchers and CAD Tools for Mechanism Synthesis", paper OS3-032, in Proc. of IFToMM 2015, The 14th World Congress in Mechanism and Machine Science, October 25–30, 2015, Taipei International Convention Center, Taiwan, 11p.

[Sardain and Bessonnet, 2004] P. Sardain and G. Bessonnet, "Forces acting on a biped robot. center of pressure-zero moment point," Systems, Man and Cybernetics, Part A: Systems and Humans, IEEE Transactions on, vol. 34, no. 5, pp. 630–637, 2004.

Deliverable Di3.2
Di3.2 – Specification of the technical Requirements for the V1 prototype
Part B – System Specification

Authors: Daniel Martin

3.1.6 Deliverable in the Context of the Work Program

The present *Di3.2 – part B – System Specification* reports the technical specifications for the V1 prototype carried out by the BOTS2REC Consortium during the first months of the Project. The system specifications are detailed in the following sections of the document and complemented with the control architecture definition and the identification of interfaces included in the deliverable *Di3.1 – Specification of the system architecture and interfaces.*

At the project start the specification of the asbestos-removal-tasks and technical requirements, initial trials with a robotic mock-up and preliminary simulations will lead to the design and realization of a first version (V1) of the robotic system with a single robotic unit. Using the V1 version of the robotic system, the automated removal of asbestos contamination (operational process) and the according process control system will be developed in several iterations in a testing rehabilitation site (without contamination). During regular workshops, the system will be tested and the central process control system and all subsystems will be advanced by implementing more ambitious functionalities and considering the "lessons learned" and updated requirements. The ongoing developments and updated technical requirements will lead to the design and realization of a second version (V2) of the robotic system with 2 robotic units. The V2 version of the robotic system will be the final prototype used for further developments and the final demonstrations and benchmarking. Using the V2 version of the robotic system, the operational process and the according process control system will be finalized to fulfil all requirements for the robotics use case.

In this report, a precise definition of the technical specifications of the BOTS2REC V1 system is illustrated in order to cover all system development aspects including the BOTS2REC V1 Robotic Unit and V1 User Interface, etc. This task is of great importance as it sets the guidelines what need to be adhered to throughout the project. An architecture and functionality of the system to be developed is provided in detail.

This document is the result of the system specifications and architecture definition phase. However, this is a planning document and therefore the final implementation may not coincide with the architecture described in version 1 of the deliverable Di3.2. Future versions will be released to reflect the up-to-date state of the implemented architecture.

3.1.7 Methods and Work Carried Out

The system specification has been conducted from the collection of partial information of different components provided by the relevant partners involved in its development, and completing it from the point of view of the integrator for a complete and reliable definition of the system.

Starting from conceptual information and/or preliminary, periodically discussion has been organized during this first phase of the project in order to evolve the system V1 concept. This information has been supplemented and detailed based on the progress of discussions and analysis done in between.

Different drafts of the specification have succeeded each until reaching this deliverable describing the system completely.

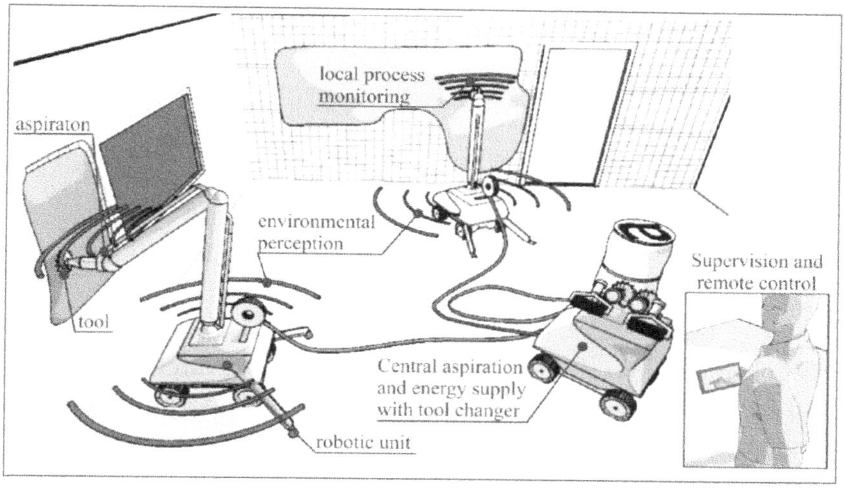

Figure 3.19. Sketch of site with two robotic units during the automated removal of asbestos contamination.

From the point of view of system integration, it has carried out the systematic gathering of information of the elements that forms the V1 system provided by the partners and the information that resulted during different discussions. With this information it has been continuously identified the main open topics to be discussed at subsequent meetings and have been completed those aspects that are part of the whole system and do not apply to a specific component.

In addition, regular discussions have been going on in parallel with the definition of requirements, a fact that has enabled the continuously compliance assessment of these requirements or revise those that could not realistically be met.

3.1.8 Specification of the Technical Requirements for the V1 Prototype – System Specification

3.1.8.1 Introduction: The BOTS2REC System

The proposed robotic system (see Figure 3.19) used for the automated removal of asbestos contamination will consist of multiple mobile robotic units that perform the asbestos-removal-tasks autonomously.

Each unit consists of a mobile platform and robotic arm with different abrasive tool. The combination of optical and radar sensor systems will allow the environmental perception and local monitoring of the asbestos-removal-tasks. Supported by a user interface, the operator can select different areas on a virtual representation of the rehabilitation site and assign asbestos-removal tasks. Based on the user input, the Central process control system plans the task allocation and collaboration between the mobile units and their movements and trajectories autonomously.

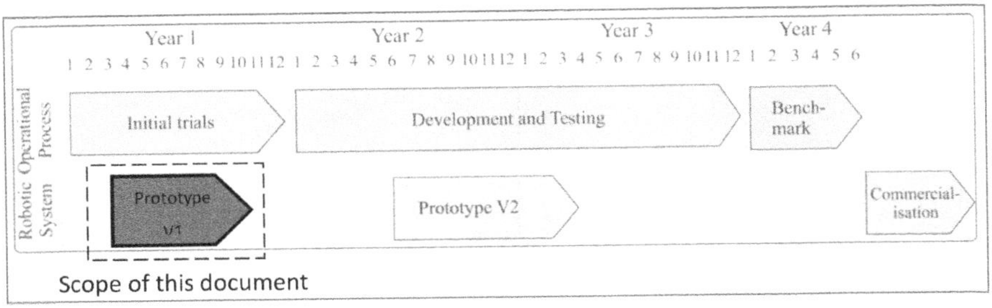

Figure 3.20. Timeline for the overall approach and methodology.

The user interface allows the permanent supervision of the automated process and optional remote control. All robotic units are connected to an aspiration system and energy supply to remove the waste from the rehabilitation site and supply sufficient energy for the robotic units.

3.1.8.1.1 The project phases

Based on the general requirements of the construction and demolition industry and the specific requirements of the robotics use case described above, an operational process and robotic system for the automated removal of asbestos contamination is being developed and implemented in different subsequent phases, as show in Figure 3.20.

At the project start the specification of the asbestos-removal-tasks and technical requirements, initial trials with a robotic mock-up and preliminary simulations will lead to the design and realization of a first version (V1) of the robotic system with a single robotic unit, which specification is the main scope of this document. In general terms, the V1 prototype will be an experimental platform, used by the consortium during regular workshops in a testing rehabilitation site, to test and improve step by step the system and its functionalities forecasted for the project, which will be joined and implemented in the V2 prototype. The ongoing developments and updated technical requirements will lead to the design and realization of a second version (V2) of the robotic system with 1 or 2 robotic units. The V2 version of the robotic system will be the final prototype used for further developments and the final demonstrations and benchmarking, performed under real world conditions.

3.1.9 The Robotic System V1

The focus of the **Robotic System V1** is placed on developing **a single robotic unit**, composed mainly by a mobile platform and a robotic arm. This V1 of the robotic system will not include the central unit, therefore, the support devices (such aspiration) will be included or attached to the single robotic unit.

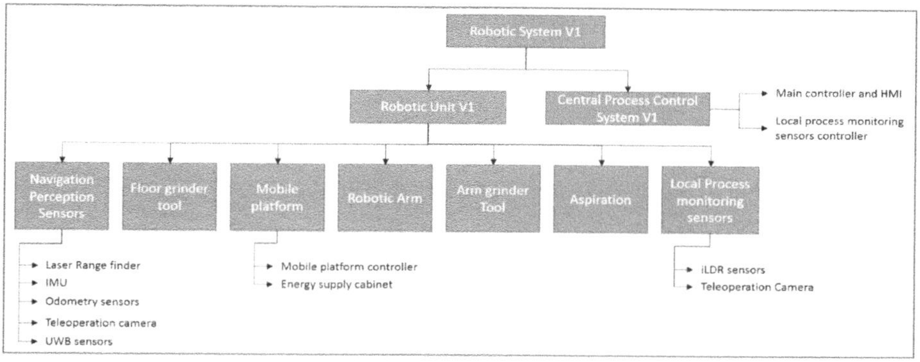

Figure 3.21. General system architecture.

The Figure 3.21 shows the main devices that compose the Robotic System V1:

- 1 Robotic Unit (V1)
- Simplified Central Process Control System

And accordingly, the Robotic Unit V1 will become an experimental platform with all of the components required for the asbestos removal tasks. The main components of this unit are:

- Mobile Platform
- Aspiration
- Supply cable automated reel
- Robotic Arm
- Tool for wall and ceiling grind
- Tool for floor grind
- Process Monitoring Sensors (PMS)
- Navigation and Perception Sensors (NPS)

The detailed Specification of the Robotic Unit and its components is included within this section 3.4 of this document and its related sub-sections.

3.1.10 The Robotized Asbestos Removal Process

3.1.10.1 The robotized procedure

The scenario of the asbestos removal tasks done by human workers allows to understanding how a robotic-system can replace human workers. Considering also that not all tasks the human workers do currently can be automated, for instance the seclusion or the cleaning, three successive periods or phases can be determined in the robotic-system process, as reported in Deliverable *D1.1 Report with a detailed specifications of the asbestos-removal tasks and possible tools* and complemented in the current specification document.

(a) Preparing asbestos removal (done manually, before robotized removal process)
   - Identification of asbestos (from D1.1)
   - Estimation of the quantities and the spotting (from D1.1)
   - Estimation of the dust level (in comparison with test-sites) (from D1.1)
   - Adapted seclusion (regarding the estimated dust level and the spotting) (from D1.1)

     Optionally the seclusion will be adapted by adding an additional (soft) layer under the foil in order to reduce the damage of the foil from the robot wheels.
   - Installation and test of the ventilation system (suction and smoke test) (from D1.1)
   - Robot assembly (platform, arm, aspiration, tools, ...)
   - Plug the robot to the external supply generator (placed on the same floor of the worksite)
   - The standard operation would be to start creating a map of the flat. This can be done very fast in manual mode. There is no need to autonomously create a map.
   - The operator introduces the removal jobs into the system through the HMI interface

(b) Removing asbestos with the robotized system
   - The robot starts the removing task autonomously or teleoperated
     - When the bag is full, it is cleaned in the decontamination lock for the waste and put in a cleaned bag that will be put in a big bag with a sign. This operation will be done manually by the operator.
     - If required any tool change or replace any part of it (such disks, etc.), the operator will stop the robot and perform the required action. Then he/she will resume the robotic system operation.
     - If there's a failure of the robotic system, the system will stop and the operator will come to solve the failure. Then he/she will resume the robotic system operation.
   - Materials incorporated in hard materials shall not be removed by the designated robotic system: asbestos cement, structural panels, pipes

(c) After asbestos removal (done manually)

- Visual inspection (from D1.1)
- Measurement of the dust level (from D1.1)
- Removal of the seclusion (plastic only used once) (from D1.1)
- Decontamination of the not renewable parts of the robotic unit and removal of renewable gaskets.
- Confinement of the robotic unit into the container solution.

3.1.10.2 Removal process conditions

3.1.10.2.1 Asbestos grinding process parameters

Regarding the removal process parameters, the system specifications collected within this document are based on preliminary inputs. Therefore, the final V1 specifications will be updated after initial trials done with a mockup robotic system (T1.2 and the resulting deliverable *Di1.1 – Technical report with protocols of the trials and specification of the process parameters*) providing information about parameters for a robot guided use of the tools, such normal and lateral forces, optimal process speed, etc.

- The thickness of the product that needs to be removed is between less than a 1mm (for cement rendering for instance) till 2cm (for ceramic tiles for example).
- As required for V1 system, the robot has to be capable of grind walls, ceiling and the floor.
- Taking into account that the same kind of tool will be used for grinding walls/ceiling and the ground, a 80N – 100N normal forces will be considered for the arm grinding tool, as well as for the floor grinding tool.
- About the disc position, the unit will be able to place the tool/disc in different positions against the wall, according to:
  - Disc inclined for rough operation, in firsts steps of grinding process
  - Disc completely parallel to the wall for fine operation, in finishing steps.

3.1.10.2.2 Environmental conditions

Joined to the *D1.1 – Report with a detailed specifications of the asbestos-removal tasks and possible tools* there is a table that shows the different products containing asbestos that this project will cover (Annex "Classification"). This table has been established based on the tool of INRS (Institut National de Recherche et Sécurité / National Institut of Research and Safety), and provides an orientative dust level regarding the material and tool used for the removing task.

In the project, it will be assure the level 2 for the dust maximally:

- Level 1: less than 100 f/l
- Level 2: between 100 f/l and less than 6.000 f/l

The aspiration system of the robotic system (see section 3.4.4) will respect this limit.

On the other hand, in order to keep the level of asbestos particles in the air as low as possible, it might become 's necessary to spray a mixing of water and soap (or surfactant) in order to keep the level of asbestos particles in the air as low as possible in the case that aspiration is not able to maintain the proper levels of dust. In that case, the next specification will be add to taken into . Taking this into account:

- Removal has to be done in the wet environment (from D1.1)
- Robot systems has to withstand water (from D1.1)
- Robotic Unit will needs to have a spraying unit (from D1.1)

Other important aspect regarding the environment where the robot will be operating is the fact that the floor will be covered, in most cases, by a common plastic of a 150 μm of thickness to protect the building ground. This issue has to be also taken into account combined with the wet environment.

3.1.10.2.3 Operation in corridors

According to the requirements stated in Di3.2 part A, the system must be capable of to operate in corridors of the 700mm of width. This fact, together with other restrictions related control and sensors precision entails that the envelope of the robot must not exceed 600mm of width to operate properly.

In order to be able to turn in corridors of that dimensions, a noticeable reduction of the robotic unit dimensions would be needed. This would hinder making that not able to place the placing of all the required components within the allowed space of the platform. Nonetheless, considering that the situation of 2 corridors of 70cm is not frequent, it will be considered corridors of 700mm next to other of 800mm for the design, which are the most frequent (80-90%). Additionally, in small corridors or places with difficulties for autonomous navigations the system will be teleoperated.

Table 1 contains the results of the tests performed in different bent corridors. It is indicated if the platform could turn in the corridors with the complete robotic unit V1 system, or with a simplified robotic unit with less protruding elements. More details are also provided in the deliverable *D2.1 Preliminary simulations of the robotic units.* In the table below there are the conclusion from that study.

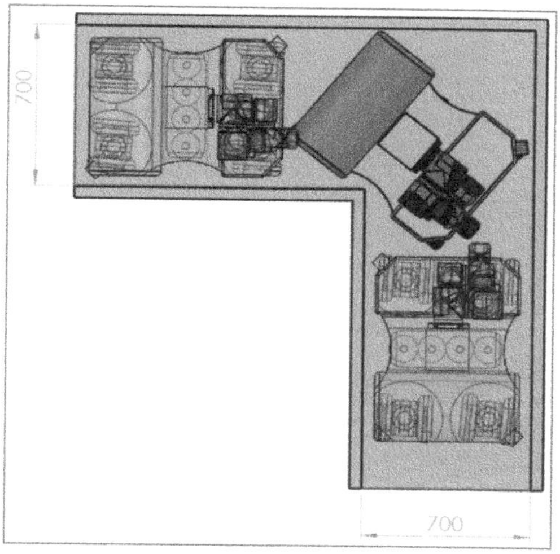

Figure 3.22. Operation in corridors.

Bent corridor Crossing	With robotic unit V1	With simplified robotic unit V1
700x700 corridor	Not able	Able
700x800 corridor	Not able	Able
700x900 corridor	Able	–
800x800	Able	–

3.1.10.2.4 Automated detection of asbestos

For the V1 of the system, the identification of the asbestos contaminated areas will be done manually by visual inspection, as indicated in section 3.3.1.

As well as the initial identification, the verification of the asbestos contamination removal will be done by visual inspection by the operator, and reprogram the system if additional operations are needed.

The automated detection of asbestos by computer vision or other technologies will be analyzed for system V2 even that it is out of the scope of the project.

3.1.10.2.5 Check of disc abrade

It will be done manually, by visual inspection, as done in the manual operation.

3.1.10.2.6 3D scanning of the worksite

For V1, a simplified 3D representation of the worksite will be generated (i.e. by adding the height coordinateds…) after 2D mapping performed by the moving the MP around. On board photogrammetry technologies will be investigated during the V1 testing phase.

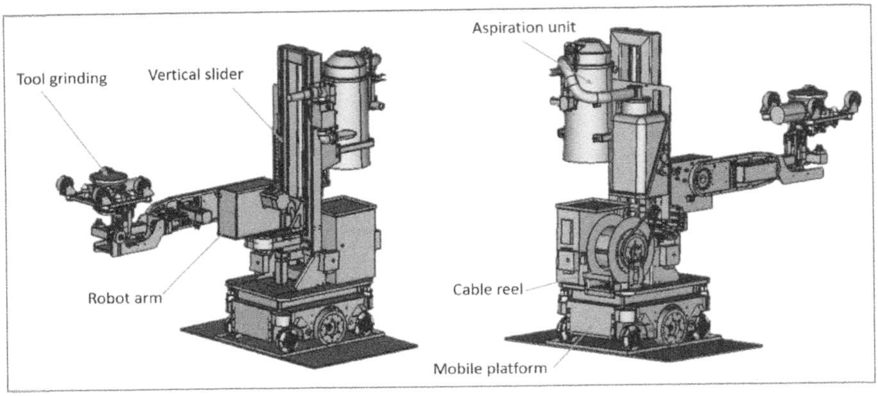

Figure 3.23. Robotic Unit V1 CAD.

For V2 it will be considered to rent a professional 3D vision system. Other technologies, such as on board photogrammetry will be considered in order to simplify and reduce system costs.

3.1.11 Robotic Unit V1 Technical Specification

The overall architecture of the Robotic Unit V1 (RU-V1) is mainly composed of a Mobile Platform (MP), a Robotic Manipulator Arm (RMA) with a single attached Process Tool (PT), including two groups of sensors: Navigation and Perception Sensors (NPS), and Process Monitoring Sensors (PMS). Figure 3.23 shows the main components of the RU-V1. Attached to this document there has been included a table which details the different components and devices that conform the unit V1 and summarizes its main specifications.

The V1 system doesn't consider the use of the Central Supply Unit (CSU) to support the RU-V1. Consequently, in this first version the aspiration unit will be on board of the mobile platform avoiding the need of an automatic pipe reel. Nonetheless, the unit will require an active cable reel for the energy supply and communications, which is described in more detail in section 3.4.8.

As standard procedure to reduce the dust, the environment might be constantly kept wet with a water sprayer. Consequently the robot and ground can be exposed to water constantly. For these reason, the robot will be dustproof and a certain level of waterproof.

3.1.11.1 Mobile Platform

The mobile platform of the robotic units will be based on existing and proven mobile platforms from Robotnik Automation. The platform design will be adapted for the versatile use on the rehabilitation site.

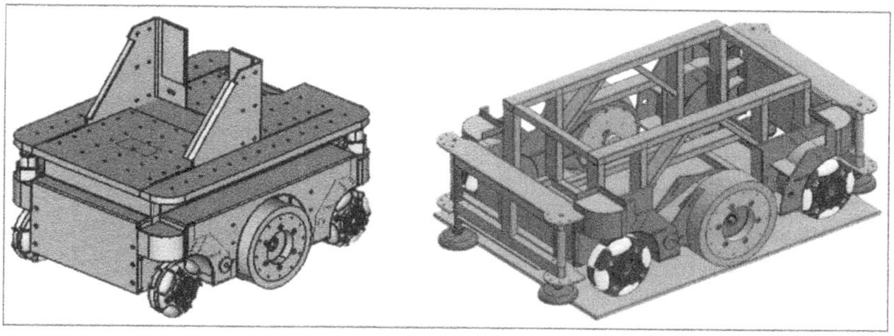

Figure 3.24. Mobile Platform CAD model (left) and tubular structure (right).

The table below includes the mobile platform main specifications:

Type/Model	Robotnik Omnidirectional with 4 wheels swerve drivee
Ground Control software	ROS
Dimensions	Width: 581mm Length: 738mmm Height: 350mm
Weight	95 kg
Payload Weight Capacity	230 Kg
Operational range/altitude	3 m
Safety mechanism	2x Onboard Emergency Stop button 1 x Remote stop 2 x Non-Safety Laser Scanner (There is a protocol that ensures robot is not moving near workers)
Payload Power Capacity	80A/48V(peak) , onboard DC/DC converters will provide different voltages
Power Type	Lithium LiFePO4 batteries 48V 15 Ah 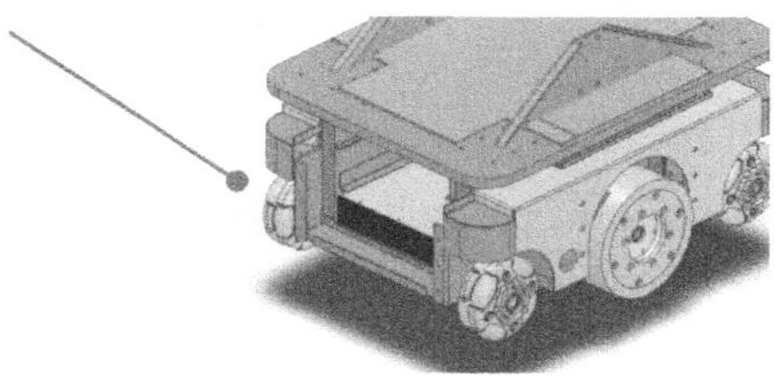
Autonomy	2 hours using only mobile platform,1, 1 hour using arm. Work is done connected to AC
Charger	input 230VAC; Charging time 1 hour (onboard charger)

Locomotion system	Differential with 2 traction wheels + 4 castors (dual wheel or omniwheel), communication with Control computer using CANOPEN
Maximum speed	1.7 m/s (will be limited)
Minimum turn radius	0, will be able to turn on the spot
Point of contact with the floor	4 omni-wheels in the 4 corners of the footprint, 2 rubber-band wheel in the lateral side
Crossing obstacles capability	Step: 2 cm Ditch: 10 cm
Materials of main parts	Structural parts: Aluminum and Steel Covers: aluminium Wheels: Rubber
Shapes and surfaces geometry	preferably flat or convex to facilitate cleaning
Surfaces processing/finishing	painted or untreated raw
IP protection grade	IP65
Attachment of the arm	The platform will include a vertical slider so the robot arm can be lift to reach a workspace of 3m height. Attachment will be defined using standard ISO screws and mounting flange (attachment between slider and MP will not be ISO flange)

The mobile platform will incorporate different brakes in order to block the platform motion when the emergency stop is triggered. A manual unblocking (brakes off) of the robotic unit must be possible in order to move the robotic unit by pulling/pushing.

3.1.11.1.1 Mobile platform control

The mobile platform controller will be an Intel NUC (Intel i5 processor, 16 GB RAM, SDD disc). More details of the mobile platform control architecture can be found in section 3.4.9 of this document.

3.1.11.1.2 Stabilizers

The first prototype will help to study the tool forces effects in vibration and stability off the platform. Into the mobile platform there will be attachments for passive stabilizers, which will be placed by hand in different configurations to test the results

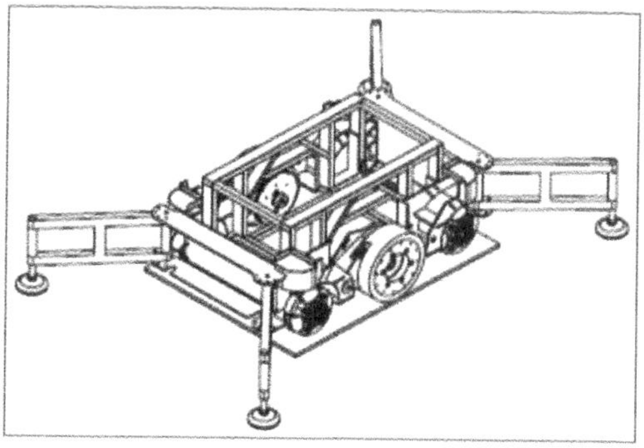

Figure 3.25. Manual stabilizers.

of simulations. After that mechatronic Stabilizers will be designed (if required) for V2 prototype. A support structure to place a stabilizer from the platform to the ceiling will be also forecasted.

In order to provide information to the control system of the presence of this element (i.e. to allow/restirict wider movement that may introduce instability if stabilizers are not present) it will be introduced some limit switch on the stabilizers.

3.1.11.2 Robotic Arm

In order to move the tool to process the different surfaces it will be necessary to place an automated manipulation solution. For the system V1 the manipulation solution will be based on a customized robotic 6DoF arm attached to a Vertical Slider system.

This robotic arm is built with brushless motors with their drivers. Further information for the last update of the robotic arm on Work Package 5 (WP5).

The table below includes the robotic arm main specifications:

robot arm	33 kg
robot shoulder	21 kg
reach	1605mm
payload	18 kg (21 kg with passive compliance)

		Voltage	Nominal power	Nominal speed
Link 1	Base roll	48 V	0,12 kW	3000 rpm
Link 2	SH-pitch	48 V	0,58 kW	3000 rpm
Link 3	EL-pitch	48 V	0,34 kW	3000 rpm
Link 4	w-roll	48 V	0,17 kW	3000 rpm
Link 5	w-pitch	48 V	0,17 kW	3000 rpm
Link 6	w-yaw	48 V	0,17 kW	3000 rpm

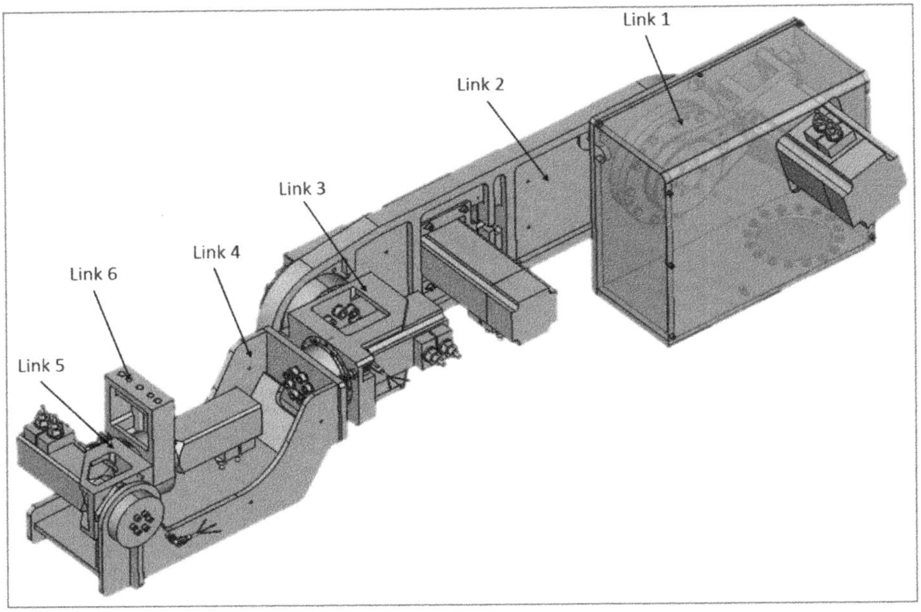

Figure 3.26. Robotic Arm Model.

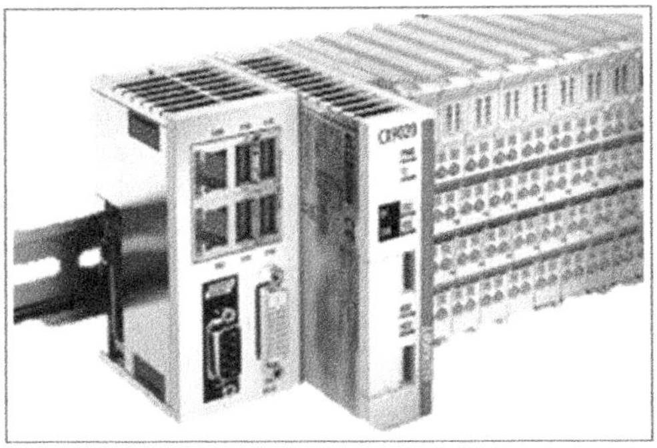

Figure 3.27. Robotic Arm Controller.

3.1.11.2.1 Robotic arm Control

The robotic arm control will be based on a distributed process based on 2 different CPU:

- A Beckhoff CPU CX9020 with 10 I/O attached cards will be the low level controller part for the robotic arm.
- An Intel NUC "PC" to perform the high level control: path planning, task management, etc.

Processor	ARM Cortex^{TM}-A8, 1 GHz
Flash memory	256 MB MicroSD (optional expandable), 2 x microSD card slot
Internal main memory	1 GB DDR3 RAM
Persistent memory	128 kB NOVRAM integrated
UPS	1-second-UPS integrated (optional)
Interfaces	2 x RJ 45 (Ethernet, internal Switch), 10/100 Mbit/s, DVI-D, 4 x USB 2.0, 2 x optional interface
Diagnostics LED	1 x power, 1 x TC status, 2 X bus status, 2 x flash access
Clock	Internal clock with battery backup for time and date (battery replaceable)
Operating system	Microsoft Windows Embedded Compact 7
Control software	TwinCAT 2 PLC runtime or TwinCAT 2 PLC NC PTP runtime
Power supply	24 V DC (-15%/+20%)
Current supply	I/O terminals 2 A
Max. Power loss	5 W (including system interfaces)
Dielectric strength	500 V_eff (supply/internal electronics)
Dimensions (W x H x D)	84 mm x 99 mm x 91 mm
Weight	approx. 590 g
Operating/storage temperature	-25 °C ... +60 °C / -40 °C ... +85 °C
Relative humidity	95 % no condensation
Vibration/shock resistant	Conforms to EN 60068-2-6 / EN 60068-2-27/29
EMC immunity/emission	Conforms to EN 61000-6-2 / EN 61000-6-4
Protection class	IP 20

3.1.11.2.2 Vertical Slider

In order to place the industrial arm in different height positions, an automated vertical slider will be implemented and adapted to be attached to the mobile platform and able to be controlled trough the robot control I/O.

weight	73 kg (including robot yaw motor)
dimensions	1440x300x86
stroke	890 mm
Supply voltage	48 V
Power	0,150 kW
Nominal speed	2000 rpm

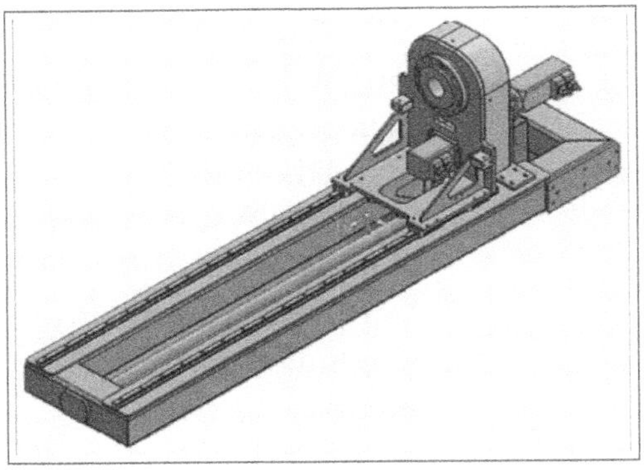

Figure 3.28. Vertical slider.

The slider will have the enough height in order to allow the robotic unit reach the ceiling height requirement at the same time to be able to meet the height limitation of the robotic unit.

3.1.11.3 Tools

The V1 robotic unit will be equipped with 1 tool, to be placed in different locations, in order to perform the required asbestos removal task for this previous version. Taking into account that the V1 system will be used as a test platform for further developments, it has been only required as mandatory that the unit will grind walls, ground and ceiling of the asbestos site, leaving the removal of tiles and other materials for being dealt in V2 prototype. Therefore, the V1 unit will be equipped the following:

- grinding tool attached to the arm's end effector, for grinding walls, ceiling and floor (see 4.3.1)
- it will study how to attach a grinding tool to the lower part of the mobile platform, for grinding the floor (see 4.3.2)

Using these tools, as checked in D1.1 the maximal dust level will be inside the **level 2: between 100 f/l and less than 6.000 f/l**.

Even the tool exchange is no predicted, any tool exchange to be performed in the V1 system will be done manually by the operator. Therefore, no active tool changer units will be implemented on this prototype version.

As indicated in section 3.3.2.5, the disk abrading will be performed by visual inspection to be carried out by the operator as done in the manual process.

The tool to be attached at the arm end effector will be a manual grinder tool adapted for being held and operated by the robot. This tool will be used to grind walls and ceiling and will be linked to the vacuum aspiration system described in section 3.4.4.

The necessary supporting structure will be developed and any required modification to the original tool will be performed. On the adaption of the tool will be taking into account the aspiration head and its pipe.

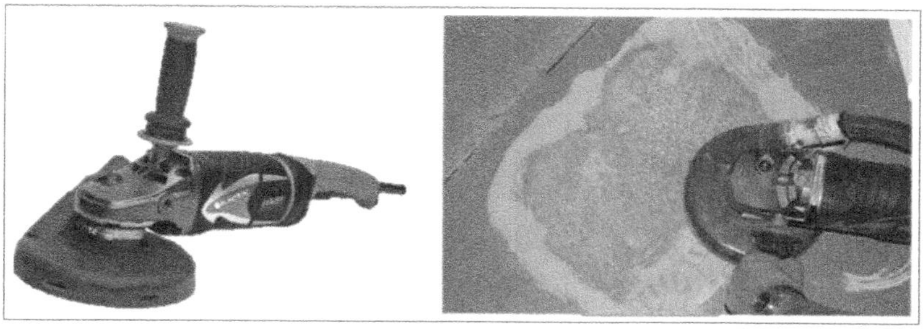

Figure 3.29. Manual grinder tool to be adapted to the Robotic Arm.

The tool is available and will be tested in two sizes. The main specifications of the tool will be similar to the ones in the tables below, corresponding to the manual tools:

Type/model	BLASTRAC BGV-180AV
Power	2600 Watts (performance 97%)
Supply Voltage	230 Vac – 50Hz – (2P+T)
Rotation Speed	6600 rpm
Weight	5,46kg (without cord)
Disc diameter	Ø180 mm
Dimensions	495mm / 185 mm / 325mm
IP protection	n/a
Acoustic pressure:	n/a

Type/model	BLASTRAC BGV-125AV
Power	1,3 kW
Supply Voltage	230 Vac – 50Hz – (2P+T)
Rotation Speed	11000 rpm
Weight	3,2kg (without cord)
Disc diameter	Ø125 mm
Dimensions	495mm/185 mm/325mm
IP protection	n/a
Acoustic pressure:	n/a

This tool will be adapted and sealed in order to reach the operation and environmental conditions required for this V1 unit. A force/torque sensor will be attached at the end effector of the robot arm, although there is the option to have a passive compliance actuator.

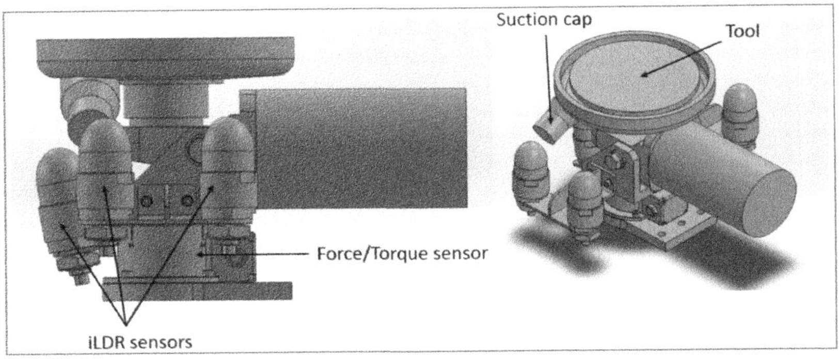

Figure 3.30. Sketch of the adaption of manual grinder tool to the Robotic Arm.

The tool will be controlled through the I/O cards of the arm controller placed on the Robot Arm Cabinet B1, and with the help of auxiliary relays and other protection devices to be placed on the Robot Arm Cabinet B2 described in section 3.4.7.

For grinding the floor, the same tool will be able to be attached directly to the mobile platform in order to perform some tests. The platform will include any required automated mechanism for its proper operation during locomotion of the platform and its normal operation.

3.1.11.4 Waste/Dust Aspiration

Taking into account the minimum aspiration pipe diameter, around Ø35mm, the RU-V1 will include an aspiration placed on the mobile platform in order to avoid the difficult of managing this kind of pipes and the need for an active pipe reel of high dimensions.

This will suppose that the mobile platform should carry with the aspiration system additional load. In addition, the aspiration system must be as compact as possible taking into account the reduced space available in the mobile platform and the whole robotic unit. The consequence of this is that the aspiration system must be splitted in half for a better assembly and save more space.

As will be done with the different tools, the aspiration system will be based on a commercial aspiration device which will be adapted to be integrated on the mobile platform.

The main specifications of this aspiration/ dust collector will be similar to the one in the table below, which corresponds to the commercial device:

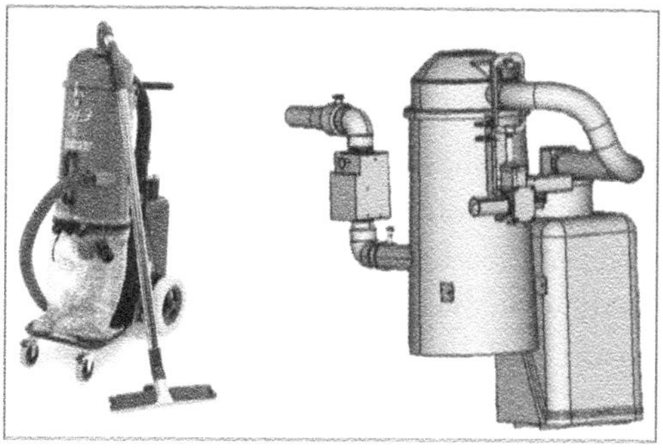

Figure 3.31. Aspiration system device S13 from BLASTRAC, parts of the CAD model.

Type/model	S13
Dimensions (W/L)	680x380x1200 mm
Weight	27 kg
Power	1200 W
Supply Voltage	230Vac single phase (50Hz)
Cleaning	Manual impulse
Bag capacity	15 l
Aspiration	210 m3/h
Pressure	22 kPa
Main filter	99,5 % (1,5 m2)

The aspiration system will be controlled through the I/O cards of the arm controller, with the help of auxiliary relays and other protection devices to be placed on the robot arm cabinet B2 (see section 3.4.7). This system will include almost an input to switch on/off the system and an output for alarm (i.e. bag full, filter blocked ...)

3.1.11.5 Decontamination, containment and transportation

It must be taken into account that the V1 prototype will not operate in real contaminated worksites. Therefore, the decontamination and containment measures are introduced in this V1 for testing purposes.

3.1.11.5.1 Decontamination and containment

As mentioned in the previous section, the robotic unit won't work in contaminated areas so there will be not decontamination process, just cleaning process.

The actual current decontamination procedure de-contamination requires 2 different protection actions for contaminated devices. The devices use are protected with a gasket for operating inside the contaminated area which is removed before the the second protection action, cleaning the device in a shower.

Regarding the RU-V1, in real contaminated worksites, it will use the following protection measures:

- The arm and the vertical slider will be covered by a renewable gasket, as done in most of dangerous nuclear operations, for instance.
- Regarding the Mobile Platform, it will be specially prepared and sealed in order to be directly washable.
- All external sensors will be prepared and sealed in order to be washable directly.
- The tool will be sealed and/or able to withstand operating and decontamination shower conditions.

3.1.11.5.2 Transportation

As pointed out the beginning of this section, the V1 unit will not operate in real asbestos contaminated sites. On this basis, it can be asserted that no special confinement or containment conditions must be considered for this V1 prototype. Nevertheless, the necessary protection measures against asbestos contamination will be considered and implemented in V2 prototype.

The necessary measures to ensure the safety and proper transportation of the V1 unit and all of its components from development and integration sites to testing sites (and viceversa), such protection boxes, pallets, etc., will be analyzed and put into practice.

3.1.11.6 Energy supply

The power supply of the robot will be 230Vac (P+N+T). It will be connected to an external power supply placed outside the contaminated area (i.e.: a diesel generator) in order to get the enough power which cannot be extracted from the building main supply. This external power supply must provide 230Vac single phase for V1 and 400Vac tri-phase for V2, and a power over 10kW.

The connection between the RU-V1 and the external power supply will be done by and extended cable through the cable reel solution described in section 3.4.8.

In the attached document *PowerSupply-Layout_v1.pdf* there's a list of the different consumer devices and all ac-dc/dc-dc converters of the design. The mobile platform will include different ac-dc converters for supplying its integrated components such the mobile platform controllers, the IMU sensors, the laser sensors, etc.

More details are also provided in the deliverable *Di3.1 Specification of the system architecture and interfaces.*

The power required by the different RU-V1 elements are described in the attached document: *Equipment List.*

It must be taken into account that a simultaneous operation of all devices is an improbable situation. Therefore, to determine the main supply cable section it will be considered the scenarios described below:

Scenario	Description	Power	Calculated cable section
A	Mobile Platform (1kW) + Tool Grinder (1,62,6kW) + Aspiration (1,3kW2kW) + battery charger (1,6 kW) + Sensors and Others (10,5 kW)	5,96,4 kW	4 mm^2 heating 4 mm^2 voltage drop (4,93,0%) for a 25m cable
B	Arm (21,7kW*) + Tool Grinder (1,62,6kW) + Aspiration (1,23kW) battery charger (1,6 kW) + Sensors and Others (10,5kW)	6,6 kW	4 mm^2 heating 4 mm^2 voltage drop (5,6%) for a 25m cable

*Considering nominal power of each arm motor. A factor of simultaneity must be applied for a more approximate calculation.

A scheme of the power supply is provided in the attached document: *PowerSupply-Layout_v10.4.pdf*

3.1.11.7 Electric Cabinets

The robotic Unit will include different electric cabinets where all the electric and electronic devices will be located in. The cabinets on the unit will be:

- Mobile platform cabinet A1, located inside the MP, it includes all the MP control and power supply elements, and the rest of control devices: router, NUC controllers, etc.
- Mobile platform cabinet A2, located on the exterior part of the Robotic Unit, will allocate theiRPU and the signal conditioning of the force/torque sensor.
- Robot arm cabinet B1, located inside the MP, will allocate the Beckhoff controller, the DC/DC converter, and the safety auxiliary relay board.
- Robot arm cabinet B2, located on the exterior part of the Robotic Unit, behind the vertical slider, will allocate the arm motor drivers, the DC/DC converter and the safety relay board.
- Battery charger box, located on the exterior part of the Robotic Unit.

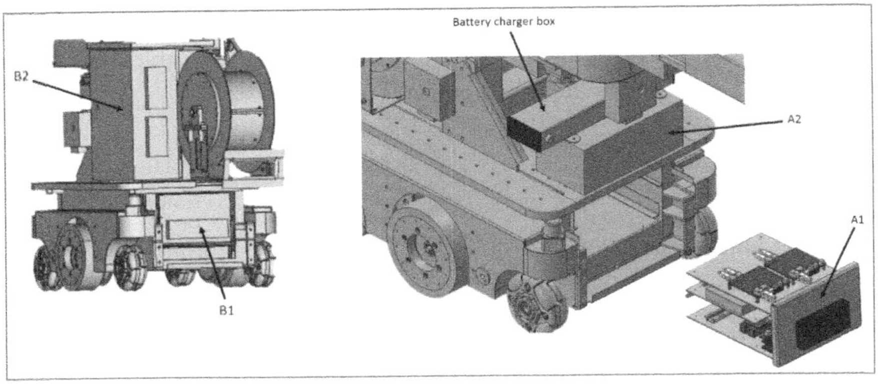

Figure 3.32. Cabinets.

3.1.11.8 Cable reel

To manage the length of the cables, it will be necessary install motorized reel, similar as the one shown in Figure 3.33.

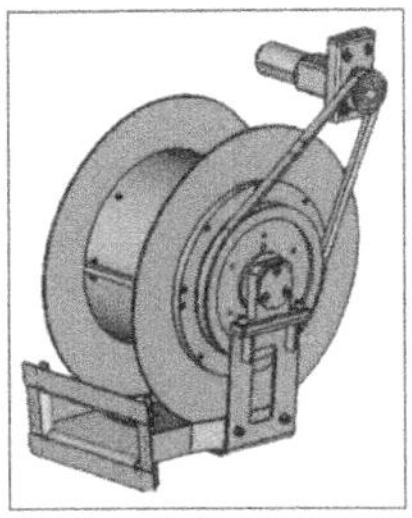

Figure 3.33. Motorized reel custom by TLABS.

This cable reel will incorporate an active control based on the odometry of the mobile platform. The main specifications of this device are listed in the table below:

Type/model	Custom by TLABS
Dimensions	Ø400mm x 150mm (only collector drum)
Cable capacity	25m 1 x (3G 4mm2 + 6x1mm2)
Weight	14.5 kg. (aprox., without cable)
Power	57,6 W (max torque)
Nominal current	2,4 mA (max torque)
Supply Voltage	24 V
speed	64 rpm (no load), 45 rpm (at max. torque)
Finished surface	Polyurethane painting resistant in corrosive environments
IP protection	IP-65

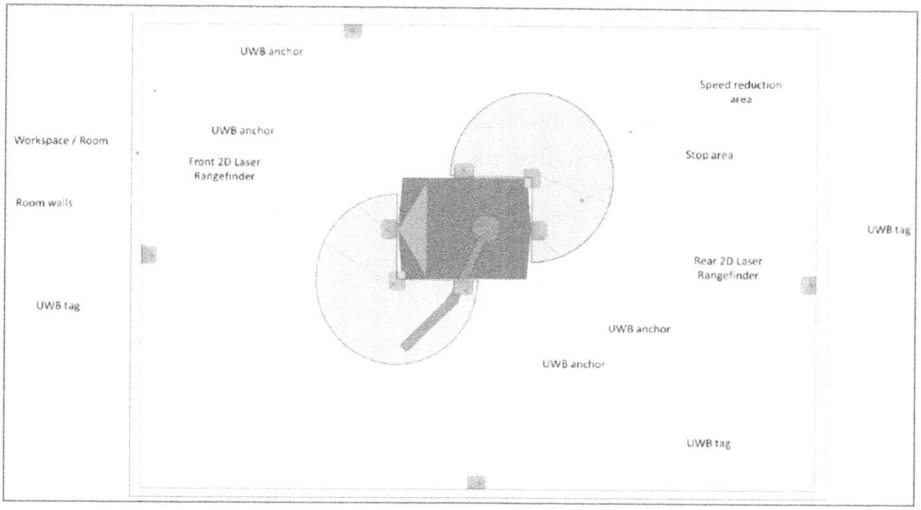

Figure 3.34. General overview of environmental perception sensors.

3.1.11.9 Navigation and Environmental perception sensors set

A set of sensors for perceiving the environment and positioning the robotic unit will be placed on the robotic system. With this purpose, for the environmental perception and navigation, the V1 system will include:

Sensor	Purpose
1 x UWB System, e.g. 4 anchors on the platform and min. 4 tags in the room	Localization
2 x Camera	Backup in case navigation fails or teleoperation mode
2 x LIDAR (in opposite corners of platform)	Localization/Mapping
1 x IMU	Localization
1 x Odometry	Localization
1 x Infrared	Collision avoidance

In Figure 3.34 General overview of is possible to see a sketch with the location of the above mentioned sensors.

3.1.11.9.1 Laser Scan sensors

As described in Figure 3.34: General overview of 7, 2 Laser scan sensors of 270° coverage will be placed on 2 opposite corners of the platforms in order to cover the whole area around the robotic unit. This laser scan sensors will be used with the following purposes:

(a) to avoid collisions with objects or other robotic units
(b) to build a map and recognize the robotic unit position (SLAM or similar), mainly for navigation.

Taking into account that the MP displacements are displacements are omnidirectional, the location is in two opposite corners so the 4 sides of the robot are covered.

Figure 3.35. Location of Laser scan when it is used for safety and Hokuyo UST-10LX laser scan sensors.

The laser scan sensor mounted on the mobile platform will be a Hokuyo UST-10LX (see Figure 3.35) or similar. Its main specifications are:

Supply voltage	12VDC/24VDC (operation range 10 to 30V ripple within 10%)
Supply current	150mA or less (during start up 450mA is necessary.)
Light source	Laser semiconductor (905nm) Laser class 1 (IEC60825-1:2007)
Accuracy	±40mm
Repeated accuracy	$\sigma < 30$mm
Scan angle	270°
Angular resolution	0.25°
Start-up time	Within 10 sec (start-up time differs if malfunction is detected during start up)
Input	IP reset input, photo-coupler input (current 4mA at ON)
Output	Synchronous Output, photo coupler open collector output 30VDC 50mA MAX.
Interface	Ethernet 100BASE-TX
LED display	Power supply LED display (Blue): Blinks during start up and malfunction state.

Ambient temperature	$-10°C$ to $+50°C$
Ambient Humidity	Below 85%RH (without dew, frost)
Storage temperature	$-30°C$ to $+70°C$
Storage Humidity	Below 85%RH (without dew, frost)
Shock resistance	196m/s2 (20G) X, Y and Z direction 10 times.
Protective Structure	IP65
Material Front case	Polycarbonate, Rear case: Aluminium
Surrounding intensity	Less than 15,000lx
ROS Driver to acquire laser data	SLAM algorithms available
Detection range	0.06m to 10m (white Kent sheet) 0.06m to 4m (diffuse reflectance 10%) Max. detection distance : 30m

3.1.11.9.2 IMU

The IMU mounted on the mobile platform will be an L3GD20H chip, integrated on Pixhawk controller. Its main specification are:

Supply voltage	2.2V to 3.6V
Supply current	5mA
Operating temperature range	-40°C to +85°C
Value output data	16 bits
Temperature output data	8 bits
Interface	I2C/SPI
Measurement range	±245, ±500, ±200 dps (user selectable)
Sensitivity	8.75, 17.50, 70.00 mdps/digit (user selectable)

Pixhawk specifications:

Processor	32-bit ARM Cortex M4 core with FPU 168 Mhz/265 KB RAM/2 MB Flash 32-bit failsafe co-processor
sensors	MPU6000 as main accel and gyro ST Micro 16-bit gyroscope ST Micro 14-bit accelerometer/compass (magnetometer) MEAS barometer

Power	Idela diode controller with automatic failover Servo rail high-power (7 V) and high-current ready All peripheral outputs over-current protected, all inputs ESD protected
Interfaces	5x UART serial ports, 1 hig-power capable, 2 with HW flow control Spektrum DSM/DSM2/DSM-X Satellite input Futaba S.BUS input PPM sum signal RSSI (PWM or voltage) input I2C, SPI, 2x CAN, USB 3.3V and 6.6 ADC inputs
Dimensions (WxHxL)	50 x 15.5 x 81.5 mm
Weight	38g

Figure 3.36. Pixhawk controller.

3.1.11.9.3 Encoders

Odometry requires a method for accurately counting the rotation of the robot wheels. A standard method for doing this is to instrument the wheels with encoders. The Mobile platform includes encoder 15T Thru-Bore eEncoders. Its main specification are:

Input voltage	5Vcc fixed voltage for communication 4.75 to 24 Vcc max for temperature up to 85°C 4.75 to 24 Vcc for temperatures between 85°C to 100°C
Input current	100 mA max (65 mA typical) with no output load
Output format	Incremental

Output types	– Open collector – 20 mA max per channel – Push-Pull – 20 mA max per channel – Pull-Up – open collector with 2.2K? – Pull-Up 20 mA max per channel – Line Driver – 20 mA max per channel (meets RS 422 at 5 Vcc supply)
Marker(index)	Once per revolution. 190 to 10000 PPR: gated output A 1 to 189 PPR: Ungated
Max. frequency	Standard frequency response is 200 kHz for PPR 1 to 2540 500 kHz for PPR 2541 to 50000 1 MHz for PPR 50001 to 10000
Noise immunity	Tested to BS EN61000-6-2; BS EN50081-2; BS EN61000-4-2; BS EN61000-4-3; BS EN61000-4-6; BS EN500811
Max shaft speed	8000 rpm
Bore size	8mm
Weight	100g
Operating temperature	−20° to +85° C standard models −40° to +85° C for low temperature option −20° to +100° C for high temperature option −20° to +120° C for extreme temperature option
Storage temperature	−25° to +85° C
Humidity	98% RH non-condensing
Vibration	10 g @ 58 to 500 Hz
Shock	10 g @ 11 ms duration
Sealing	IP50 standard; IP64 available

Figure 3.37. Encoder used in the mobile platform.

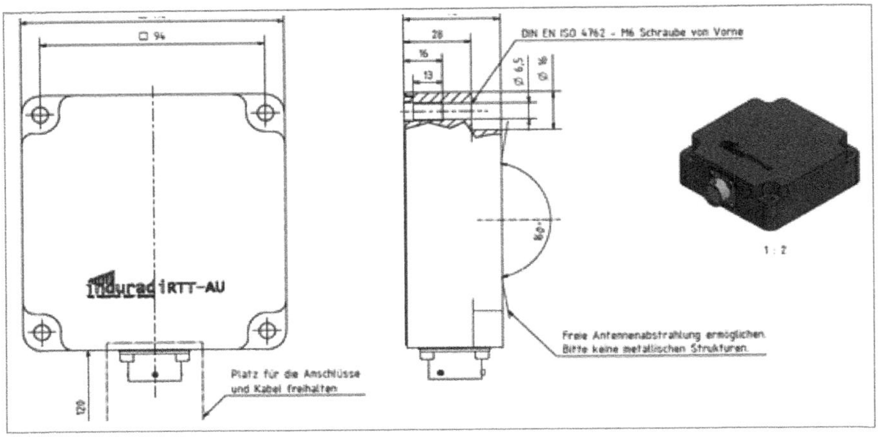

Figure 3.38. UWB anchor design.

3.1.11.9.4 UWB positioning sensors

The UWB sensor provides the robotic unit position based on different measures between the anchors placed in the mobile platform and the tags placed on the walls of the different rooms. Therefore, 4 or 5 devices (anchors) of the UWB positioning system will be placed on the different sides of the robotic unit, as depicted in Figure 16. If required, more devices will be added to the robotic unit in order to ensure the proper operation of the system and getting the required positioning accuracy.

The Two technology can be implemented is based on : ToF (Time of Flight) based measurements even or TDoF (Time Difference of Flight) measurements can be considered if necessary.

This positioning system will require also to place different emitter tags inside the room, also depicted in Figure 3.38. It will be required almost at least 4 emitter tags per room, even there is no upper limit of nodes.

In the table below there is a summary of the main characteristics of this sensor system (T=tag, A=anchor):

Weight (kg)	T,A: ~0.3kg
Voltage Range	T: Battery 8 V, A: 65 V
Power consumption	T,A: ~1W
Rechargeable battery	7V to 60V, 1200mAh
Data types/ formats	output will be platform pose data format not yet defined
Hardware interface	CAN
Communication speed	110kBit/s to 6Mbit/s

Frequency range	6 to 7 GHz
Connector	8 pin Souriau (M12)
Update rate	~10Hz
Resolution	mm
Accuracy	>=5cm
Measurement range	~200m
On-board pre-processing features	Pose calculation
Anchor size	80x75x56mm

3.1.11.9.5 Teleoperation Cameras

In order to perform the more complex removal jobs, the robotic unit V1 will incorporate 2 RGB cameras to help on the remote teleoperation from the Simplified Central Process Control System and HMI in the outside of the contaminated area.

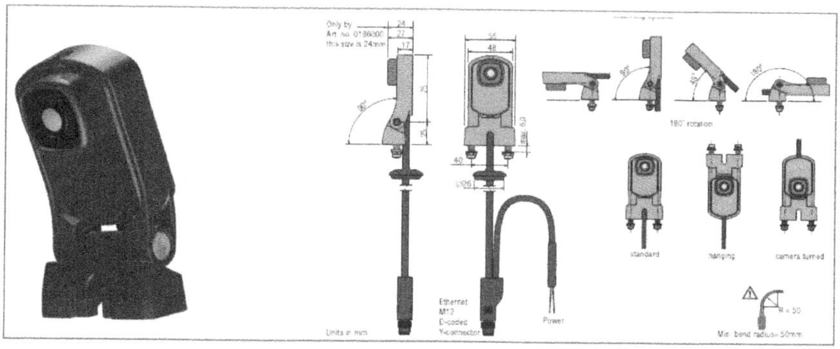

Figure 3.39. Teleoperation Cameras – ORLACO Camera EMOS.

The system will include 2 Ethernet cameras ORLACO Camera EMOS with the following specifications:

Sensor element	High resolution 1/3" CMOS rolling shutter, 1280H x 960V
Field of View	120° version
Video	image 60fps at 720p SDR, 45fps at 960p SDR image processing and 56fps at 720p HDR, 43fps at 960p HDR image processing
Protocols	RTP (MJPEG RFC2435 and h.264 RFC6184) over UDP, PTP and gPTP, ISO17215-2014, IEEE1722 (AVB)

IP address	DHCPv4 or static IPv4 (IPv6 possible)
Latency	<100ms depending on hardware processing platform (50ms demonstrated on Orlaco hardware)
Light sensitivity responsitivity	5,5V/lux-sec, Low light feature (<0,1Lux)
High dynamic range	>115dB
Compression	MJPEG (H.264 expected in Q1, 2016)
Interface	100Mb/s (Fast Ethernet)
Image processing	Color and gamma correction, 3D noise correction, edge enhancement, digital WDR, advanced contrast enhancement, auto white-balance, auto exposure control, mirroring, flipping, photometric and geometric lens distortion correction
Power input	12/24V/DC. Below 7V: camera is non-functional. Above 33V the overvoltage protection is activated. This overvoltage-protection is deactivated below 30V. Powercircuit is protected up to 80V/DC. Outputs are Short Circuit Protected.
Power consumption	<2W
Connector	Cable 6 wire: 2x twisted pair for data and 2 wires for power supply. Connector 4-pin M12 d-coded for data and open wires for power supply.
Housing	Anodised aluminium, black, UV resistant, light fastness >8, corrosion proof according IEC 60068-2-52 salt mist, cyclic.
Ingress protection	IP68 according to IEC 60529 (up to 10 m under water), IP69k according to DIN 40050-9
pressure cleaning/washing	with water: 14-16L/min. 80°C and 100 bars flow
Lens glass	Chemically hardened, toughed, tempered float glass: 5 to 7 times stronger than ordinary glass, protected against acid: class 2-3; DIN 12116, passed flying stone test
Shock constancy	Shock and vibration resistant for usage on trucks, cranes, fork-lifts, maritime applications, machinery

Camera bracket Material	glass reinforced plastics, test: 50 Nm at -40°C to +85°C
Weight	0,24kg. including cable, bracket and mounting material. 0,30kg. in standard packing
Operating temperature	-40°C to +85°C
Storage temperature	-40°C to +125°C

The cameras will be placed on each side of the slider-arm structure. With the cameras in this position an operator will be able to see most of the surrounding area of the robotic unit.

In order to avoid dust problems, it has been considered a dustproof and waterproof camera (IP68) to work in contaminated environments as mentioned in previous sections. Cleaning will be done manually.

3.1.11.10 Local process monitoring and teleoperation sensors set

A set of sensors for local process monitoring and teleoperation will be placed on the arm end effector. With this purpose, the V1 system will include:

- **iLDR sensors** for measuring the removal process

3.1.11.10.1 iLDR sensors

For local process monitoring it will be developed a sensor system consisting of 4 5 iLDR sensors (120 GHz frequency), for 1D distance measurement in one direction are will be developed. This technology is especially robust under dusty environmental conditions which will fit the prototype requirements.

This iLDR sensors will be located as described in Figure 3.40. This configuration will maximize the grinder disk reach and at the same time will allow to:

- Monitor the processed material depth
- Get the Tool orientation referenced to the wall in real time

Data processing will be done by a computer installed within the Control CabinetA1 (see section 3.4.7)

Data transfer from the tool to the computer will be done by USB2.0 cables through and USB-Hub which will be used for supplying the sensors with 5Vdc. The accuracy will depend on the number of targets found. Single targets (single peak) can be found more accurately. This sensors are not affected for the process vibration

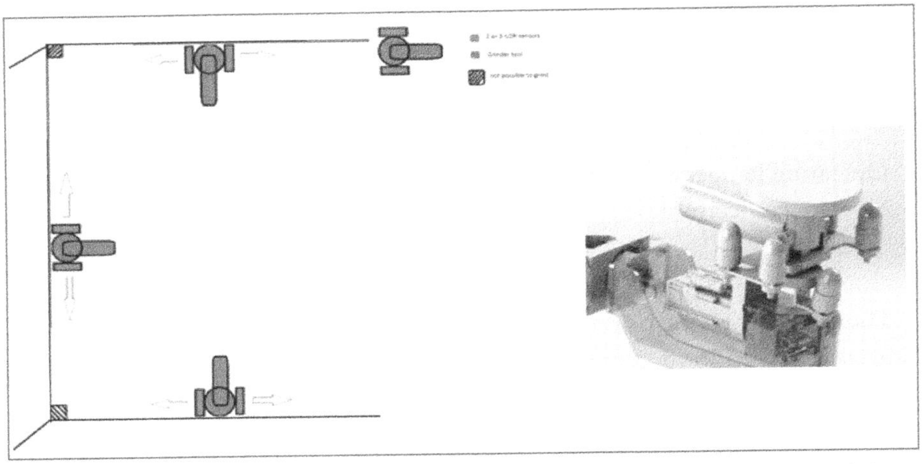

Figure 3.40. iLDR Sensors Configuration (left) CAD model iLDR sensors (right).

due to the incorporated IMU sensors. In the table below there is a summary of the main characteristics of this sensor system:

Frequency	119GHz to 126GHz
Max. radiation power	0dBm (1mW)
Operation principle	FMCW (Frequency modulated continuous wave)
Lens diameter	45mm
Lens length	about 35mm
Lens material	PTFE
Power supply	5V/2.5W – a high power USB connection can be used; optional nominal 24V (7V....65V), Power over Ethernet (802.3af/at)
Interface	USB, optional Ethernet 100BASE-TX, IEEE 802.3 Clause 25 with increased power requirements
Output data	Raw data IQ-sampled
Measurement rate	up to 1000
Single measurement duration	about 220μs
IP protection	IP67 or better probably IP69/K
Temperature range	−40...+85°C
Dimensions	60x60x60mm
Weight	350 gr

3.1.11.11 Emergency stop and safety during operation

For the system V1 preliminary prototype the machine safety area will be outlined with the red ribbon and near the laptop used for system validation there will be an electrical box with emergency button and system reset button. Similar devices are depicted in Figure 3.41.

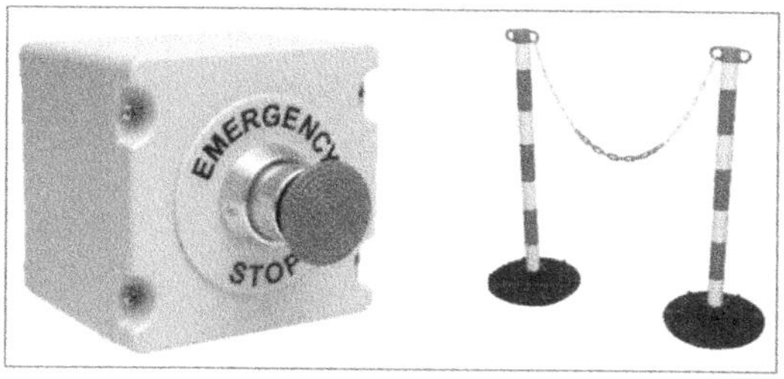

Figure 3.41. Work area safety measures.

Other emergency stop devices will be installed in the robotic unit, in a visible an easy accessible location. Once the emergency stop is triggered, mobile platform and arm brakes will be triggered in order to ensure a fail-safe behavior of the system. Also, it is also planned to could be use an emergency barrier stop, placed outside of the contaminated area, this signal will also be able to stop the operation of the robotic unit. Below a figure of the test site with the safety elements is shown.

More details can be found in Di3.3

3.1.11.12 Communications and Control Architecture

In order to the operator will be able to send asbestos-removal-tasks or receive some feedback about the work of the robot, it is necessary create a communication network.

A Wi-Fi router will be installed on the robot to establish communication with the operator from a not contaminated area.

All the controllers of the robot (arm controller, mobile platform controller, UWB controller, iLDR controller) will be connected to the router Wi-Fi via Ethernet cable. Below, is detailed with the way to connect each senor or actuator to its own controller.

3.1.11.12.1 CPCS

Taking into account that the V1 system is set for research and testing purposes in the earlier phases of the system development, a simple version of the Central Process Control System (sCPCS) will be developed just with essential functions.

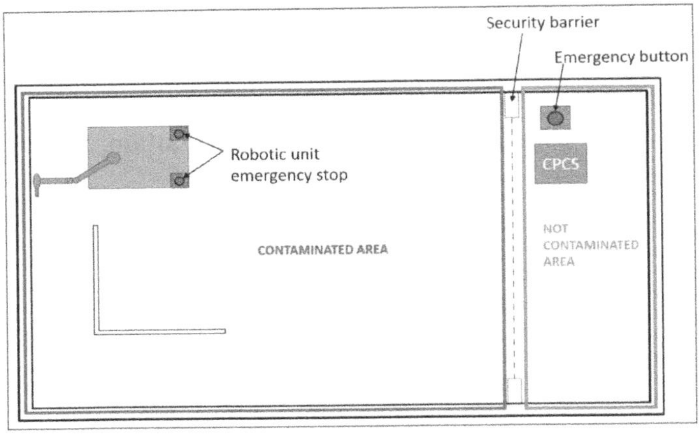

Figure 3.42. Test site with safety elements.

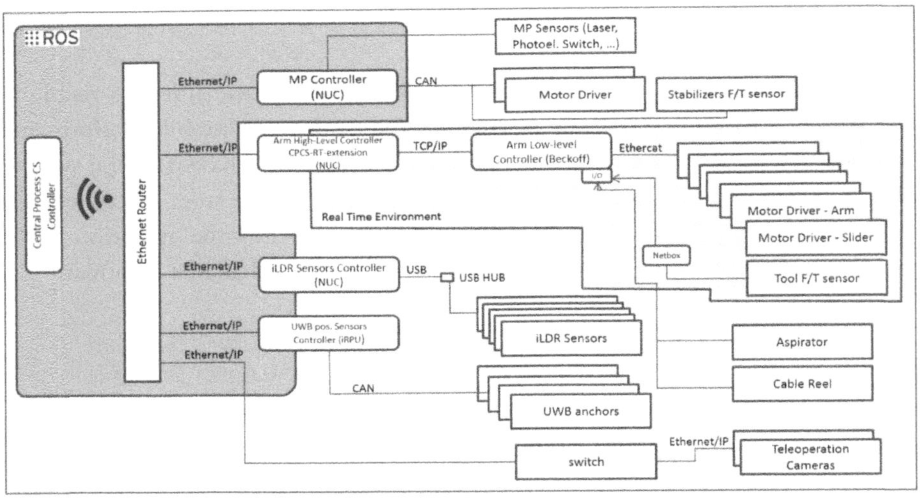

Figure 3.43. Control architecture.

As pointed in section, the CPCS will be further developed during the regular tests and modifications to achieve its final version for system V2.

The essential functions that the CPCS must have for V1 are:

- Teleoperation through a computer screen and a standard keyboard
- Automated execution of simple grinding tasks with simple geometries (such square areas).
- Monitoring of process parameters
- Alarms and warning management

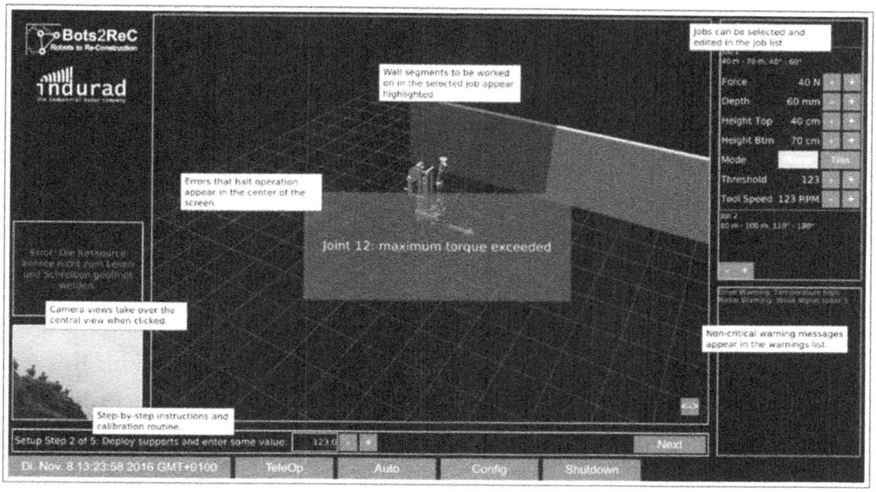

Figure 3.44. Preliminary HMI (running on laptop and desktop PC).

The CPCS will be set in a portable computer to be placed next to the robotic unit or outside of the working area, depending on the tests being carried out.

The computer which will host the control system (CPCS) will be able to operate in soft real-time. The main control loops (kinematics, start/stop signals, alarms, etc.) will be executed on Beckhoff controller and on a TLABS NUC with RT-Preempt for real-time.

3.1.11.12.2 HMI interface

The HMI will be set in the same screen as the controller, both in the same laptop or desk PC. It will be developed and implemented a multifunction HMI interfaces between the autonomous system and the operator. The HMI will control each operation of the different phases of the complete process from the initial operator task programming.

It will be adapted to be introducing the asbestos removal jobs location, based on a previous 2D/3D map and its parameters.

It must have the controls for planning and supervision of the platform motions in each process and security parameters: free supervised motion for general displacements of the platform, planned motion for getting navigating inside the worksite and planned motion for grinding process.

The operator must be able to supervise the removal process, and teleoperate the system through different live images coming from the cameras placed on the platform, modify the control strategies and parameters, and plan actions of the robot, between others. Also the operator must be able to supervise the removal jobs process and read in real-time the process variables.

The HMI will include safety parameters and alarms that show the detection of danger situations like a worker detected into the danger zone by the perception system, collisions or occlusions of the computer vision systems and problems on the platform or its components.

3.1.12 Outlook and Future Work to be Carried Out

The definition of the main technical specification of the V1 system and the devices that compose it has been done, according to the Work Program. With this work, it has updated the detailed information of each part of the system according to the design phase works.

A sufficient level of detail has been provided in order to get a detailed description of the robotic unit V1. Thanks to the information collected and from the tests to be performed, it will help for define and design the robotic unit V2.

3.1.13 Summary and Conclusion

In this deliverable the technical specifications for Robotic Unit V1 has been defined, which will become the basis for the further references in the becoming phases of the project.

It is therefore concluded that aim of the definition of system technical specification has been achieved and the work has been carried out according to the work program.

Deliverable Di4.1
3D CAD drawings, manufacturing drawings and list of part for V1

Authors: Miguel Moreno, Tim Detert

3.1.14 Deliverable in the Context of the Work Program

This internal deliverable is an output from *Task 4.1 "Design-adaption of the mobile platform"* to show the Design of the first prototype (V1) of the Bots2Rec platform.

It uses the input from previous deliverables "*Di3.1 Specification of the system architecture and interfaces*" and "*Di3.2 Specification of the technical requirements for the V1 prototype*".

This deliverable contains the CAD files, CAD drawings, part lists and electrical drawings necessary to manufacture the first version of the mobile platform, as a part of Milestone 1 (Overall system layout and design of the first prototype and central system finished) Following to this document is the manufacture of the Mobile platform, its test, and integration with the rest of subsystems included in First prototype.

3.1.15 Methods and Work Carried Out

To reach this deliverable the platform has been dimensioned according to the basic requirements specified in task 3.1, 3.2 and 7.1.

3.1.15.1 Kinematic configuration

Different kinematic configurations have been evaluated, including different suspension system. The main requirements considered for design of the mobile platform were: (See Di3.2 for further details)

- High mobility to be capable of moving through corridors and turning 360 degrees in place
- Reduced footprint to pass through doorway and to fit inside an elevator.
- Stability to carry high Centre of Mass and a total weight of the robot of 300Kg.
- Capability of crossing small obstacles in manual operation
- Preservation of the protection of the room (plastic sheet)

The configurations preferred to meet the requirements are "Swerve Wheels", "Dual Swerve Wheels" and "Differential Drive". The next table shows them.

Another question considered for the kinematic configuration is that the surface over the platform is moving could be a plastic covering the floor, the platform needs to move over this surface without damaging it. So tests over this surface have been done with different wheels to see the damage created.

The cover material used for testing is Polyethylene film with a thickness of 37 and 125 microns (the thicker is the one recommended by Bouygues Construction). The wheeled vehicles used are Robotnik existing platforms:

- **Robotnik AGVS robot.** 2 Swerve wheels + 2 fixed wheels. Robot weight 250 Kg.

Table 3.1. Kinematic configurations

Configuration	Advantages	Drawbacks	Picture
A) Swerve Wheels. 4 directional wheel on each corner.	• High maneuverability: can move in any direction without rotating the platform • More traction due to 4 motorized wheels.	• High complexity because of the quantity of motors/servodrives. • Higher weight • To change direction of robot movement the wheel often has to rotate in the spot, what can damage the plastic cover	
B) Double Swerve Wheels. 4 directional double wheel on each corner. The 2 wheels on each directional block move differentially	• High maneuverability like (A) • Double wheel configuration avoids rotation of the wheel on the spot	• Higher complexity and weight than (A) for adding extra wheels.	
C) Differential drive 2 Traction wheels on each side + 4 omnidirectional wheels in the corners	• Low complexity • Low weight • Rarely wheels rotate on the spot (can be avoided by High Level Controller). • Good Maneuverability	• Less traction than (A) and (B) because there are 2 motorized wheels and the weight is shared with non-motorized wheels. • The robot has to rotate to change direction.	

With in-site Swerve wheel rotation it destroyes the 37 micron film in seconds and the 125 microns film in some minutes. With combined linear + rotational movement holes in the plastic only appear where there were small stones or big dust particles.

- **Castor platform.** Kinematic configuration: 2 fixed + 2 castor wheels. Weight ~500Kg

Test was made only with 125 micron thickness. There are situations where the fixed wheel rotates in place, so the results are equivalent to test with AGVS.

- **Robotnik RB1 robot.** 2 traction wheels + 3 omni wheels. Robot weight ~30Kg + payload 62Kg

Table 3.2. Test over plastic cover

Type of Movement	Castor	Swerve/fixed	Omni wheels
Linear + rotational	Damage if small pebbles and granules	Damage if small pebbles and granules	Damage if small pebbles and granules
Rotation at low radius	Damages in some minutes	Damages in few minutes	Damage if small pebbles and granules

Test is made only with 125 micron thickness. It is possible to make holes only in places where there are small stones or big dust particles. In this configuration is is not possible to make the omni wheel rotate on the spot so a test was made with one omnywheel rotated by hand, but there was no extra damage to the over surface.

Next table sumarizes the results of the test regarding type of wheel.

Considering the previous studies the selected configuration for V1 prototype is (C) Differential Drive, because it is lighter and simpler and avoids the problem of the wheel rotating on the spot.

The final design is a platform with 4 omni wheels (one on each corner) and two driven wheels at the middle of both sides of the platform. There could be alternative configurations with castor wheels instead of omni since this configuration avoids corner wheels to rotate on the spot.

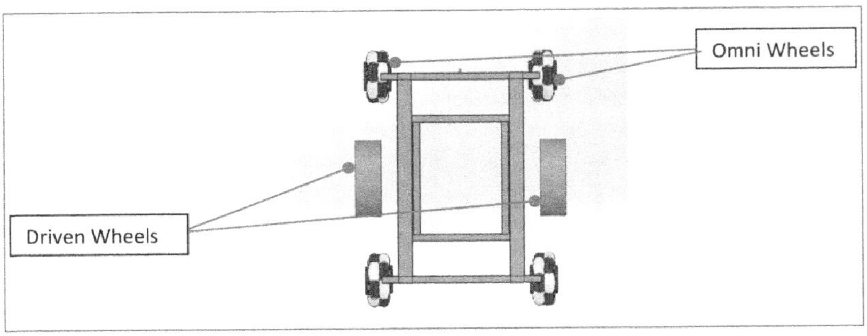

Figure 3.45. Wheel configuration.

This platform configuration needs that all omni wheels are in contact with floor, to be sure of that, at least one of them has to be capable to adapt to small height variations, so an oscillation axle has been designed.

Regarding driven wheels, they include a 500W pancake type motor with a solid rubber wheel with 220 mm in diameter.

The electrical hub has a 6:1 gear reduction, giving a max torque output of 37 Nm. The wheel is mounted in a suspension arm and, in order to be capable to test different suspension configurations for this first prototype, different

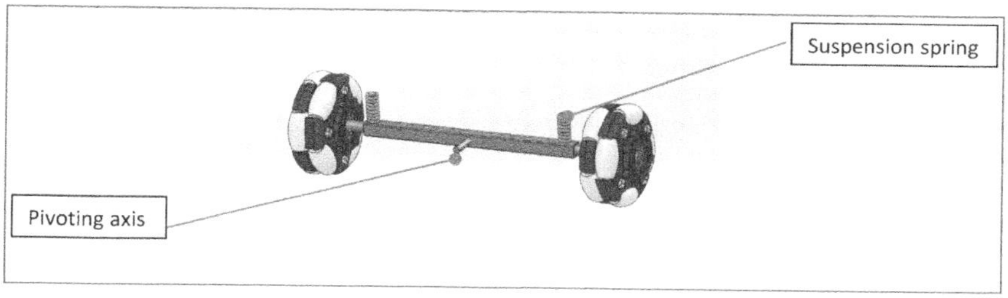

Figure 3.46. Oscillation Axle for Omni-Wheels.

air shock absorbers can be mounted. This shock absorbers must apply a force to the floor so the wheel friction forces ensure enough mechanical traction to the platform.

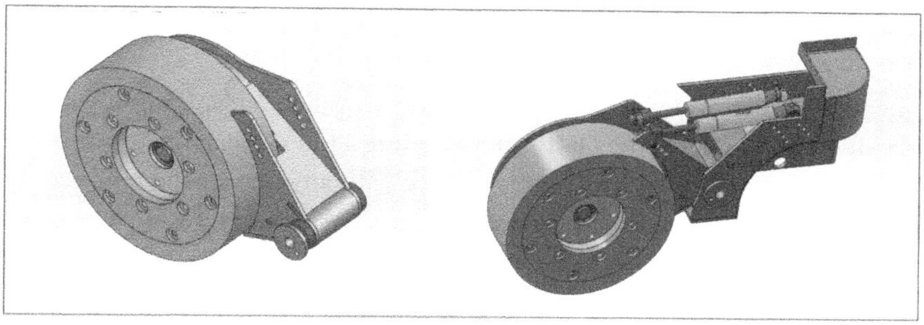

Figure 3.47. Driven wheel and its suspension.

3.1.15.2 Frame and covers

The frame of the platform is designed using rectangular steel tube. This type of structure is intended to provide enough rigidity to the platform without adding unnecessary extra weight to it. The resistance of the platform has been evaluated using finite elements analysis.

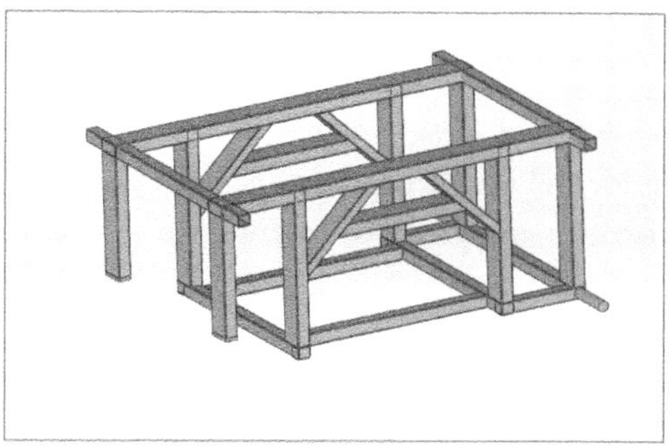

Figure 3.48. Steel tube frame.

The platform has to be easy to decontaminate so all the boxes for electronics are designed to have IP65 sealing.

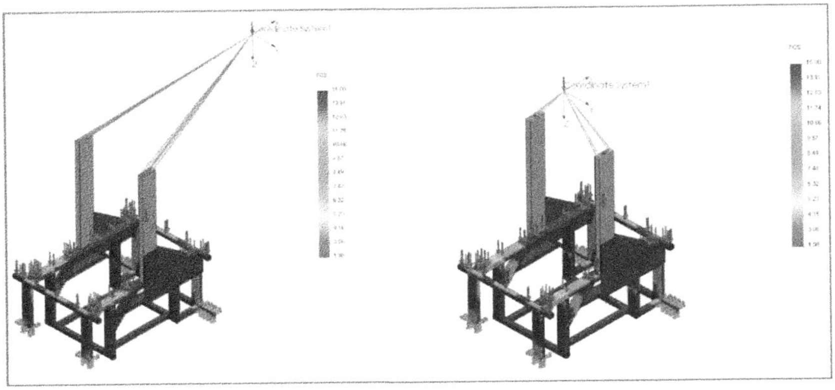

Figure 3.49. Steel tube frame Finite Elements Analysis under different loads.

The mobile platform will mount two Hokuyo UST Lasers, mounted on rear-left and front-right corners of the platform, providing a full vision angle around it. This lasers need to have full view without colliding with any part of the robot so, an air gap is needed in the lateral covers of the robot to allow laser beam to pass. Lasers are installed under top cover, in order to protect them from direct dust.

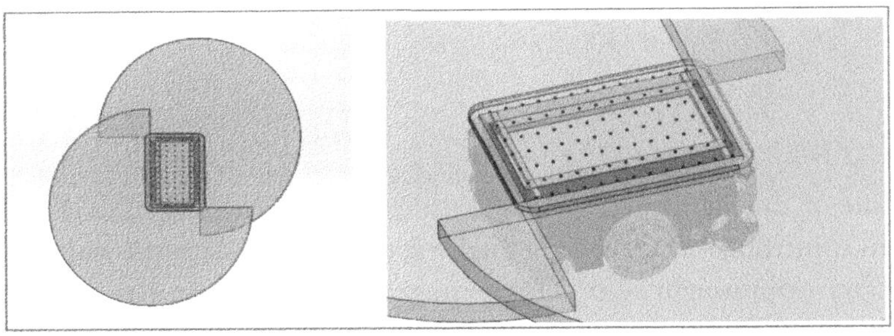

Figure 3.50. Laser range.

3.1.16 3D CAD Drawings, Manufacturing Drawings and List of Part for V1

The mobile platform for the first prototype has been designed.

The results are shown in attached documentation:

- 3D CAD models
- 3D CAD drawings
- Electrical drawings
- List of parts

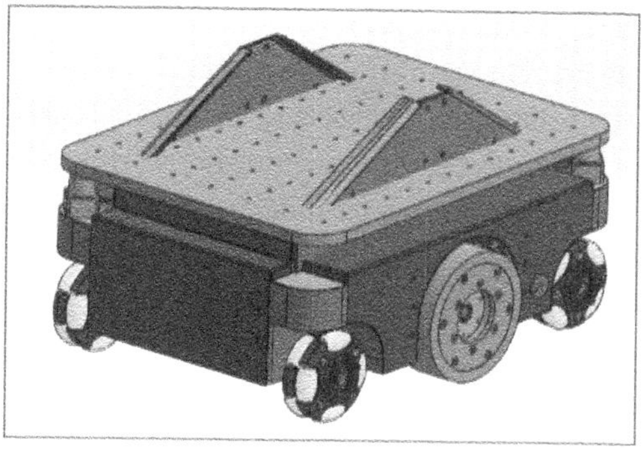

Figure 3.51. Mobile Platform.

3.1.17 Outlook and Future Work to be Carried Out

This deliverable shows the result of the work done during tasks 4.1 in order to create the first version of the platform. This work has been done in parallel with the work performed to design the rest of the subsystems, including the coordination work made in the integration workpackage WP3, so this WP is interconnected with the others.

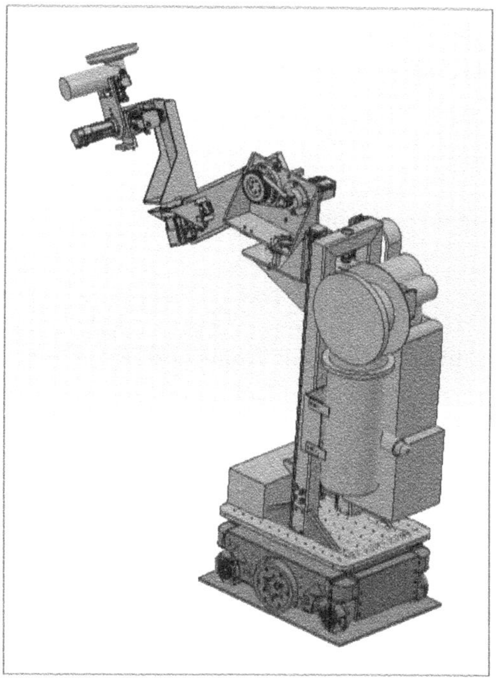

Figure 3.52. Mobile Platform in Bots2Rec first prototype.

Some problems have been encountered due to delay in decisions regarding requirements but they are not significant. Another delay has come from the problem encountered in arm definition: there was no commercial model able to fulfill the requirements and a new design has been done, so this affected the integration with the mobile platform and produced a delay of 1 month in integration of this subsystems with the others. We think this delay, as it is the first version, is not affecting the success of the project.

3.1.18 Summary and Conclusion

This deliverable achieved its objectives and provided a detailed design of the platform. With this design it is possible to manufacture a first mobile platform prototype able to fill the technical requirements and to be integrated with the rest of subsystem to build Bots2Rec V1 system.

Next steps are to manufacture the platform, test it with "dummy" weight to validate its performance, and integrate it with the rest of the subsystems. This project include several design cycles to future error corrections or improvements are expected for the next prototypes.

Deliverable D4i.3
Di4.3 Mobile platform V1 manufactured and ready for subsystem test

Authors: Miguel Moreno and Tim Detert

3.1.19 Deliverable in the Context of the Work Program

This internal deliverable is an output from *Task 4.4 Production and assembly of the mobile platform* to show the manufacture of the mobile platform for the first prototype (V1) of the Bots2Rec platform. It uses the input from previous tasks 4.1 4.2 and 4.3 and the information from deliverable *Di4.1 3D CAD drawings, manufacturing drawings and list of part for V1.*

The core part of this deliverable is the manufactured subsystem prototype ready to be tested. This report contains the explanation of the manufacturing process. Following to this deliverable is the testing of the Mobile platform individually, and then its integration with the rest of subsystems included in First prototype.

3.1.20 Methods and Work Carried Out

To reach this deliverable it has been required the input from task 4.1, 4.2 and 4.3 and the information in deliverable *Di4.1 3D CAD drawings, manufacturing drawings and list of part for V1*. As a result of the work in these tasks the procurement of some materials started before Di4.1 was finished.

The gathering of materials have been done from regular suppliers of Robotnik as most of the electronics components selected were already working inside robots of Robotnik product range.

To show the manufacturing process it has been divided between Mechanis and Electronics.

3.1.20.1 Mechanical Construction

Starting from the drawings of deliverable *Di4.1 3D CAD drawings, manufacturing drawings and list of part for V1*. The manufacturing process of different parts has been performed at Robotnik facilities but also at the facilities of different local Machining Workshops.

3.1.20.1.1 Frame

The frame has been designed to support high payload: the weight of the arm, tool, slider, aspiration system and extra elements.

The frame of the platform was designed using rectangular steel tube, it was assembled by Arc Welding. This type of structure was selected due to:

- High MOI (moment of Inertia) with low weight.
- Easy gathering of the material, construction and modification

The resistance of this chassis was evaluated using finite elements analysis, as shown in Di4.1, and has been tested as described in D4.1.

Once finished frame has been treated with a passivation process to be protected against rust and painted.

The heaviest component that the frame has to support is the slider, that's carries the weight of the arm to the platform. To hold this slider and transmit the weight and torque to the structure, two support brackets have designed.

These brackets are made of folded carbon steel (ST 37) sheet, the 5mm sheet is folded to provide rigidity and to add two different mounting surfaces.

3.1.20.1.2 Wheels

There are two types of wheels: four omni-wheels in the corners and two traction wheels in the sides, as shown in the next drawing.

Figure 3.53. Welding process details.

Figure 3.54. Frame before surface treatment.

Figure 3.55. Frame after being blued and painted.

The four omni-wheels have an inside bearing to provide free rotation around its central axis so they provide free movement in X axis, and Polyurethane (PU) rollers along its external circumference to provide free movement in Y axis. Omni wheels

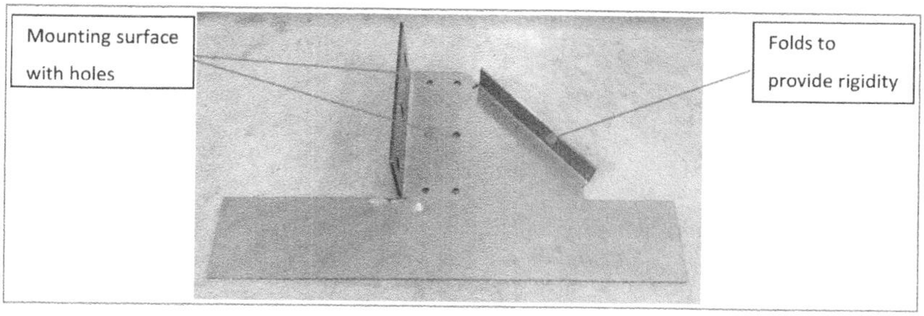

Figure 3.56. Supporting bracket for the slider.

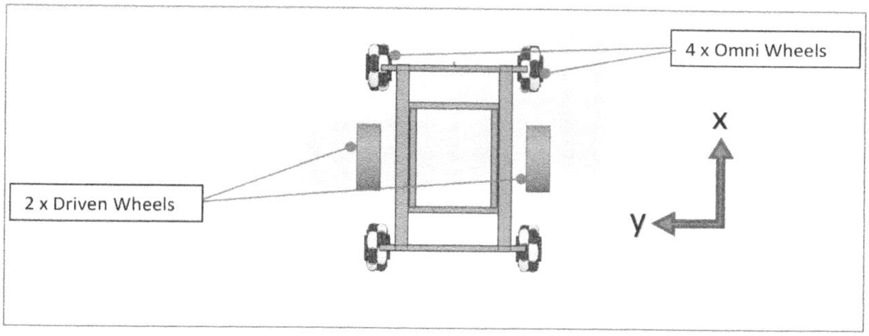

Figure 3.57. Wheel configuration and robot axis.

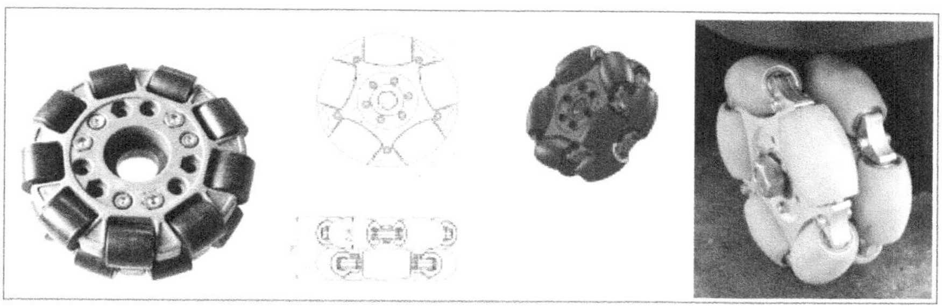

Figure 3.58. Regular size Omni-Wheel, and Bots2Rec Omni-Wheel.

have been ordered directly to a regular supplier but with a size bigger than other Robotnik robots to allow higher step crossing.

Traction/driven wheels have been developed by Robotnik and have a built-in gearbox and motor inside to provide up to 1000W of power, and an encoder to measure and control wheel movement.

The outer moving part of the when is a lathe-machined iron hub with a PU coat.

Driven wheels are mounted in a suspension system consisting on a Rocker Arm made of steel and a spring. The wheel inner hub is a lathe machined cylinder soldered to the rocker arm.

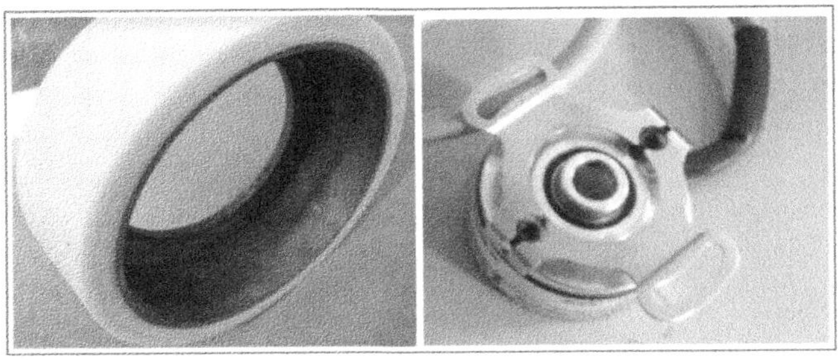

Figure 3.59. PU outer wheel and encoder of the driven wheel

Figure 3.60. Plinth used for soldering rocker arm to Lathe machined motorwheel hub.

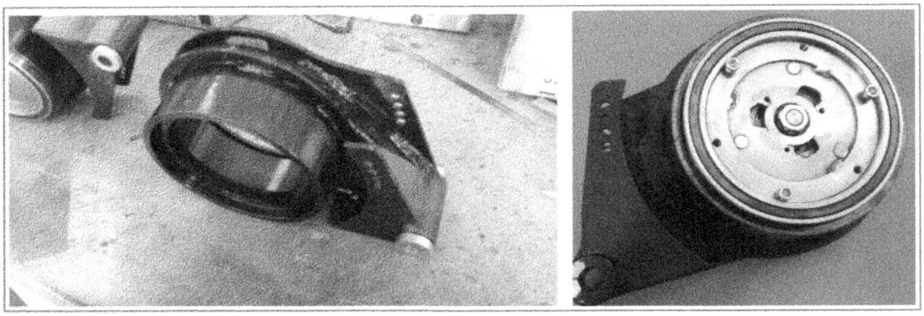

Figure 3.61. Hub and Rocker arm finished with Bronze bearing inserted.

The rocker arm is attached to the robot frame with a bronze bearing joint and is pushed to the ground by gas springs. The springs can be mounted in different positions to allow different forces, depending on the robot load.

3.1.20.1.3 Covers

The covers are designed to protect the platform avoiding external elements to enter the platform. It has been decided that electronics are mounted inside sealed boxes to avoid contact with dust or water so external covers are not needed to be IP65 but they have to be easy to decontaminate.

The top cover is a ST37 steel 4mm plate. Its corners are rounded to facilitate rotation in narrow corridors. As the robot is a first prototype, an array of M8 threads

Figure 3.62. Rocker arm.

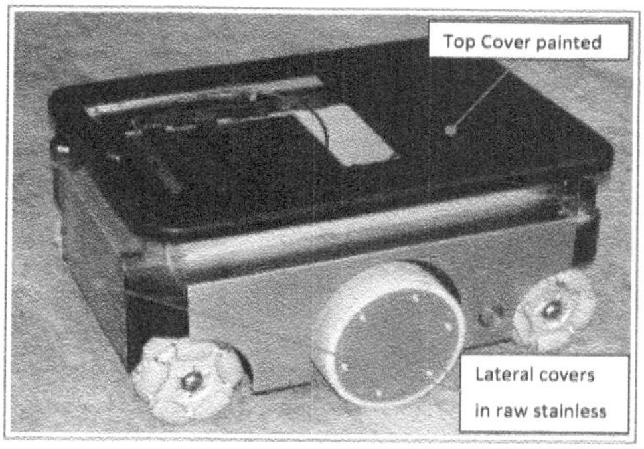

Figure 3.63. Covers.

are done in the top to facilitate attaching different elements, these holes will be sealed in future versions to avoid water splashes to enter.

The lateral covers are made of 1 mm stainless steel sheet folded. They are four independent covers to facilitate independent access to different parts of the mobile platform: front and back cover to electronics boxes and right and left covers to driven wheels and suspension system.

Driven wheel access has also been protected with the same type of stainless steel cover.

3.1.21 Mobile Platform V1 Manufactured and Ready for Subsystem Test

The mobile platform for the first prototype of Bots2Rec has been manufactured and it is ready for testing.

General parameters of finished platform have been measured:

Length: 760 mm
Width: 595 mm

Figure 3.64. Mobile platform V1 manufactured and ready for subsystem test.

Height: 355 mm
Weight: 95 kg

3.1.22 Outlook and Future Work to be Carried Out

This deliverable shows the result of the work done during tasks Task 4.4 Production and assembly of the mobile platform.

The manufacturing processes are not different from the ones used in other Robotnik mobile robots, except from the structural frame, as this kind of steel tube frame is normally not used, and so has occupied more time than expected but not adding a delay to the process.

The manufacturing process has suffered different problems that cannot be considered unusual for a prototype development process.

- Some delay on the decisions about the interfaces between the mobile platform and other subsystem have delayed the process of manufacturing. Specifically the position and configuration of the arm was defined later than expected, and the use of the inner space of the platform for the electronics has been changed during design and manufacturing process.
- It have been a delay due to problems of communication with omni wheels supplier, it is due to the fact that there is a very small number of providers of this kind of material and are based in China. It is expected that this will not

be a problem in future versions as it is planned to have stock of this type of wheel once it is validated by the complete robot tests.

- The covers attachment design has suffered small redesign due to redesign of V1 prototype redesign of cable routing.

3.1.23 Summary and Conclusion

This deliverable achieved its objectives and provided the manufacture process of the mobile platform. It has been shown the construction details and the solutions used to pass from the CAD and manufacturing drawings to a working prototype.

Next steps are to test the platform, to validate its performance, and to integrate it with the rest of the subsystems.

3.1.24 Robotic Arm

Deliverable Di5.1

Customization of the robotic arm 3D CAD drawings, manufacturing drawings and list of part for V 1

Authors: Francesco Becchi, Tim Detert

3.1.25 Deliverable in the Context of the Work Program

The Scope of this internal deliverable is to present the CAD design of the robot arm developed within the scope of WP5. The deliverable is an output from Task 5.1 “Design of the robotic arm first prototype (V1). It uses the input from previous deliverables “Di3.1 Specification of the system architecture and interfaces” and “Di3.2 Specification of the technical requirements for the V1 prototype”.

This deliverable contains the CAD files, CAD drawings, part lists and electrical drawings necessary to manufacture the first version of the robotic arm, as a part of Milestone 1 (Overall system layout and design of the first prototype and central system finished). The following section introduces the guidelines defined in the arm design trough first conceptual design. The third section details the reference final design.

3.1.26 Methods and Work Carried Out: Main Design References

The original arm layout was based on two main links (upper arm, forearm) with a 2 dofs shoulder, single dof elbow and 3 dofs wrist. The link length was defined so to match the reach requirement defined in the system specifications (2D simplified models were used for such purpose). Local motor+gearhead positioned on each

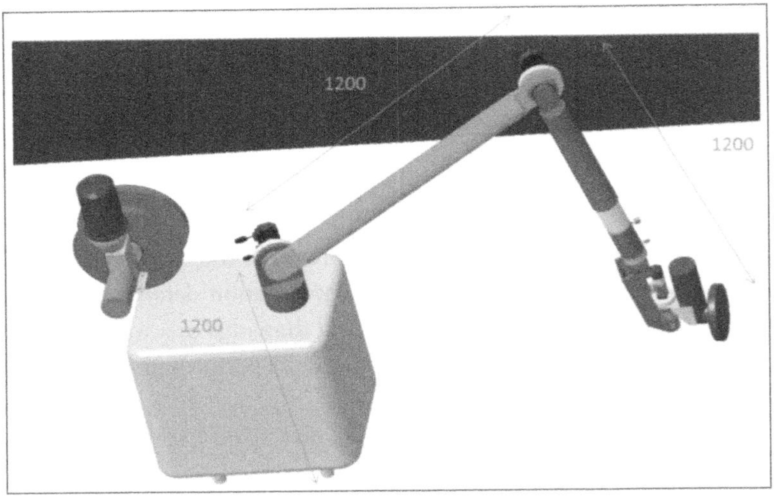

Figure 3.65. First arm conceptual design.

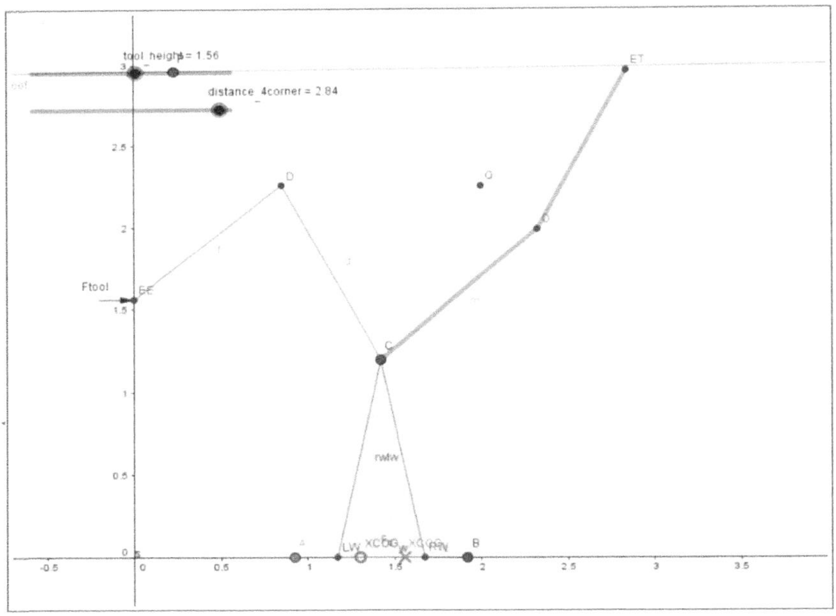

Figure 3.66. First arm 2d reach verification.

joint was the reference solution. An axial compliant slider was also included in the forearm to allow tool compliant control.

The operational requirement in limited/narrow spaces as corridors took to an alternative design where the articulated links of the robot arm were significantly reduced but the needed vertical reach was ensured adding a vertical slider.

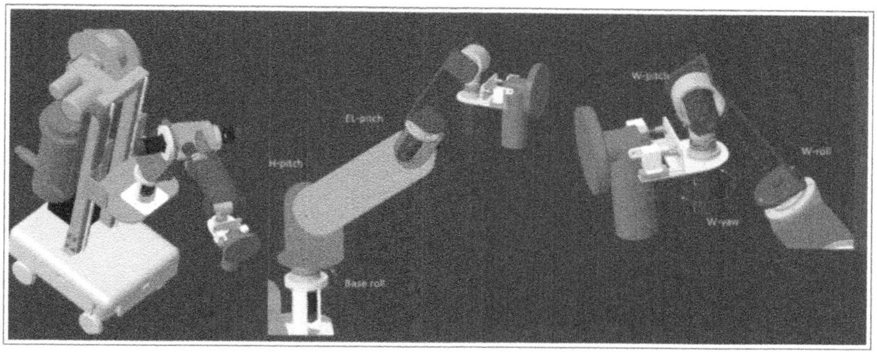

Figure 3.67. Second solution conceptual design.

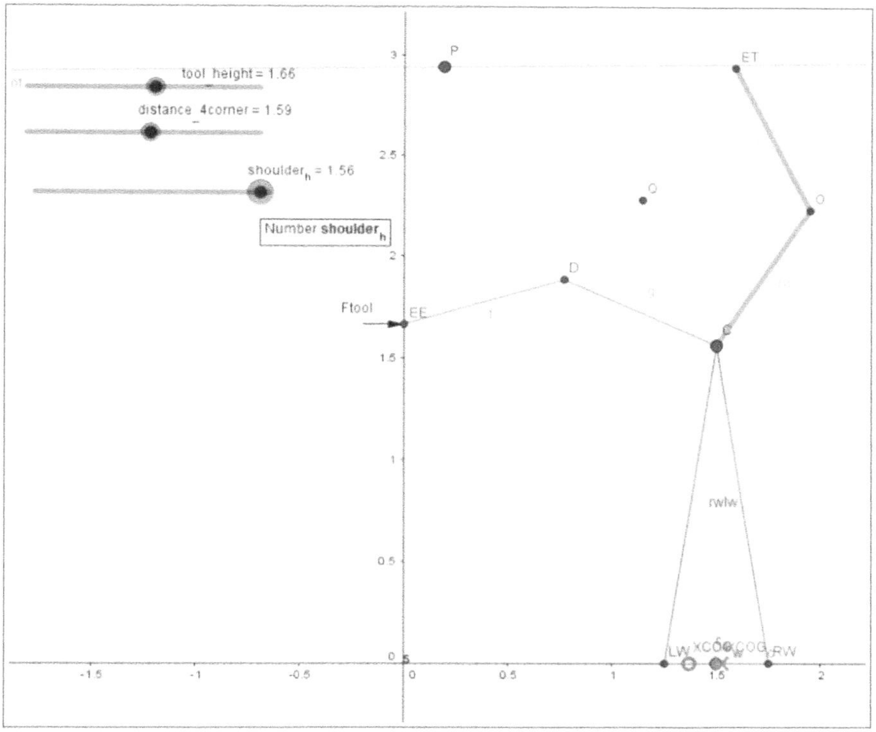

Figure 3.68. Second solution 2d reach verification.

3.1.27 Final Design

This section details de final design of the robot arm. In line with the updated conceptual design introduced in the previous section the reference design is based on a 6 dofs serial arm integrated on a vertical slider frame. All system dofs are controlled in closed loop with a brushless DC motor.

The following pictures give an overview of the robot arm including the slider and identify each DOF.

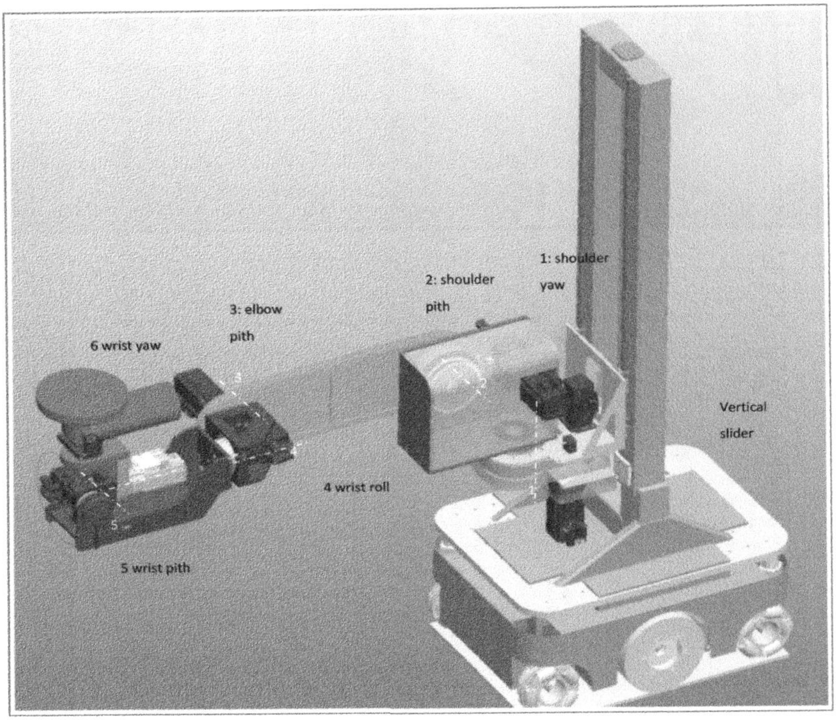

Figure 3.69. Overall view of the robot arm final design.

Each dof is based on a Harmonic driver gearbox with integrated crossed roller bearing. The Gearbox and supported bearing were sized according to the following load conditions:

- Robot arm own weight;
- Robot arm configuration: fully extended (as shown in the above picture);
- Safety factor on tool weight 2 (10 kg);
- Tool/wall-ceiling interaction force 80 N.

The table that follows summarize the load condition considered in the design.

NOTE: coding standard used in the design is based on TLABS internal QA rules. For quick reference the BOTS2REC project ID is 0000120 (internal job). Parts are identified with the "P" letter followed by a serial, sub-assembly with a "G" and top level assembly with "A".

Quick Overview	Gearhead			Torque				Bending			
	Harmonic drive model	Max output Torque (Nm)	Max bending torque (Nm)	Arm own weight (Nm)	Tool interaction force (Nm)	Total (Nm)	SF	Arm own weight (Nm)	Tool interaction force (Nm)	Total (Nm)	SF
YAW UNIT	HF UC-17-50-2UH	54	64	24.72	8.8	33.52	1.61	24.72	14	38.72	1.65
PITCH UNIT	HF UC-17-50-2UH	54	64	34.53	12.4	46.93	1.15	34.53	15.6	50.13	1.28
ROLL UNIT	HF UC-25-50-2UH	98	156	41.45	8.8	50.25	1.95	127.53	32	159.53	0.98
ELBOW UNIT	HF UC-32-160-2UH	372	313	158.22	48.8	207.02	1.80	158.22	57.6	215.82	1.45
SHOULDER	HF UC-40-100-2UH	617	850	366.89	112	478.89	1.29	129.36	120	249.36	3.41
BASE ROLL	HF UC-40-100-2UH	617	850	0.00	112	112.00	5.51	467.58	120	587.58	1.45

3.1.27.1 Vertical Slider

The vertical slider is based on a welded and machined stainless steel frame. Two parallel 15 mm four races guideways with 2 cart per rail support the main arm frame. The displacement of the vertical slider is made with a rotating trapezoidal screw driven by a brushless motor. The translating brass nut is integrated in the robot arm support frame.

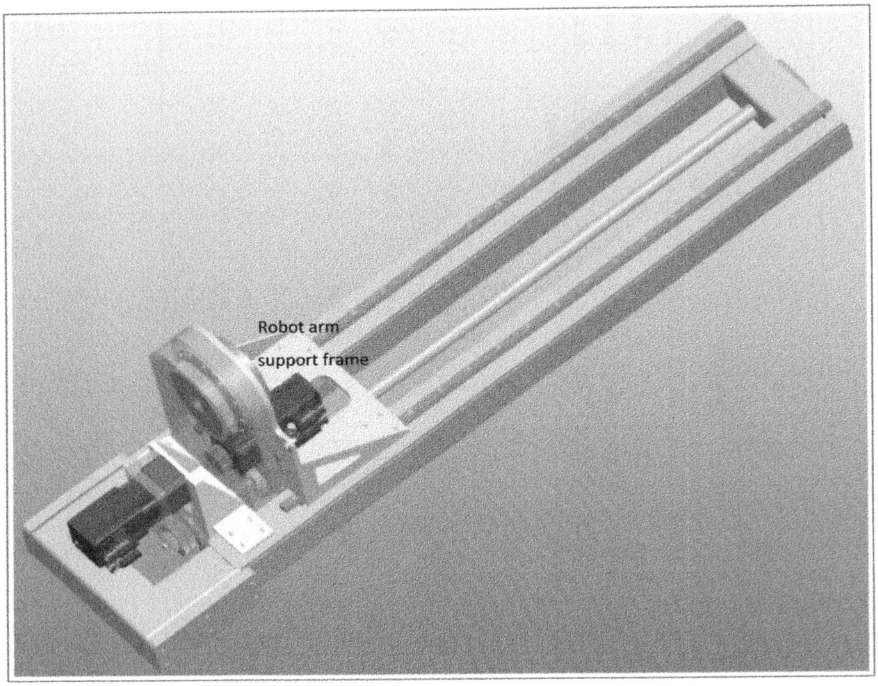

Figure 3.70. Vertical slider (T000120G001).

The support frame is made out of a welded 6000 series aluminum alloy machined on its main mating reference surfaces.

As visible in the detailed view that follows the shoulder yaw actuation is also integrated in the support frame as follows:

The HFUS gearbox is assembled on the frame so to support on its output flange the main shoulder frame. The driving motor is positioned vertically behind the shoulder frame and connected to the HFUS input flange with a timing belt transmission.

NOTE: all the timing belt transmission considered in the design are protected with a dedicated carter for safety reason.

The two red block visible in the picture are the adjustable end of stroke of the shoulder yaw. An inductive sensor is also included to perform motor calibration.

Figure 3.71. Robot shoulder support frame (T000120P002).

Figure 3.72. Shoulder yaw actuation.

3.1.27.2 Main Shoulder Frame

Following a similar arrangement than the main arm support the main shoulder assembly is made out of a welded and machined aluminum alloy frame that integrates HFUS harmonic drive and its driving brushless motor.

3.1.27.3 Robot Arm Elbow

The robot elbow assembly is based on HFUC 32 harmonic drive with integrated brushless motor on the input flange. The elbow pitch is connected to the shoulder pitch with a AL6082T6 machined aluminum alloy frame (T0012P008). The elbow output supports with a machined interface frame a similar actuation unit with an HFUC25 and its integrated motor used as wrist roll.

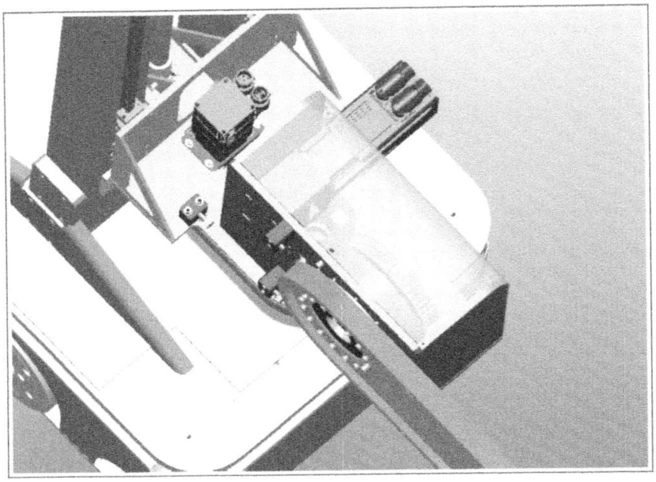

Figure 3.73. Main shoulder assembly (T0000120G006).

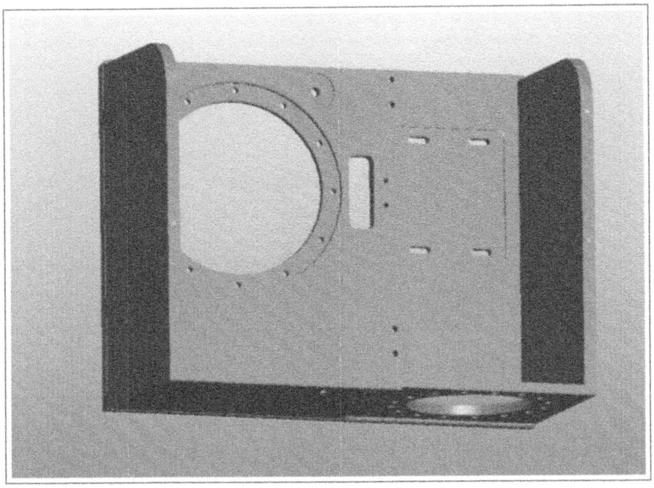

Figure 3.74. Robot shoulder main frame (T000120P060).

3.1.27.4 Robot Forearm/Wrist

The robot forearm structure is based on a special "cradle" aluminum frame designed for an optimized tool attitude control on vertical walls. The distal end of the forearm supports the last two wrist degrees of freedom both based on HFUC-17 harmonic driver gearhead.

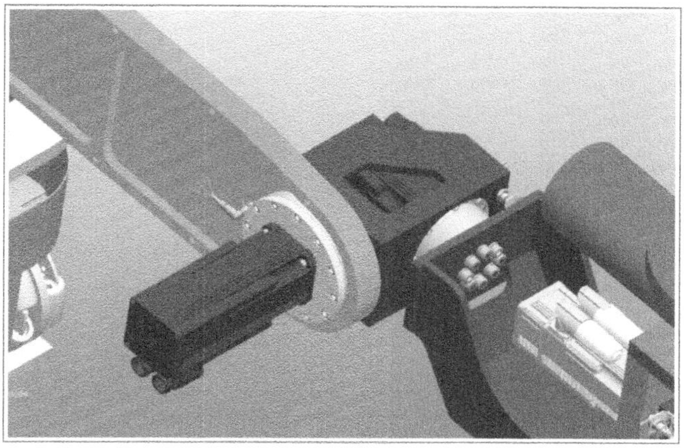

Figure 3.75. Elbow pitch joint and wrist roll.

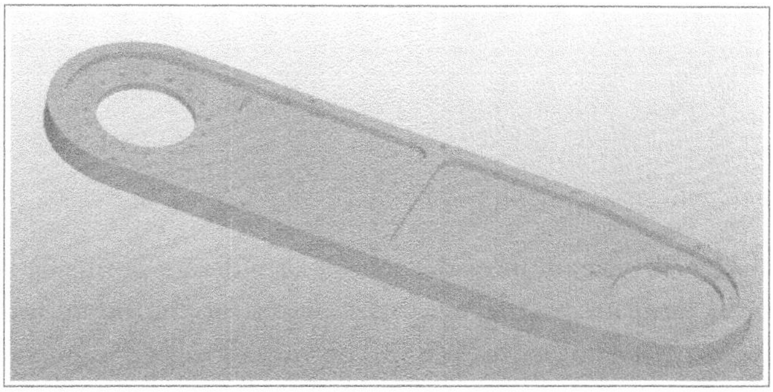

Figure 3.76. Upper arm frame (T000120P008).

Figure 3.77. Forearm wrist layout.

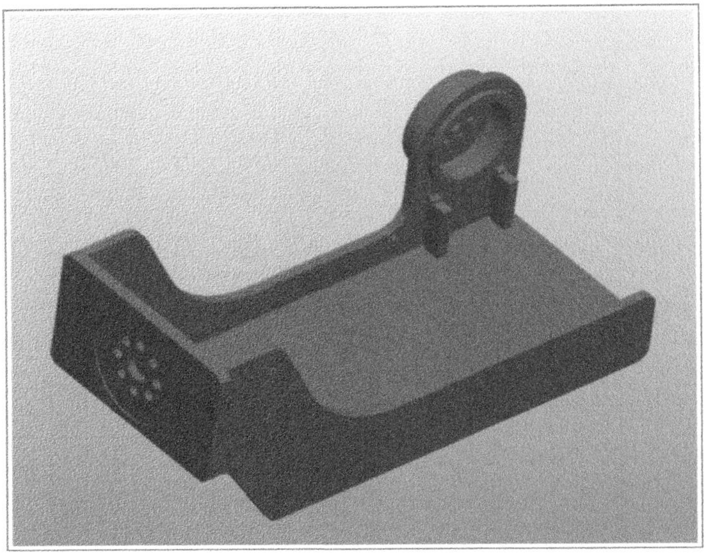

Figure 3.78. Forearm cradle frame (T000120P065).

3.1.28 Outlook and Future Work to be Carried Out

This deliverable shows the result of the robotic arm design performed during tasks 5.1. It has been done in parallel with the design of the other subsystems and the overall system integration in WP3. The overall design process was delayed with respect to the time plan: Difficult conceptual considerations for the robotic arm and controversy discussion led to a longer conceptual phase. No commercial model was able to fulfill the requirements and a new design had to be considered.

After finishing the design, the arm implementation has started but the targeted Deliverable Di5.3 ("Robotic arm V1 manufactured and ready for subsystem test", due 01.11.2016) and D5.1 ("Robotic arm tested and ready for integration ...", due 01.12.2016) will be delayed by 2-3 month. However, the current time plan should allow to keep the targeted Milestone 2 ("First Prototype available and functional") by end of 01/2017. The Christmas break and planned buffer during the end of the year will be used for the realization of the robotic arm.

3.1.29 Summary and Conclusion

The arm design was performed based on the technical requirements. After conceptual considerations mainly of the kinematic structure (considering workspace, reach

Figure 3.79. Arm mechanical integration (shoulder and elbow).

and payload/operational forces) the final design was implemented considering all technical specifications. The design steps were iterated in discussion with the partner and considering the relevant interfaces for the system integration. The design is finished and first steps towards the realization have already been performed. The following realization and testing is delayed due to an extensive period of the conceptual design, but the targeted first prototype of the robotic system will nevertheless be finished in 01/2017 by speeding up the arm manufacturing and using some buffer time at the end 0f 2016.

Deliverable Di5.3
Di5.3 Robotic arm V1 manufactured and ready for subsystem test

Authors: F.Becchi

3.1.30 Deliverable in the Context of the Work Program

Scope of this internal deliverable is to summarize the integration (mechanical and electrical) of the first prototype of the robot arm. The following section reports views of different stages of integration of the robot arm.

3.1.31 Mechanical Integration

All custom made welded and machined component with commercial components are assembled in the TLABS workshop following the robot arm design and documentation.

Figure 3.80. Arm mechanical integration (shoulder and elbow).

Figure 3.81. Vertical slider and shoulder pitch transmission details.

A temporary support frame made with a welded interface and a large assembled frame is also prepared to position the arm and its vertical slider support in its nominal configuration to perform electrical integration of arm motor and sensors, auxiliary subsystems , B1 and B2 cabinet in line with the robot assembly layout.

3.1.32 Electrical Integration

The two B1 and B2 arm electrical cabinets are first assembled on the cabinet back plate and then integrated with all cables and connectors in the cabinet. First cables are routed on the arm support structure following the path planned in CAD.

Figure 3.82. Complete arm structure on dummy frame structure ready for electrical integration.

Figure 3.83. First cable routing.

Figure 3.84. B2 and B1 cabinet first integration.

Figure 3.85. Arm cabling nearly completed.

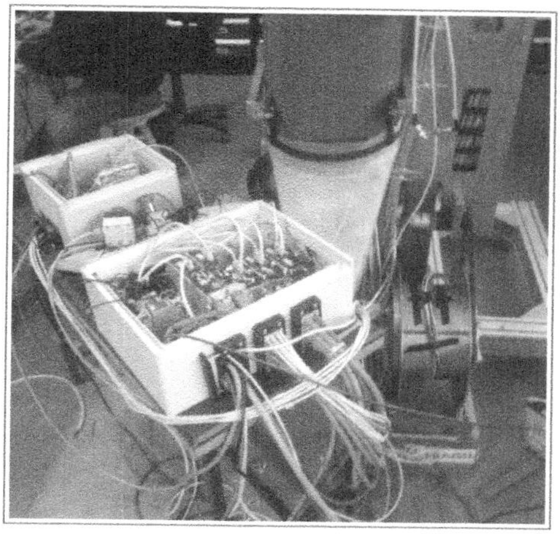

Figure 3.86. B1 and B2 cabinet in integration.

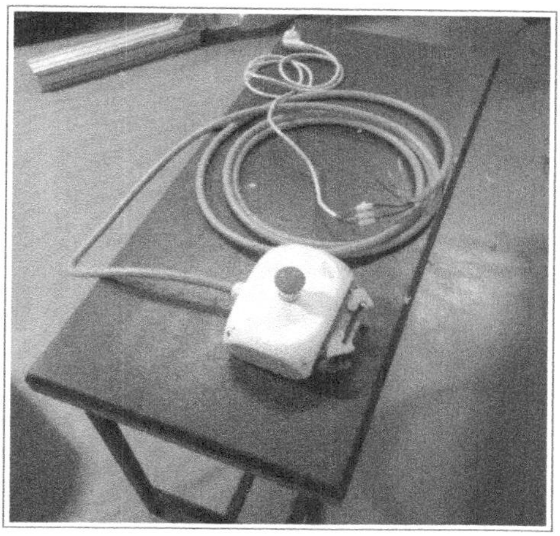

Figure 3.87. Terminal box.

Figure 3.88. Integration complete (rear view).

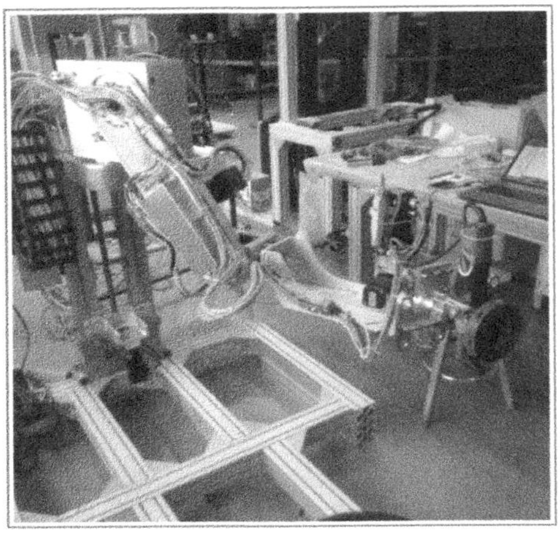

Figure 3.89. Integration complete (front view).

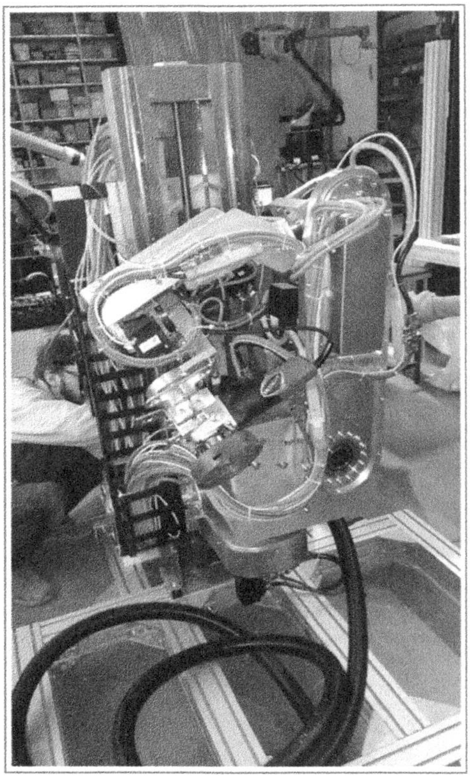

Figure 3.90. Arm in "storage" position.

3.1.33 Outlook and Future Work to be Carried Out

After completing the mechanical and electrical integration and the configuration of the different modules, a preliminary stand alone testing of the robotic arm and its functionalities is foreseen.

3.1.34 Summary and Conclusion

This internal delivery resumes with a sequence of shots made in the TLAB workshop the robot arm mechanical and electrical integration.

3.1.35 Radar Sensors

Deliverable 4.2
Sensor system for environmental perception ready for implementation

Authors: Dr. Matthias Rabel, Masrur Doostdar

3.1.36 Deliverable in the Context of the Work Program

Positioning, Navigation and Control of automated equipment relies on sensors detecting the environment. Within the Bots2Rec project a robot shall be constructed that allows remote operation and processes predefined tasks. The field of operation is asbestos removal within buildings, thus it shall work in an environment where humans hardly can work in. This deliverable focuses on the mobile part of the robot. Sensors shall be able to detect walls, obstacles, steps and holes and thus enable the mobile units to move without misadventures. Also the operators should be able to visually check the environment.

3.1.37 Methods and Work Carried Out

There is a variety of sensors available to detect objects around a robot. First, the given constraints have to be analyzed. Besides several use cases, overall, the robot shall be realizable in terms of cost during the project and afterwards in real environments. As orientation the robots cost shall be limited to about 30.000Ł. This estimate does not yet contain expensive sensor systems. Methodically, a top down approach seems to be straight forward to narrow possible options down.

First, the robot must be moved or delivered to its field of operation. This must be possible using a remote control to move the robot. Possible required sensors are:

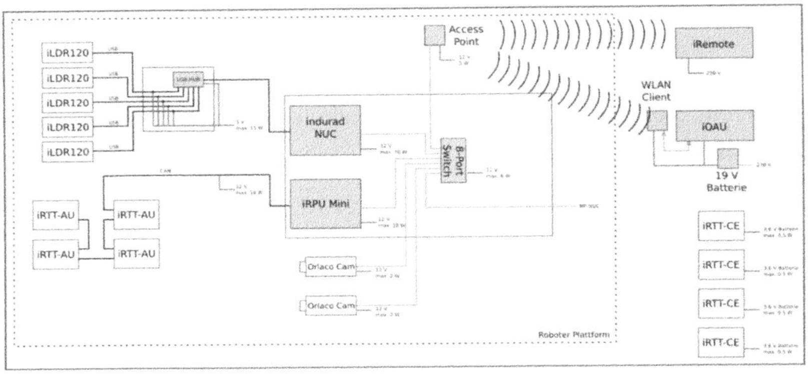

Figure 3.91. Sensor system layout, for mobile platform and robotic arm.

- Cameras so that the operator can see the surrounding
- Sensors to detect obstacles to prevent accidents, e.g., ultrasonic ranging, radars, lidar 2D or 3D, stereo vision cameras or time of flight cameras
- Indoor positioning systems, e.g. GPS-like systems working indoors; one technology known to measure distance is ultra-wide band radio ranging, measuring the time of flight of radio signals
- Movement detection sensors like acceleration sensors and gyroscopes

Second, it is worth looking out for other industries. Especially the automotive industry tries to detect the surrounding environment since several years. Gradually these systems become more and more complex. Modern vehicles have assisting systems using cameras, radars and ultrasonic devices. Infrared cameras are also being used to allow vision during foggy weather conditions.

Third, as the partner *indurad* focuses with its solutions on the mining industry with very tough environmental conditions, *indurad* is well aware that any optical system will not operate reliably. Optical sensors must be either cleaned repeatedly or protected using movable covers that can only be opened when the environment is clean. This limits possible measurement principles to radio links like UWB distance measuring or radar-based systems. Both are known to perform well in dusty, foggy and muddy environments. Nevertheless, cameras are still required to give the operator the option to see the environment.

3.1.38 Layout of the Sensor System

Figure 3.91 shows the general layout of the sensor system implemented on the current prototype V1.

Within the dashed area the mobile platform is shown. Grey colored infrastructure components are used to transfer and analyze the data. An access point is

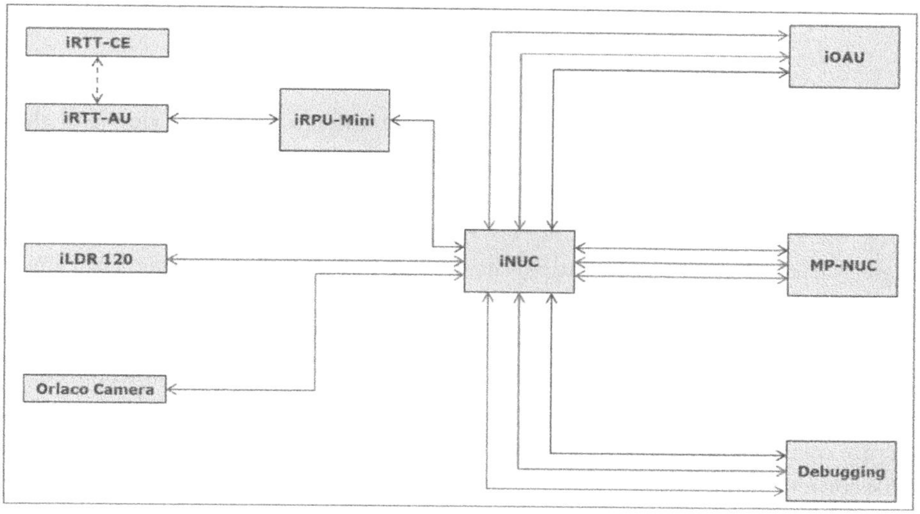

Figure 3.92. Signal and control flow.

mounted for wireless data transmission to the external system components. The sensor system for environmental perception contains:

- 2 Cameras produced by Orlaco
- 4 Devices called iRTT-AU, short for "indurad Radio Transponder Tag – Antenna Unit". It uses an ultra-wide band radio for ranging.
- 5 iLDR120 "indurad linear dynamic radar". They are mounted on the robotic arm to monitor the process locally (see Task 5.2). Together with the robotic arm these sensors could also be used to scan the environment.

Additionally to this current setup up of the sensor system, typical automotive radar, measuring a 2D segment and a 360° scanning radar, measuring distances in a 2D plane, have been evaluated preliminary.

The data flow diagram shows the data path between the main systems. iNUC[1] is the central instance of the measurement system providing the interface to the MP-NUC (mobile platform – NUC), the user interface and debug interface. Additionally there is an external iOAU "indurad operator assistant unit", an operation panel with the user interface. The iRemote is another device that allows *indurad* to remotely analyze and debug the components.

1. "NUC" is an Intel compact computer that can be mounted on the robots platform.

3.1.39 Sensors Integrated to the Current Layout

3.1.39.1 Camera System Orlaco

The camera system is used to give the operator an environmental view and shall be used to check the result of the grinding process. Deliverable 4.2 focuses on the environmental view, deliverable 5.2 focuses on monitoring the process results, while using the same cameras. The cameras are available with 120° and 180° opening angle. Variants with 90°, 60° and 30° are planned by the manufacturer. The camera allows a video resolution of up to 960p with 45 frames per second. The housing is IP68 and IP69k specified. Thus, it can easily be used in dusty environments. The camera has a high dynamic range of > 115dB and has a low light feature. Figure 2 shows the camera and a picture taken underground, with a 120° variant. The picture quality is quite good. The only area exceeding the dynamic range is at the dark corners of the image and at the light bulbs.

For now, the camera data is only intended for visual inspection. This might change later.

3.1.39.2 Ultra-Wide Band positioning system iRTT

Depending on the used definitions ultra-wide band (UWB) communication operates with a signal bandwidth of >=50MHz or >=500MHz. Bandwidths between 50MHz and <500MHz are named as wide band (WB) systems, whereas ultra-wide band uses >= 500MHz as bandwidth. The use of UWB radios is strongly limited to the intended use as described in the relevant European specifications (ETSI EN 302065). Indoor there are only power restrictions. Outdoor license free use is limited to handheld or vehicular.

UWB radios embed a fast running internal clock that is able to record timestamps on picosecond basis. Thus, the time of arrival of a message can be detected exactly. By exchanging data packets between the radio devices the time of flight can be calculated. This process is called "two-way double sided ranging", allowing to calculate the clock offset between the devices. The clock skew causing errors can be minimized. Exact measurements require a direct line of sight and ideally, multipath free environment. A theoretical distance accuracy of about 5 cm seems to be possible with a single measurement.

3.1.39.2.1 Setup and configuration of iRTT

The ultra-wide band positioning system iRTT (Figure 3.93) consists of iRTT-CE (Client Equipment), iRTT-AU (Antenna Unit) and its process control unit called iRPU-Mini. It requires additional components that must be installed at fixed locations inside the room the robot operates in. The iRTT-CE "indurad radio

Figure 3.93. Orlaco camera EMOS Ethernet; Example picture underground.

transponder tag – client equipment" is a battery operated device that can be used for this purpose.

The technology is applied within very robust housings that can be used in underground mining. Thus, there is potential to reduce its size for the second, final robot prototype 2. The antenna units are mounted on the robot, the distance between any of the antenna units shall be maximized, and also the mounting height must be variable to prevent symmetric axes or planes. E.g., if all four antenna units were mounted vertically along a line, then the system could not detect a direction, it could only detect the height and a radius. If all four antenna units were mounted within one plane then the height would be ambiguous: either the solution is above or below the plane. A way to eliminate this ambiguity is to use context information. Inside a room the robot will most likely not change its height very much. Anyway, in case of more complex rooms, the height should be detectable.

The four or more client equipment components are placed around the robot. They need to be fixed inside the room. These 4 devices can be used as anchor nodes. If the position of those four nodes is known, every node on the robot can be localized using simple trigonometric functions.

The selected mathematical algorithm is taking into account that the measurements might be wrong. Therefore an approximation based algorithm like Gauss-Newton is used. An initial localization is used to start the algorithm. The output is generated after about 50 cycles.

The results of all four measurements are evaluated against the known mounting places of the antenna units on the robot. This allows matching the robots shape into the found positions. Finally, the robots pose is determined. Additionally, an unscented Kalman-Filter implementation is used to track the position.

Figure 3.94. iRTT-AU and iRTT-CE.

Figure 3.94 shows a row of pictures with different positions of the robot. On the left, the calculated position is drawn. As additional information, the measured distances from any iRTT-CE to any iRTT-AU are drawn as circles around the anchor nodes (iRTT-CE). The large black dots are the iRTT-CE inside the room. The four colored dots are the iRTT-AU on the robot. It can be seen that the shown positions fit to the photograph.

Additionally, in Figures 3.95 and 3.96 circles have been driven and the robot was moved forward and backward. It can be seen that the measurements are repeatable. Nevertheless, the accuracy is just a rough estimation, since the robot could hardly be maneuvered in a circle with its remote control and also the algorithms are in an early stage of development

3.1.39.2.2 Perceptive Capabilities

Ultra-wideband ranging has been found to be quite sensitive in multipath environments and in environments with limited direct visibility between anchors and tags. The according situations are named LOS for "line of sight" and NLOS for "non line of sight". A number of possible design changes to improve the performance have been evaluated. Also, a system to analyze the positioning quality has been developed.

Hardware improvements

A second ranging system has been implemented using the ISM band at 2.4GHz. This allows for a higher output power of about 20dB compared to UWB. The longer wavelength allows a better visibility in NLOS situations.

An additional motion detection system is mounted to the robot platform. Measuring the acceleration will allow to calculate the motion sensor data and the iSDR data together in one model using Kalman-Filters. Furthermore, a hardware change on such a small robot would be possible due to the relatively short cables required. The anchors could be synchronized using hardware synchronization. This would allow almost synchronous operations. It reduces the measurement noise detecting the timestamps drastically.

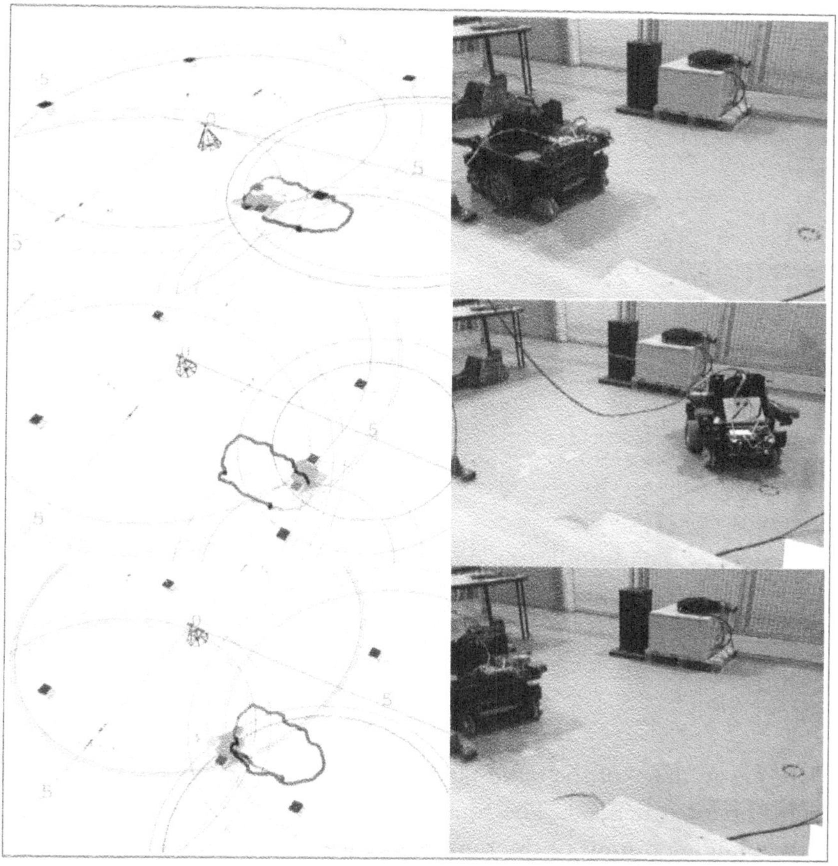

Figure 3.95. Robot position calculated/photograph.

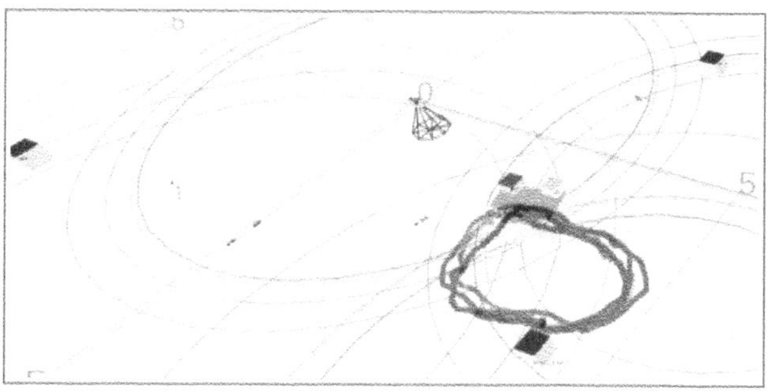

Figure 3.96. Repeatability: mobile platform driving circles.

Software improvements

Any type of ranging allows detecting a number of failures. First quality markers have been added. With every data packet these indicators are calculated, e.g., RSSI

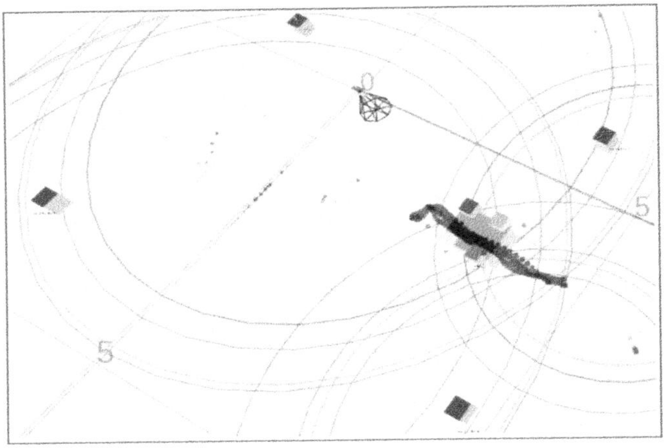

Figure 3.97. Repeatability: mobile platform driving for- and backward.

(received signal strength indicator. Second data analysis improvements have been realized to evaluate the measured sensor data, e.g., see Figures 3.97 and 3.99, both screen copies of an analysis tool. Within the analysis-tool, stored real-time data can be replayed forward, backward and with different speeds.

3.1.39.2.3 Accuracy

To verify the accuracy of the radio ranging technology a grid with squares similar to a chess board is defined. The squares are 0,5m x 0,5m. Additionally, on the robot the middle position of the robot is marked. Now, the robot can be moved exactly with its center along the chess square edges. On every node, where two lines cross each other, the robot will stop and a number of measurements are taken for statistical analyses. Figure 8 shows the white grid using the left and right 180° camera mounted on the robot. First trials have shown relatively strong deviations from the nominal positions.

In Figure 3.99 the first line shows the gyroscopes perpendicular to the robot plane. Mainly only noise can be seen. The second line shows the measured yaw value from the IMU. The third line shows the x/y coordinates of the robot and the last graph shows the deviation between the positioning and the estimated ground truth, derived from the camera pictures. It has been shown that a number of improvements must be developed, as the deviation of up to 1m is not yet acceptable. A major reason for the huge deviation is that on the second trial performed on the testing site in Valence now the mobile platform is mounted together with the manipulator, this causes numerous NLOS situations. For now there is no multipath filter enabled, although the measurement system has detection possibilities for multipath situations.

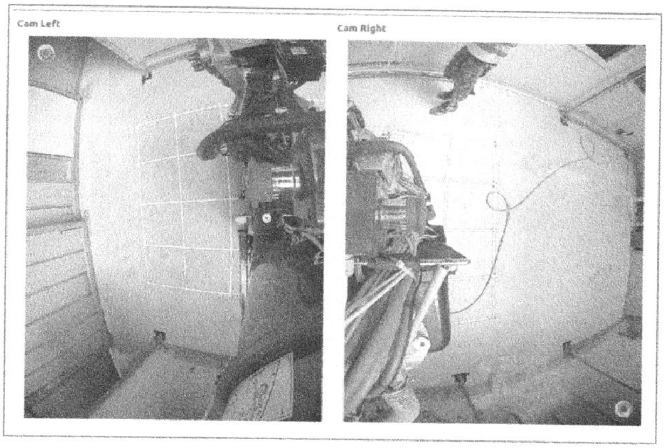

Figure 3.98. Accuracy evaluation using a grid.

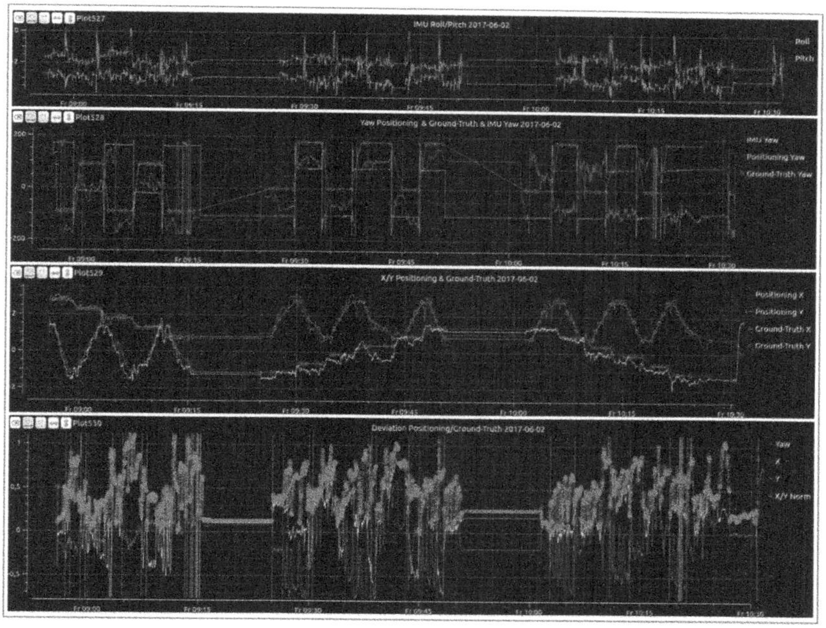

Figure 3.99. Inertial data, position data and deviation between measurement and nominal position over time.

3.1.40 Sensors to Augment Environmental Perception

3.1.40.1 360° Scanning Radar iSDR

The iSDR is a rather large device optimized to detect objects in several hundred of meters. A rotating parabolic mirror focuses the radar beam. It is a highly configurable device that allows using the whole frequency range usable for vehicular

Figure 3.100. Indurads 360° scanning radar iSDR.

applications (76–81GHz). Its beam focus is about 1.4° (+/-0.7°). Its angular resolution is also about 1.4°. Recording the peaks with a mechanical angular resolution of 0.18° might allow detecting the angles quite exact. 5GHz as bandwidth allows unambiguous distance measurements down to 3cm. Rotation rates can be applied from 0 to 15 rounds per second (Figure 3.100).

Although these values are promising and tests have shown rather good results, the device is far too large and heavy to be used on the robotic platform of the project. Due to the fact that the device might be replaced with a far smaller variant it is still worth evaluating.

The iSDR output is a list of detected targets of every single measurement, together with its measurement angle. The list is truncated by a minimum and

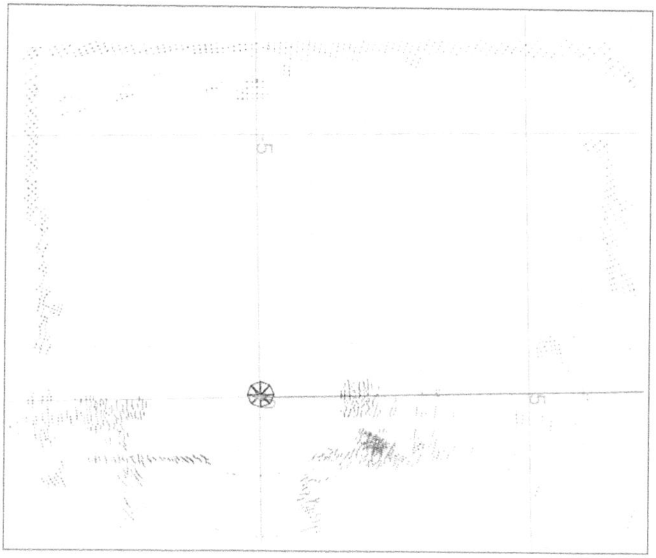

Figure 3.101. iSDR example of point cloud.

maximum distance. Also a minimal signal strength value can be configured. The detected outputs are passed with the signal strength in logarithmic scale. Figure 11 shows an example of such a point cloud. One grid square is 1m. It can be seen from the origin 1.5m down left and down right, there seems to be a circle. This is the set minimal distance, so the near multipath reflections of the antenna and random are suppressed. In the scope of task 6.2 these point clouds will be further analyzed and used for localization and mapping of the environment.

3.1.40.2 360° Scanning Radar iSDR-C

A new and far smaller scanning radar has been developed. In Figure 3.102 a new prototype is shown. In comparison to the one shown in chapter 3, the sensor has a smaller antenna. The opening angle is $+/-4°$. The estimated maximal measurement distance is about 50m. Also fast moving objects can be measured. The rotation rate is specified with max. 25 rotations per second. *indurads* product name is iSDR-C (indurad scanning dynamic radar – compact). The current development state is a fully new designed housing. Internal, the radar technology is based on the same pcbs as in the large iSDR. With numerous patches, the radar is ready to use. Further work has to be done to finalize the pcb redesign and the design of the motor control unit.

The iSDR-C is planned to be mounted horizontally, this allows to detect not only all objects around the robot. It will also allow detecting the robots movements and rotations directly from distance measurements and changes. In comparison to

Figure 3.102. iSDR-C.

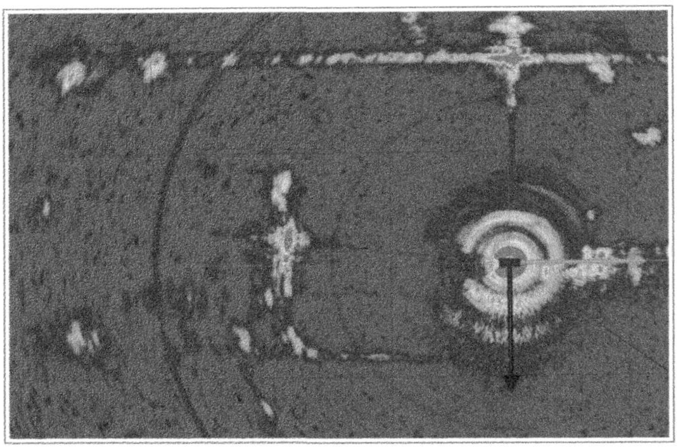

Figure 3.103. iSDR-C vertical scan with person.

speed measurements, e.g., with a Doppler radar there is no integration step required anymore. Even inertial motion data must not be used anymore. Detected accelerations must be integrated two times to calculate the movements.

Its principle operation was shown using one iSDR-C scanning a vertical plane. The iSDR-C was lying on a desk. The ceiling, floor and wall can be seen. In the middle left side a person walking through the measurement zone can be seen in Figure 3.103.

A first prove of concept was done, by reconstructing the surrounding fixed environment. The single 360° pictures of the radar are stitched together using 2D cross

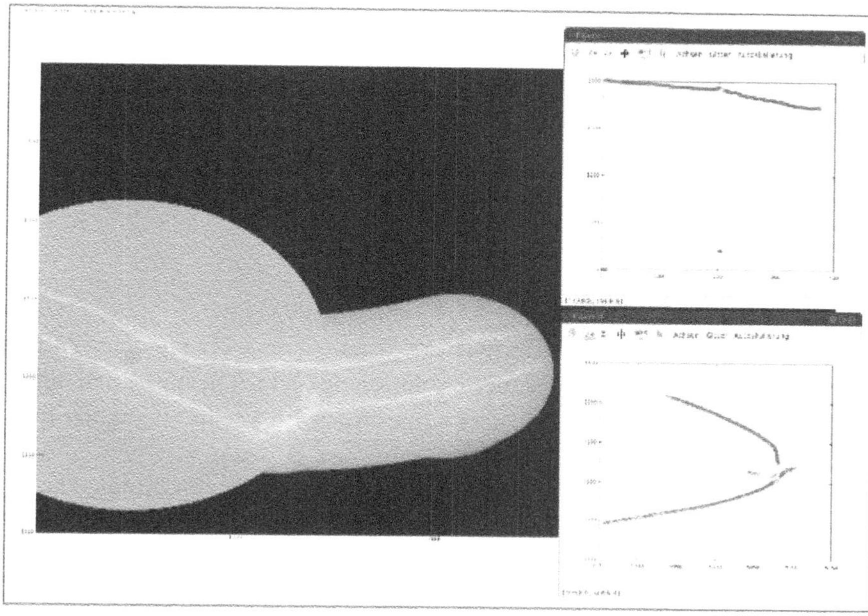

Figure 3.104. Reconstructed environment.

correlation. In Figure 3.104 such a picture is shown. The small graphs on the right show the detected angle with 2000 increments per 360° on the top, on the bottom a position is shown. (Please note the axes are scaled not equally). In Figure 14 it also can be seen that there has been obviously a wrong detection. There, the angle of the robots pose jumped to another angle and back. A complete system would use the radar data, the data of radio ranging and also the data measured by inertial measurement units. This will help to stabilize the stitching process and the speed detection.

3.1.40.3 Further Sensors

In addition to the previously mentioned sensors the LIDAR sensor technology used for the mobile platform (see Deliverable 4.1) is available for any non-dust situation. Furthermore the following sensors might be included in the future.

3.1.40.3.1 Inertial Sensor

As already stated in the subsection 4.2 additional acceleration sensors and gyroscopes might be used as additional sensor source to stabilize the UWB measurements. It allows evaluating the ranging information from measurement to measurement against the inertial sensor. Thus, invalid detection values can be identified more reliably and furthermore, using dead reckoning can be used during phases with poor ranging information. If applicable, the fusion of IMU measurements with exteroceptive sensors is in the scope of Task 6.2.

3.1.40.3.2 Lidar and 3D Camera

Lidar sensors and 3D cameras are also evaluated in Task 6.2. They might be used as initial measurement tools before grinding and maybe with a separate high precision environment sensing vehicle.

3.1.40.3.3 Automotive Radar

Typical automotive radar allows to measure distances on a 2D segment. A typical angular resolution is 2° with a distance resolution of 20cm. The horizontal opening angle is 16°, additionally there are near range areas with a reduced angular resolution of 4° and an opening angle of 50°.

The radar was primarily developed to detect other cars, e.g., for distance-keeping cruise control systems. More modern variants also allow break-assistance systems detecting objects as large as humans. The measurement rate is 10 to 15 measurements per second.

These radar types are optimized for environmental perception on a horizontal plane with comparably small required opening angles. To detect the surrounding of the robot a minimum of eight devices must be mounted, either on the robot or inside the room the robot shall work in.

In principle, the radars seem to be usable, however from experience *indurad* knows that in complicated multi-target situations the primary limit of the radars is the data rate of the CAN bus they are attached to. The bandwidth is limited, thus, only the targets with highest reflection value are transferred. This means dry wall materials like wood or drywalls hardly can be seen, but objects behind or inside them. Also, small objects might not be detected.

Limited distance resolution is another drawback. It is not easy to detect the position inside a room while moving. While standing still, statistical calculations on a number of measurements might allow sufficient position detections. In conclusion, the iSDR(-C) is preferable compared to automotive radar sensors.

3.1.41 Outlook and Future Work to be Carried Out

The basic sensor system has been developed successfully. Further work has to be done to analyze the camera system, including mounting positions, number of cameras and view angle. Also, motorized DOM cameras might be used in future to allow detailed analysis of grinded surfaces.

Since the last reporting period a major step has been done miniaturizing the iSDR. A functional prototype is now available.

The ultra-wide band system has shown to be able to detect the robots position and pose. Further work has to be conducted to verify its accuracy. First, the UWB measurement itself must be improved by taking the antenna delay into account. The antenna delay depends on the direction and on the mounting point at the robot. Therefore, calibration mechanisms have to be implemented.

The algorithms must be revised, such that all measurement results for all antenna units are used at one time and to deal with multipath effects. This gives the least square minimum optimized over all parameters. The unscented Kalman-Filter must be optimized to increase the output speed and minimize the deviation in case of single wrong measurements. Methods must be implemented to evaluate the validity of a single measurement. One of the preliminary findings is that it is better to skip possibly wrong measurements than to use them.

The whole localization system must be usable by the end-users of the asbestos removal robot. Thus, a simple method to setup the iRTT based localization system must be developed.

The final choice for a sensor system and thus the final improvements depend on the results of task 6.2, where the localization and mapping algorithms are developed. Therefore, the final adjustments of the sensors, both iRTT and iSDR, will be conducted in close cooperation with the participants of task 6.2.

3.1.42 Summary and Conclusion

Within Deliverable 4.2 a variety of sensors that can be used to detect the environment and the robots position and orientation have been analyzed. The sensors have been preselected on a nearby basis. The main criteria have been size and cost. Ultra-wide band localization is a promising cost effective solution. Cameras are required to help the operator, to document the grinding result and for users convenience. Inertial sensors can be jointly used together with other sensors. Mathematical methods like unscented Kalman filters can be used to track the robots position realizing the sensor fusion.

The work package 4.2 is not fully finished, as final design and development steps still need to be performed for some sensors. Further developments on the sensor systems are ongoing and will accompany the further developments of the prototypes. The functionality and accuracy of the UWB will be developed and analyzed further and the integration of the first functional iSDR prototype with interfaces to the robots is seen as a next major step to the completion of all systems.

A further main step is to implement a user interface to set up the UWB localization system conveniently. Ideally, this will be done using the iOAU based wireless user interface. With the touch screen the position of the anchor nodes mounted in the room shall be set in the embedded computer calculating the pose.

The delivery of this deliverable report was delayed due to the ongoing technical developments and the absence of some project members during the summer vacations. The delays are however not severe and the ongoing developments are continued, to ensure not to slow down the overall prototype development.

Deliverable 4.2 – Appendix
Control system of the mobile platform layout and ready for implementation

Authors: Dr. Matthias Rabel, Lucas Rohe, Robin Bernau, Martin Gritzan

3.1.43 Deliverable in the Context of the Work Program

Deliverable 4.2 has been completed since 2017-03-08. Within the related work package various sensors that can be used to measure the robot's position and orientation in the room have been introduced. One finding has been that the radio ranging system (iRTT) is not as precise as required to detect the accurate position. Thus, iRTT was not included in the second version of the asbestos removal robot. Instead the scanning radars are used to measure in 2D the surrounding environment. With simultaneous location and mapping algorithms the robot position can be detected exact.

During the second phase of Bots2Rec iRTT system was further analyzed and even modified to improve the system. The report covers this work.

3.1.44 Methods and Work Carried Out

This report is an appendix to Deliverable 4.2. The introduction of the radio ranging system (iRTT) which is mainly based on Ultra-Wide Band (UWB) communication is repeated for convenience and then extended with details to the hardware architecture and additional features not yet reported. A third communication system based on Phase Measurements that is embedded in the iRTT system is introduced and analyzed. Secondary a different ranging scheme based on time difference of arrival (TDOA) is introduced. Finally an improved multi-tag tracking method that uses inertial motion sensor data additional is introduced.

3.1.44.1 iRTT Hardware Platform

The following section 3.1 has been reported already in Deliverable 4.2.

Figure 3.105. iRTT-AU and iRTT-CE.

3.1.44.2 Ultra-wide band-based location detection

Dependent on the used definitions ultra-wide band (UWB) communication operates with a signal bandwidth of >=50MHz or >=500MHz. Bandwidths between 50MHz and <500MHz are named as wide band (WB) systems, whereas ultra-wide band uses >= 500MHz as bandwidth.

The use of UWB radios is strongly limited to the intended use as described in the relevant European specifications (ETSI EN 302065). Indoor there are only power restrictions. Outdoor license free use is limited to handheld or vehicular.

UWB radios embed a fast running internal clock that is able to get timestamps on picosecond basis. Thus, the time of arrival of a message can be detected pretty exact. By exchanging data packets between the radio devices, the time of flight can be calculated. This process is called "two-way double-sided ranging". With this the clock offset between the devices can be calculated. The clock skew causing errors can be minimized. Exact measurements rely on a direct line of sight and in best case multipath free environment. A theoretical distance accuracy of about 5cm seems to be possible with single measurement.

indurad has developed ranging devices using UWB. Two of them are iRTT-AU and iRTT-CE (Antenna Unit and Client Equipment). They have been developed for different purposes but are reused in a variation within the Bots2Rec project (Figure 3.105).

The technology is applied within very robust housings that can be used in underground mining. Thus, there is potential to reduce its size for the second, final robot prototype 2. The antenna units are mounted on the robot, the distance between any of the antenna units shall be maximized, also the mounting height must be variable to prevent symmetric axes or planes. E.g., if all four antenna units were mounted vertically along a line, then the system could not detect a direction, it could only detect the height and a radius. If all four antenna units ere mounted within one plane then the height would be ambiguous: either the solution is above or below the plane. A way to get rid of this ambiguity is to use context information. Inside a room the robot will most likely not change its height very much. Anyway, in case of more complex rooms, the height should be detectable.

Around the robot the 4 or more client equipment components are placed. They need to be fixed inside the room. These 4 devices can be used as anchor nodes. If the position of those four nodes is known every node on the robot can be localized using simple trigonometric functions.

The selected mathematical algorithm is taking in account, that the measurements might be wrong. Therefore, an approximation based algorithm like Gauss-Newton is used. An initial localization is used to start the algorithm. The output is generated after about 50 cycles.

The results of all four are measurements are again evaluated against the known mounting places of the antenna units on the robot. This allows matching the robots shape into the found positions. Finally, the robots pose is determined.

Additionally, an unscented Kalmannfilter implementation is used to track the position.

Figure 3.106 shows a row of pictures with different positions of the robot. Left, the calculated position is drawn. As additional information, the measured distances from any iRTT-CE to any iRTT-AU are drawn as circle around the anchor nodes (iRTT-CE). The large black dots are the iRTT-CE inside the room. The four colored dots are the iRTT-AU on the robot.

It can be seen that the shown positions fit to the photograph.

Additional in Figures 3.107 and 3.108 circles have been driven and the robot was moved forward and backward.

It can be seen that the measurements are repeatable. Nevertheless, the accuracy is just a rough estimation, since the robot could be hardly maneuvered in a circle with its remote control and also the algorithms are in an early stage.

Further work has to be conducted to improve the accuracy of the whole system. First the UWB measurement itself must be improved by taking the antenna delay into account. The antenna delay depends on the direction and on the mounting point at the robot. Therefore, calibration mechanisms have to be implemented.

The algorithms must be revised, such that all measurement results for all antenna units are used at one time. This gives the least square minimum optimized over all parameters.

The unscented Kalmanfilter must be optimized to increase the output speed and minimize the deviation in case of single wrong measurements. Methods must be implemented to evaluate the validity of a single measurement. One of the preliminary findings is, that it is better to skip possibly wrong measurements than to use them.

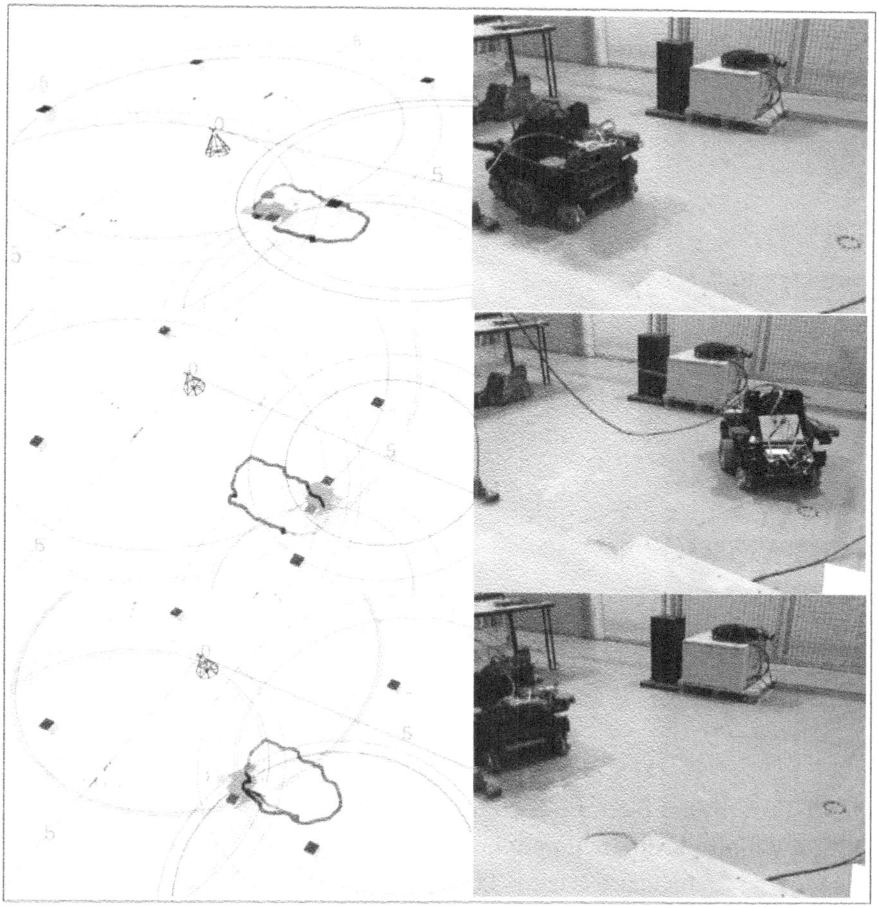

Figure 3.106. Robot position calculated / photograph.

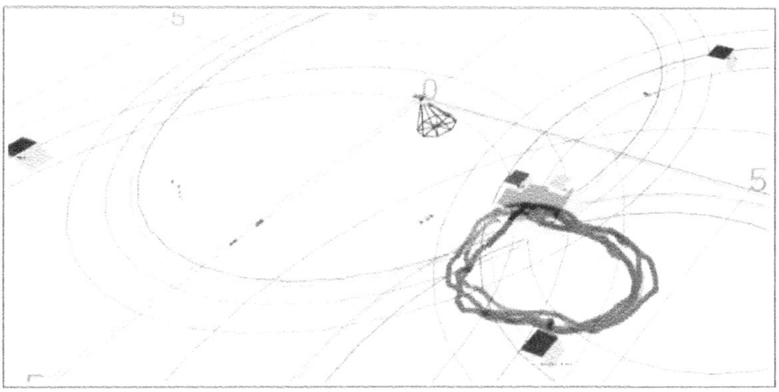

Figure 3.107. Repeatability: mobile platform driving circles.

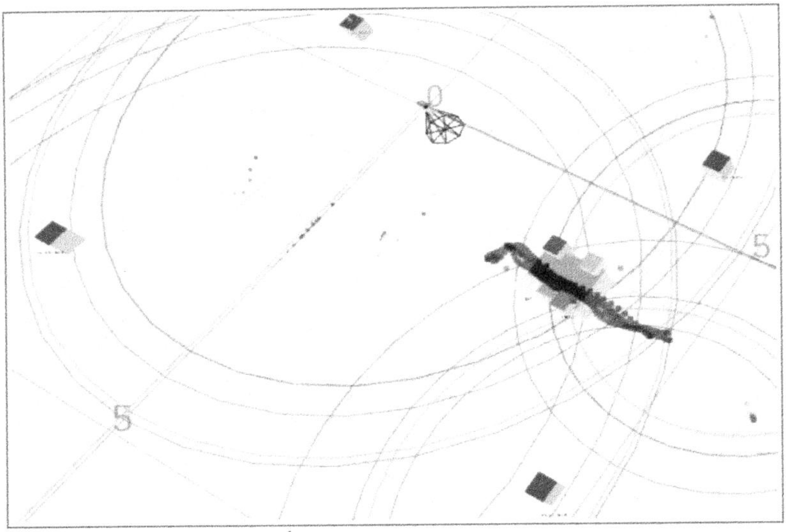

Figure 3.108. Repeatability: mobile platform driving for- and backward.

The whole localization system must be usable by the end-users of the asbestos removal robot. Thus a simple method to setup the iRTT based localization system must be developed.

3.1.44.3 Anchor System

The iRTT Hardware Platform is comprised of an indurad Radar Processing Unit (iRPU) and multiple different peripheral devices. The iRTT Antenna Units (iRTT-AU) are used as vehicular receivers. Their positions within the robot's geometry are known and form the ground truth for tracking calculations. Yet, Antenna Units can act as ranging anchors and tags alike which is further detailed within Section TODO. An iRTT-AU unites radio interfaces for the three different channels within one circuit board as shown in Figure 3.109.

One microcontroller (STM32) is used as the main processor of each Antenna Unit and serves multiple purposes. The UWB Ranging protocols are implemented within the main processor as it interfaces the Decawave DW1000 UWB radio frequency interface chip directly. The main processor computes distances for the time-of-flight (TOF) ranging scheme and forwards UWB timestamps required for accurate tracking via CAN bus towards the iRPU. Also, the main processor maintains configuration data accessible to the tracking logic through a specialized CAN-based protocol. Furthermore, it serves as a proxy for communication via the Sub-1 GHz radio.

The Sub-1 GHz channel is used for device presence detection and meta data communication between remote iRTT devices only. The robot's local iRPU can

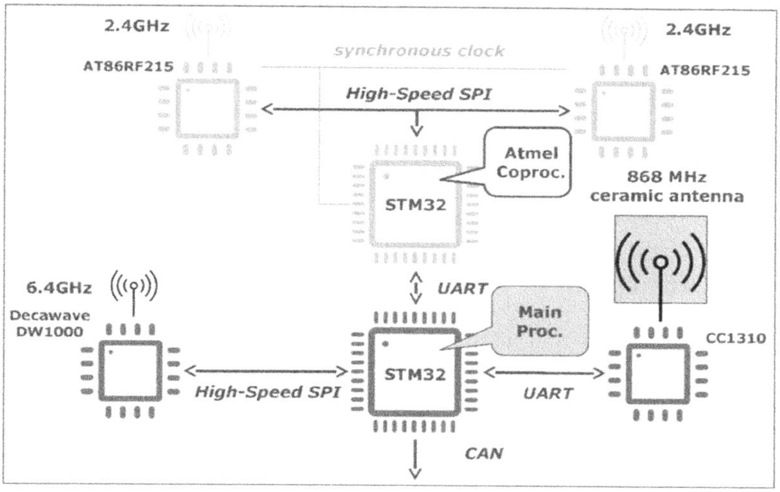

Figure 3.109. Sensor system layout, for mobile platform and robotic arm.

issue commands towards remote devices via the CAN bus which are forwarded through the main processor towards the dedicated coprocessor which broadcasts those messages towards all iRTT devices in the vicinity. A Texas Instruments CC1310 System-on-Chip is used for this purpose as it incorporates a user-programmable ARM-based processor alongside the actual radio interface.

The two ISM 2.4 GHz radio interface chips need strict timing requirements to be fulfilled for accurate measurement results, especially for the antenna diversity measurement scenario. A second STM32 is introduced as the dedicated ISM coprocessor. It services a specialized SPI interface towards the two radio interface chips to achieve the timing requirements. A serial connection to the main processor is maintained to exchange commands and results.

3.1.44.4 Inertial Sensor Extension

As stated in the report Deliverable 4.2 in the subchapter 3.1, a 6-axis inertial measurement unit has been connected to the extension interface of the radio ranging anchor hardware.

The additional acceleration sensors and gyroscopes are used as additional sensor source to stylize the UWB measurements. It allows evaluating the ranging information from measurement to measurement against the inertial sensor. Thus, invalid detection values can be detected more reliably and further more using dead reckoning can be used during phases with poor ranging information. Chapter 5 describes an experiment with a combined setup that uses UWB and the acceleration sensors.

3.1.44.5 Tag System

The iRTT Client devices serve as ranging tags of known position within the robot's working area. Much like the Antenna Unit the client unites all three radio interfaces on one PCB, however, with focus on the comparably smaller size and power consumption required for operation independent of mains supply and without hard-wired connection to any supervising iRPU. The structure is shown in Figure 3.110.

Within the client the same Texas Instruments CC1310 also applied as the iRTT-AU's Sub-1 GHz coprocessor is used as the main processing unit which interfaces all RF chips directly.

For UWB the Client only serves as a Tag and does not compute any results. Hence, the required logic is much simpler compared to the Antenna Unit's implementation. Same holds true for the ISM ranging where only a single radio interface is applied for the sake of both device size and battery life. This also reduces the timing requirements' strictness compared to the Antenna Unit allowing the Client to serve both RF chips through the same serial interface.

An inertial measurement unit is introduced for orientation detection on the Client devices. The orientation information can be used to optimize tracking based on the known antenna geometry.

3.1.44.6 Time of Flight (TOF)

The so called Three-Way-ToF method visualized in Figure 3.111 enables measurements of distances between transponder tags and stationary reference anchors. For each distance measurement three data packets are exchanged: (1) tag → anchor, (2) anchor → tag, (3) tag → anchor.

This results in six timestamps which can be used to determine the signal propagation delay between the tag-and anchor pair. Using the speed of light, the delay values can be converted into distance information. This method does not require

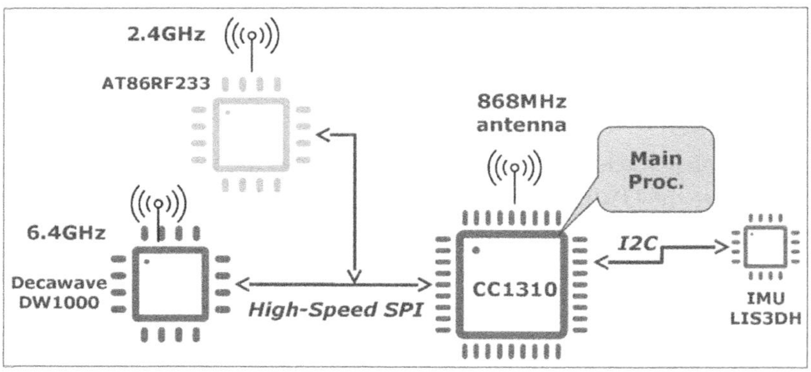

Figure 3.110. Signal and control flow.

the clocks of the devices to be synchronized. Anyway, the assumption is that the clocks are running stable. In fact, the phase noise which is in principal an inherent part of any clock source is a major source of errors.

The graphical representation of a single ToF in 3D space is a sphere with radius r = distance and center position p = position of the anchor. For the positioning of a tag in that 3D space at least four ToF measurements from different anchors are required in order to get an unambiguous point of intersection. In Figure 8 three small circles and some more large circles can be seen. The blue cube is drawn at the point where all circles have their intersection. In static measurement situations the result is very stable.

In case the tag moves, the distances between multiple anchors and one tag must be acquired within short time. This is achieved by an optimized channel access algorithm. Anyway, the sequence of measurements dependent on the tag speed will always lead to an error. Robust tracking algorithms must be deployed.

The initiating handshake packet is transmitted by the tag as a broadcast message. Each anchor is assigned a time slot to respond within when it receives such. For each responding anchor the tag stores response timestamp information and publishes all these in the final message of the handshake. This scheme enables up to 24 anchors

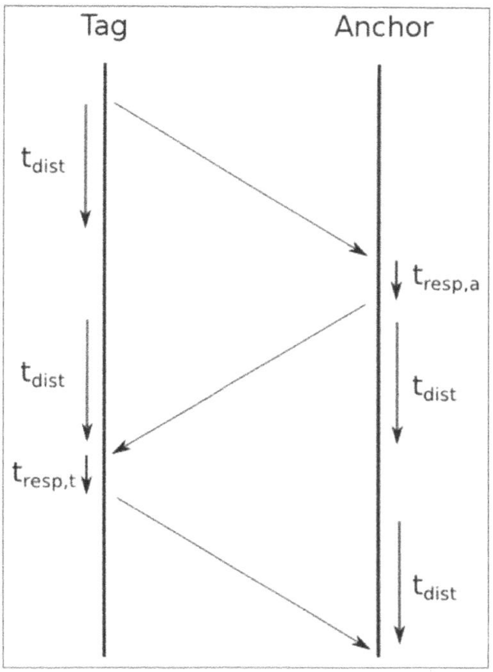

Figure 3.111. Time-of-Flight (ToF) Three Way Ranging.

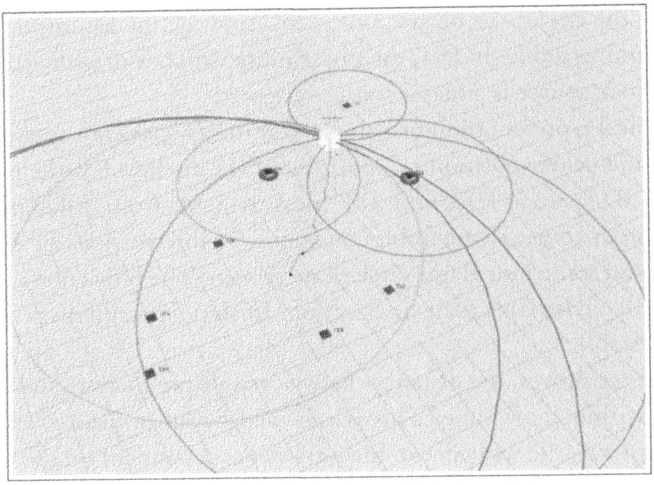

Figure 3.112. Visualization of Time of Flight (ToF) as distance circles between anchors and tag (blue block).

to measure distance to a single tag within one set of handshakes, taking only about 30 ms of air time.

Especially the ToF measurements rely on the radio interface chip to accurately measure the frames' timestamps. These measurements can, however, be impaired by radio frequency packet collisions. Such collisions must be avoided as much as possible. The probability of a collision occurring is reduced by means of configuring the anchors to each respond in a different time slot and by the tags modifying their handshake initiation period at random. These two precautions are found to contribute largely to the quality of ToF measurement results.

3.1.44.7 Time Difference of Arrival (TDoA)

The second method, TDoA, is based on differences of the propagation delay between the tag and a number of different anchors. The tag sends out broadcasts which are received by the anchors at different points in time (Time of Arrival, ToA). As long as the clocks of the anchors are synchronized the differences of the receive timestamps (Time Difference of Arrival, TDoA) can be determined. Due to the speed of propagation (speed of light) of the electromagnetic waves this method demands an error in synchronization smaller than a nanosecond.

The synchronization is based on dedicated data packets (Time of Send and Arrival, ToSA) that are sent out via radio waves from the synchronization reference anchors with a sufficiently high rate of 3 Hz. This way a time delay and drift of the clocks can be calculated and the clocks of the different anchors can be translated into a global time system.

Alternatively, the reference clocks for all UWB radio interfaces can be driven from a single clock source. Additionally, a synchronization pulse can be distributed among all devices to reset their timestamp counters to zero at about the same time. A small inaccuracy is still introduced by the different electrical lengths between the central clock distribution and the individual devices and has to be calibrated out during the subsequent tracking algorithm.

A TDoA measurement can be represented as a 3D hyperbola whose inflection point lies in between the receiving anchors and corresponds to the tag position. For the positioning in 3D space at least five hyperbolas are required.

The iRTT system uses both methods in order to compensate the disadvantages of each method with the advantages of the other: The ToF method does not scale well with many participants (transponder tags). A measurement of one tag with 24 anchors lasts approximately 20 ms during which time no other tag can initiate a measurement. The advantage is the direct result which is consistent and does not need an additional complex processing afterwards.

For the TDoA method the clocks need to be actively synchronized and two timestamps are necessary to obtain one TDoA measurement. The processing and the underlying communication protocol are more complex. The key advantage of the TDoA method is the scalability. A TDoA broadcast takes less than 1 ms can be received from many anchors at the same time. This enables setups with a large number of anchors and tags.

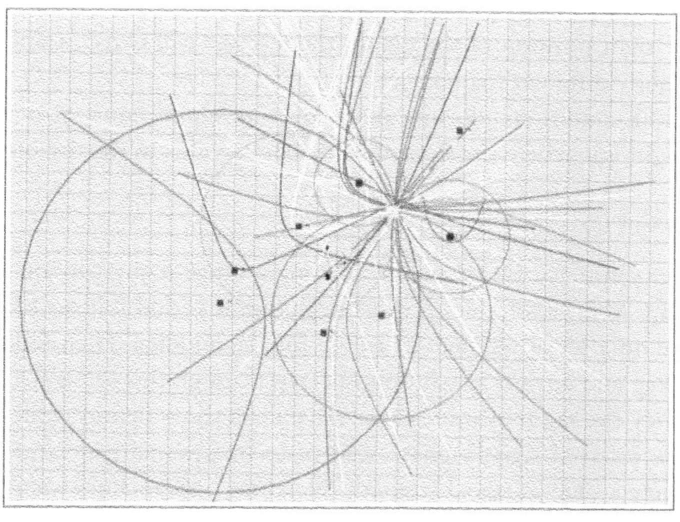

Figure 3.113. Visualization of TDoA measurements (hyperbolas) and ToF measurements (circles).

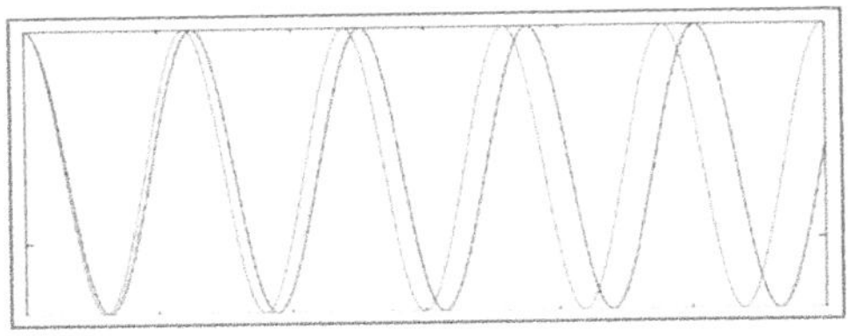

Figure 3.114. Small frequency differences lead to a changed phase position in the receiver.

3.1.45 Phase Measurement System

This ranging system operates within the 2.4 GHz ISM (industrial scientific medical) band. The IEEE 802.15.4 channel 26 at 2.48 GHz is used for negotiation purposes with QPSK modulation scheme while the actual ranging uses a frequency sweep across the whole ISM band from 2.405–2.483 GHz.

3.1.45.1 Phase Measurement Method

A phase measurement unit (PMU) in the radio chips measuring phase angle on constant frequencies. An equal spaced number of frequencies are stepped through and the differences are calculated and averaged. The resulting value is a measure for the distance. At a given distance the multiple of wavelengths filling up that distance is different across frequencies. The ambiguity can be resolved measuring round trip times in the communication. Typical at a round 120m the phase angle roles over the first time. Figure 10 visualizes this effect. The horizontal axis is the distance, the vertical axis is the phase angle. When the sender is on the far-left and the receiver is on the far-right position of the graph, the receiver will detect a different phase angle for slightly different frequencies.

Under ideal conditions plotting phase over an equidistant set of frequencies yields a straight line where the slope directly corresponds to the distance.

3.1.45.2 PMU Ranging Scheme

The ranging is initiated by an anchor sending a range request to any tag using a standard IEEE 802.15.4 frame. A short negotiation period follows where tag and anchor agree on a set of frequencies to use for the ranging. Then follows a phase where both tag and anchor synchronously step through this set of frequencies. At each frequency the tag transmits a sine wave for a short period while the anchor measures the received signal's phase relative to its internal generator along with other parameters such as signal strength. Next, tag and anchor switch roles, with the anchor transmitting the sine and the tag measuring before commencing to the next frequency.

After finishing this phase measurement set the tag transmits its measurements back to the anchor, again using standard IEEE frames. The anchor is able to compute a distance from these and its own data.

The double-sided process is necessary because in practice it is not directly possible to measure the arriving signal's phase relative to the transmitter but only to the receiver's internal reference. The divergence between the two are calibrated out by comparing the phase samples of the same frequency across both tag and anchor receivers.

In practice, non-line-of-sight conditions, timing inaccuracy and external disturbance results in the phase samples not forming a straight but rather, a point cloud. To still obtain the distance a straight is subsequently fitted through the point cloud and outliers are eliminated.

3.1.45.3 PMU Ranging with Antenna Diversity

The iRTT Antenna Unit incorporates two separate interface chips which are clocked synchronously. This allows to use passive antenna diversity to overcome further disturbances.

Only a single RF chip is used for communication with the remote device in both tag and anchor mode. During the local device's receiving turn of the frequency stepping procedure the second chip's phase measurement unit is synchronously sampled, yielding an additional frequency-phase point to the point cloud.

This method ensures that the ranging scheme itself does not need to be altered compared to the single-receiver case and allows both iRTT clients with only one interface and Antenna Units with two interfaces to act as ranging targets within the scheme.

3.1.45.4 PMU Ranging Experimental Setup

In order to evaluate the accuracy and the reliability of the newly developed distance measurement system based on radio phase differences in the 2,4 GHz ISM band, two kinds of experiments were carried out on the V2 Robot at the IGMR in Aachen:

Straight traversal of the robot front first and rear first in a constant speed towards an iRTT-CP-Tag. Distance measurements with from all AU's to the tag with 1Hz rate.

Rotation of the robot basis on in front of an iRTT-CP-Tag with constant angular rate and fixed distance towards the CP-Tag. Distance measurements with from all AU's to the tag with 1Hz rate.

Both experiments were conducted inside the IGMR-laboratory with the back part of the V2 Robot, which was steered manually with a wireless gamepad. The

Figure 3.115. Mounting positions and orientation of iRTT-AUs and CP-Tag.

iRTT measurement system consisted of four sideward facing and one upwards facing iRTT-AU-V2, mounted on a wooden board which was fixated on top of the robot and in parallel to its basis as shown in Figure 3.110.

Figure 3.115: Mounting positions and orientation of iRTT-AUs and CP-Tag shows the schematic setup of the iRTT measurement system on top of the board. The four iRTT-AU's S0A1 S1A1, S2A1 and S3A1 were mounted at each corner of the board with their backs facing to the center. The S4A1 was also used as a

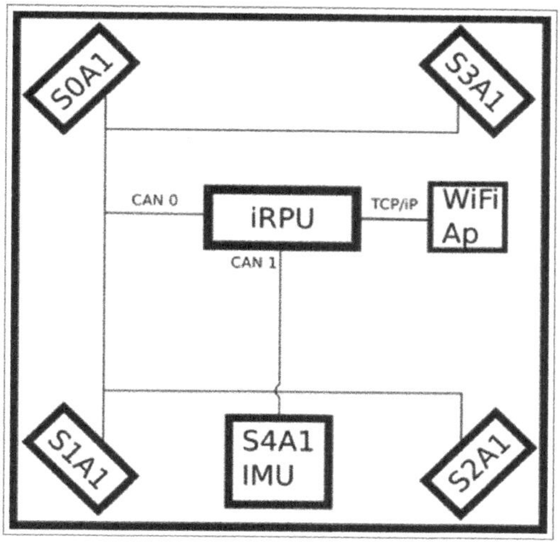

Figure 3.116. Schematic setup.

reference anchor for TDoA measurements and provided additional inertial sensor data. Therefore, it was mounted facing upwards to simplify yaw-angular rate measurements.

Distance measurement streams were provided via indurads proprietary stream protocol based on one shared high-speed CAN-bus connection to a central processing unit called IRPU. Due to significantly higher sample rates for the IMU measurements, a second CAN bus was established for S4A1 to reduce bus utilization.

3.1.45.5 Result and Discussion

Figure 3.112 shows the distance measurements from tag b6d43d to all installed AU's based on the technologies UWB Time-of-Arrival (UWB ToF) and phase differentials (PMU).

The graph depicts the straight traversal of the robot starting at 1 m distance towards the 7 m mark (distance to front facing AU's) and back again

It is clearly visible that the improved UWB-ToF ranging outperform the newly developed PMU based ranging both in accuracy and reliability. While there is a visible distance trend especially for the front facing two AU's S0A1 and S3A1, the measurement scatter of the other three averted AU's is too severe to be used in any positioning algorithm.

Even though the 2,4 Ghz ISM band allows for a higher output power of about 20dBm compared to UWB with -14dBm and its longer wavelength should provide easier NLOS detections, there is currently no reliable measurement quality indicator offered from the PMU chip to reliably detect multi path measurements and the overall error variance of the system is too high.

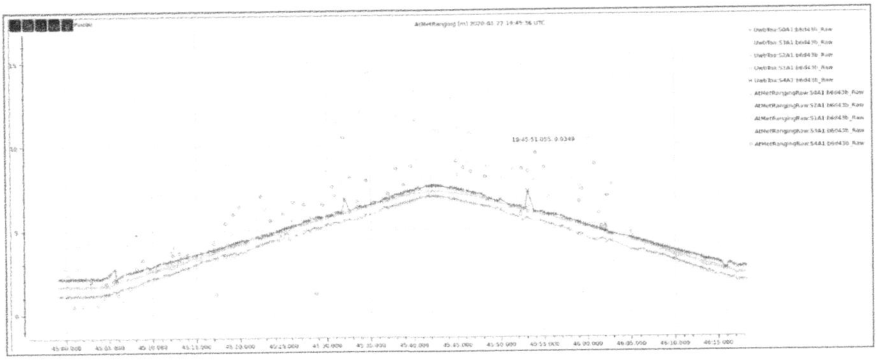

Figure 3.117. Front facing straight traversal all AU's.

However, the signal processing chain of the PMU system might be further improved with appropriate signal filters. Furthermore, the system might be used as an occasional support to resolve positioning ambiguities of the UWB ToF-system.

3.1.46 Multitag Tracking

The pose tracking system is based on a set of indurad iRTT-AUs (M) which get installed onto the robot (p_M) and a set of iRTT-CE tags (N) which are installed in the working area and get measured in accurately in the rooms coordinate system (p_N). Additionally, an inertial measurement unit is used for tracking movement dynamics.

The following sequence describes the main steps realized to get the robots pose. Figure 3.118 show this graphical:

- The tags periodically initiate distance measurements (d_{MN}) with all anchors installed on the robot. This resulting set of distances is used for estimating the rooms pose R (x,y,z, roll, pitch, yaw) relative to the robot (or vice versa). With corresponding object function of form:

$$f(R) = min_R \left\{ \sum_{MN} ||p_M - R * p_N||_2 - d_{MN} \right\}.$$

- Simultaneously the tags emit broadcasts which get received by the robot's anchors. Since the anchor clocks get synchronized via air, a TDoA (Time Difference of Arrival) can be deduced. Using the speed of light, the TDoA can be converted into a distance difference of arrival (DDoA). For pairs of receiving anchors at position p_{aM} and p_{bM}we can estimate the tags pose R using the object function:

$$f(R) = min_R \left\{ \sum_{(MxM)N} (||p_{aM} - R * p_N||_2 - ||p_{bM} - R * p_N||_2) - DDoA_{abMN} \right\}$$

- We combine samples of both measurement methods in a common Unscented Kalman Tracker state. During the update step, either objective function is used for comparing the data set with the Kalman trackers state.
- Measurements with a higher probability of being NLOS are down-weighted by applying a higher measurement noise to them. Additionally, we reduce the Kalman Trackers innovation by down-weighting measurements which deviate too much from the current state (Robust Kalman).
- A Madgwick filter is used for combining and integrating the inertial measurements units (IMU) accelerator and gyroscope measurements to an absolute orientation of the robot. We duce orientation changes from the filters result and apply those to the Kalman trackers state as control input.

3.1.46.1 Multitag Tracking Results

The most recent robot setup was tested during a trail at IGMR institute. A set of six iRTT-Tags got installed at the rooms edges and a two reference squares (1 m x 1 m, 3 m x 3 m) were measured in accurately relative to the rooms local coordinate system.

The system was configured to conduct ToF/TDoA measurements with a sample rate of 3 Hz and IMU measurements with 100 Hz.

During pre-tests the TDoA subsystem has been evaluated. Even though the synchronization is functional and stable, the provided accuracy is not sufficient for a narrow anchor configuration.

The system is subject to a typical synchronization error of ten Decawave time units which leads to a TDoA measuring error of approx. 10 cm.

The systems pose estimation proved to be reliable in the chosen measurement configuration. The maximal position error was approx. 20 cm from the reference position and orientation estimation was stable thorough the trail showing a maximal orientation estimation error of approx. 20°.

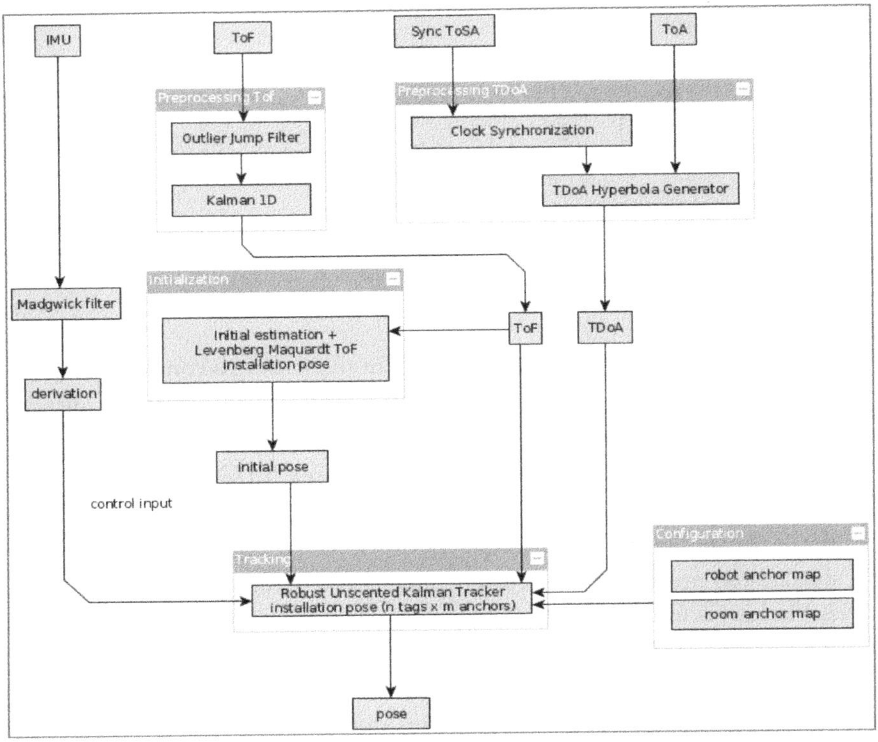

Figure 3.118. Tracking: data flow and algorithms.

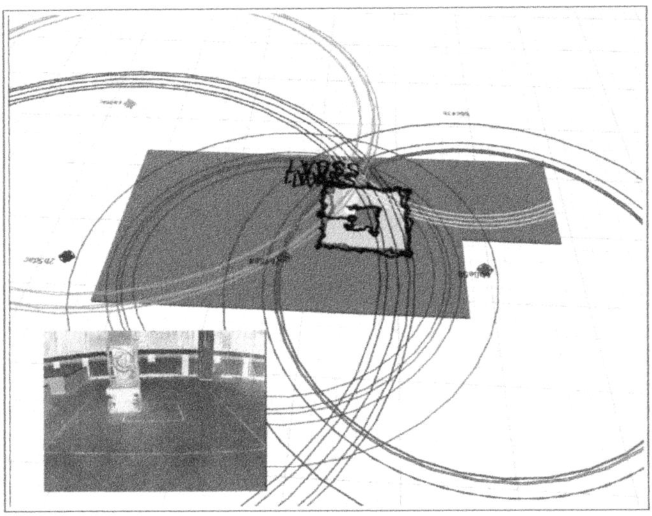

Figure 3.119. Pose Tracking: estimated trajectory (black path) and corresponding distance measurement circles.

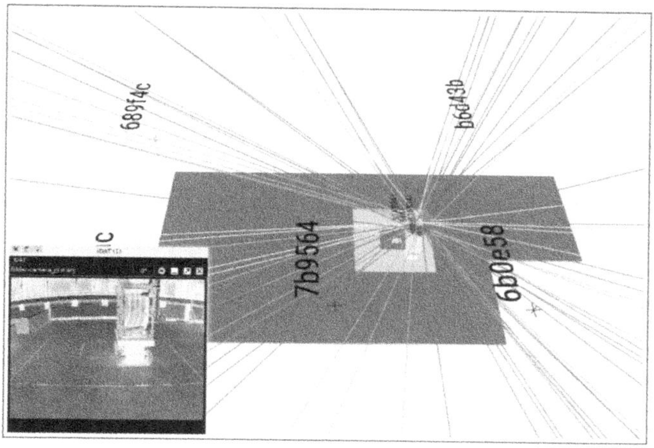

Figure 3.120. TDOA: measured TDoA visualized in plane z=2 m. Sending iRTT tag id is encoded as hyperbola color.

3.1.47 Summary and Conclusion

In this Appendix to Deliverable 4.2 we have elaborated different possibilities to improve radio ranging accuracy. A phase measurement scheme at 2,4GHz has been analyzed and tried. Far too high measurement deviations and also a rather slow measurement system that limits its practical use to static systems must be reported. Thus, the system will not be followed up further.

Another trial with over the air synchronized UWB anchors shows a far more stable location detection compared to the results in Deliverable 4.2. The UWB measurements further have been stabilized with acceleration measurements from a 6-axis inertial motion sensor. These results are promising, the combined TDOA an IMU system will be improved in future, with hard wired synchronization lines removing the different anchors clock jitter almost entirely.

Deliverable D5.3
Sensor System for local process monitoring implemented and functional

Authors: Dr. Matthias Rabel

3.1.48 Deliverable in the Context of the Work Program

Scanning the surface before, during and after treatment is a key technique for optimized work processes. Before treatment, the work area is scanned as reference and to optimally assist the user who makes the selection. During processing, the area is monitored to document the progress. Events like sudden spalling of large areas, or the detection of disturbing objects shall be detected to optimize the work process during the treatment. Visual process monitoring by camera might possibly not be usable during the treatment due to the amount of dust generated and will be supported by radar technique to overcome this issue. Similar to deliverable 4.2 the same camera system is preliminary used on the prototype 1 of the asbestos removal robot.

3.1.49 Methods and Work Carried Out

The asbestos removal robot is moved using the mobile platform. It is placed in front of a wall or other surfaces where grinding has to be done. The grinding disc can then be used to operate on a selected region of the surface. Since the robot can never be placed with an accuracy of a few millimeters and the mechanics has also tolerances, a measurement system is required to detect the actual geometric situation. This is required to allow a controlled grinding process. The grinding disc must be oriented exactly parallel to the surface at any time. Thus, a measurement system is required that is insensitive against dust.

Several systems are discussed:

- A radar based system that can be used on multiple positions around the disc to measure the distance to the wall.
- The second sensor is a force measuring component placed in between the robot's arm and the robot's grinding machine. Linear and rotational forces are measured with this device with strain gauges. There is no experience yet with the force feedback sensor, it is currently replaced by a dummy component.
- The third system, a camera based system, shall only be used to check the grinding results. It is implemented and functional, however, the real usage is still under development.

The radar based system has been developed from scratch during the first project year. Its concept is described in the further chapters.

3.1.50 Sensor System for Local Process Monitoring

Figure 3.123 shows the general layout of the sensor system implemented on the asbestos removal robot prototype 1. Within the dashed area, the mobile platform is shown. The manipulator arm is mounted on the robot platform. Up to 5 iLDR120 "indurad linear dynamic radar" shall be mounted around the grinding disc. Additionally, two cameras might be placed on the robot to evaluate the grinding result.

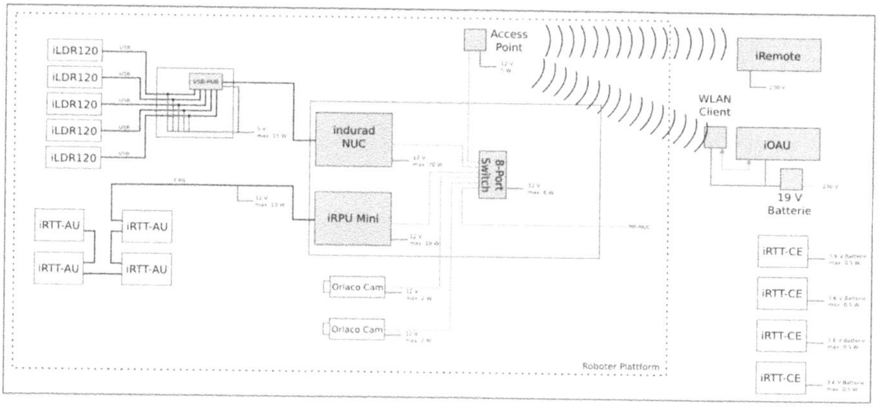

Figure 3.121. Sensor system layout, for mobile platform and robotic arm.

Figure 3.122. Orlaco camera EMOS Ethernet; Example picture underground.

Several other components are shown. These are used on the mobile platform. The sensor system is further described in Deliverable 4.2. The shown indurad NUC intel computer is used to connect the radars with USB 2.0. The data is processed and the results can be fetched via the graphical user interface iOAU (indurad operator assistant unit) or the robot control processing unit. The iOAU is also used to calibrate the sensors.

3.1.50.1 Camera System

The camera system is used to give the operator an environmental view and shall be used to check the process result of grinding. The work conducted in deliverable 5.2 focuses on monitoring the process results. Within deliverable 4.2, the focus is on the environmental view using the same cameras.

Figure 3.123. iLDR120 with FE (left) and DSC (right) board.

The cameras are available with 120° and 180° opening angle. Variants with 90°, 60° and 30° are planned by the manufacturer. The camera allows a video resolution of up to 960p with 45 frames per second. The housing is IP68 and IP69k specified. Thus, it can easily be used in dusty environments.

The camera has a high dynamic range of > 115dB and has a low light feature. Figure 3.122 shows the camera and a picture taken underground with a 120° variant. The picture quality is quite good. The only areas exceeding the dynamic range are the dark corners of the image and at the light bulbs.

The full usage and functionality will be evaluated with the first real grinding tests including the operational GUI. The concept will be revised after the first results.

3.1.50.2 1D Radar System

To be able to measure the distance from the tool to the processed wall, five iLDR120 sensors should be mounted next to the cutter tool. The requirements are as follows:

- Small package dimensions
- Light weight
- Dust resistant
- Easy to mount on the arm
- Only one cable for power supply and data transmission
- A small beam width in the antenna diagram

By using a highly integrated 120 GHz radar chip, the dimensions of the sensor can be minimized. This chip offers a bandwidth of up to 7 GHz from 119 GHz to 126 GHz. This allows operation in the so-called industrial, scientific and medical frequency band (ISM) that starts from 121GHz and ends at 122 GHz. The silicon-germanium based chip includes mixer, power amplifier, two patch antenna arrays for transmitting and receiving, low-noise amplifier and IQ down converter.

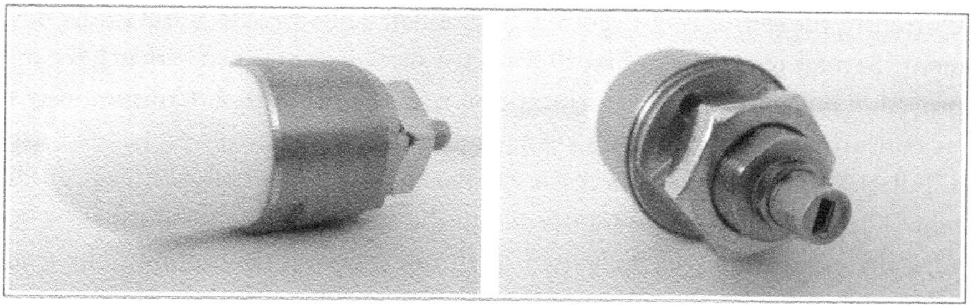

Figure 3.124. iLDR120 sensor housing.

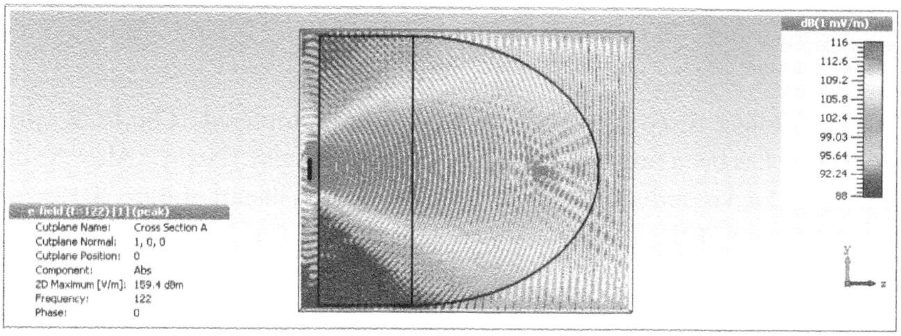

Figure 3.125. Simulation model and electric field magnitude [dBm].

Two printed circuit boards (PCB) have been designed and manufactured as depicted in Figure 3.123. The frontend (FE) circuit contains all components for frequency generation including the phase-looked-loop (PLL). The reference signal is generated by a programmable PLL. In the receiving path, multiple amplifiers and filters prepare the signal for the analog-digital conversion on the digital signal controller (DSC) circuit. On this second PCB, which is directly connected to the frontend circuit, the DSC is used to process the received signal. The 512k x 16 bit SRAM allows the storage of several radar measurements before transmitting the data via a serial USB interface. This USB interface is used as power supply of the sensor as well.

The sensor casing in Figure 3.124 consists of a PTFE lens and a stainless steel body. To mount the sensor on the arm easily, a thread has been integrated at the lower end. There is only one M24 nut required to fix the sensor. A Mini-USB-M12 connector ensures a dust-resistive connection of the sensor.

The DSC and FE circuit will be placed into the lens, which has been developed to narrow the beam width of the chip antennas. For the present application, a beam-width of $+/-25°$ is not sufficient and has been improved to $+/-1.8°$. The lens has been constructed with a microwave simulation tool and converts a spherical wave into a plane wave. An antenna gain of 32 dBi has been reached. The simulation model and the electric field magnitude can be seen in Figure 3.125.

Currently, the sensor is not yet fully functional. The software is still under development. Several tests on the first PCBs have shown several issues that have to be improved. The overall power consumption must be optimized. Continuously the radar consumes about 3.5W. This is far too much to be usable embedded inside of a Teflon lens. Thus the concept is further developed to embed a heat guiding system. The heat will then be transferred to the steel parts of the housing.

Also the electronics will be revised to shrink power consumption. The goal is a maximum power consumption of 2.5W. This equals a standard USB port that provides 500mA at 5V supply voltage.

3.1.51 Realization

During the project three design cycles have been completed. The PCB (printed circuit design) has been improved to solve production related issues. Also the design has been improved to reduce the power requirements. Till end of June 2018 we have received Version 3 of the sensors.

- Version 1: has only partly been working, we could only test basic functions and static radar functions that did not allow a distance measurement.
- Version 2: this version has been working. The radar has been tested and has been installed at Bots2Rec Prototype Number 1. The following subchapters are summarizing the results.
- Version 3: Now of writing this document only a first evaluation has been done. The radar has been working out of the box. New power management features are working in principle. During the measurement 500mA as current is drawn, turning off all radar related components the current is 250mA. Dependent on the amount of calculations and the measurement rate. The DSP might also be powered down partly, thus further power saving is possible. The design additional has a strongly improved design to reduce thermal resistances. That allows an easier thermal flow towards the 1D-radar mount.

3.1.51.1 Radar Tests

One of the first tests have been distance tests. Therefore, a linear axis with a stepping motor has been used to move a reflector in front of the radar. The distance of the maximum signal peak has been evaluated. Since the linear axis is moving with a constant speed, any measurement is ideally on a linear line. The following plot shows a distance in [m] on the vertical axis, on the horizontal axis the measurement number. The measurements up to now might jitter several milliseconds. In future firmware versions, using an already existing hardware timer will solve this issue. The speed has been 12,5mm/s.

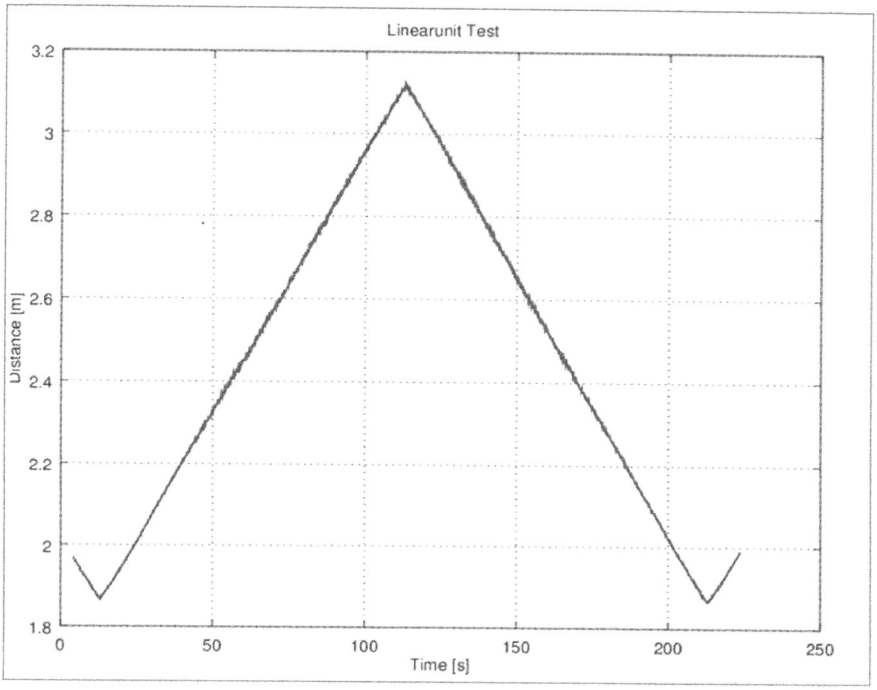

Figure 3.126. Linear Axis Test, Overview.

Differentiating the graph the result should show a horizontal line with a positive or negative offset. The following picture shows the result. The measurement rate has been 25 samples per second. The averaging length is 20 measurements. This results in 10mm steps in the following graph.

Obviously, the measurements are not yet linear and there is measurement noise. The following picture shows the samples 500 to 2000 more in detail. The deviation is +/-1,5mm over the length of 1,2m.

One main reason causing noise is the programming of the radar chip and the AD convertor. In Board revision V2R1 the operation is non-coherent. The second reason is that the reflecting target has been a trihedral reflector. It should be a sphere instead, as it is a point reflector. Another source of error is the measurement quite near to the linear axis, thus multipath will influence the measurement slightly. The direct and the multipath are causing targets close to each other. Due to the minimal distance, those two peaks cannot be separated.

During discussion of the results we have found, that the PCBs in the mounting fixtures are not entirely fixed. Thus as the linear axis is using a stepper motor and the steps cause a stuttering movement, including that the mounting fixture is not completely stiff, the dynamic movements and vibrations cause as well a noisy result.

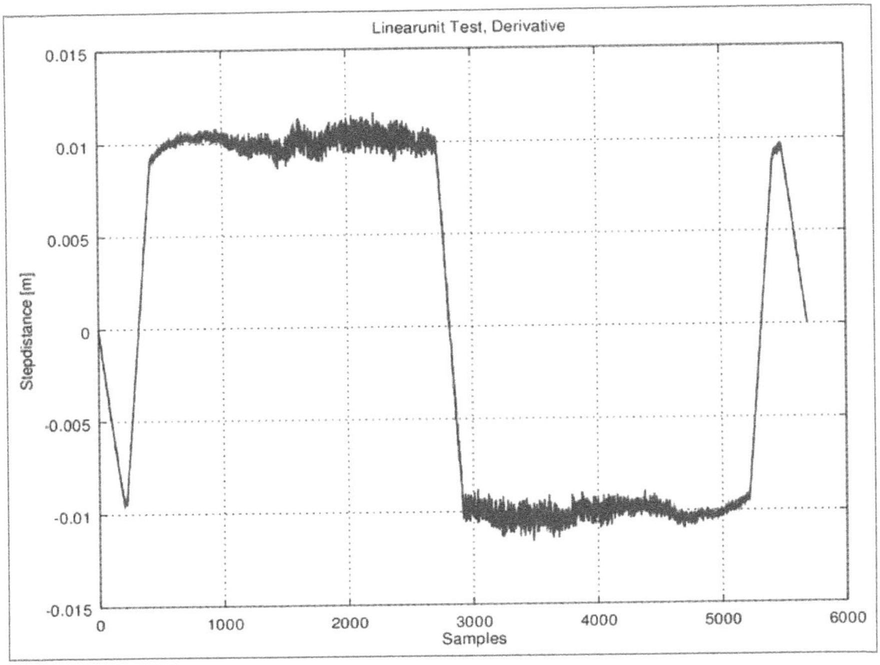

Figure 3.127. Difference from measurement to measurement.

Another experiment that should give us an idea of the effective antenna diagram has shown, that there is some deviation of the maximum peak in the antenna diagram that should be perfectly aligned. The deviation is $-2{,}1^{\circ}$ around the view axis, and 7.6° to the side. the This error was expected, but could not be simulated in advance, as we do not have enough detailed knowledge of the materials used in the radar chip. Since the transmitting and receiving antennas are separated there are two antenna diagrams. We need to find the point behind the lens, with the resulting maximum peak along the axis of the antenna. This only can be realized in experiments.

3.1.51.2 Installation on the handling arm on Prototype 1

4 of the radars are installed on the asbestos robot prototype V1. The integration has been realized using USB cables that are connected to a USB hub, which provides every radar the supply voltage of 5V and maximum current of 500mA. The USB hub is connected to an embedded computer, the NUC PC on the robot. The software executed on the NUC PC receives the raw data, calculates the FFT and searches the maximum peak in configurable distance limits. The following pictures shows the mount.

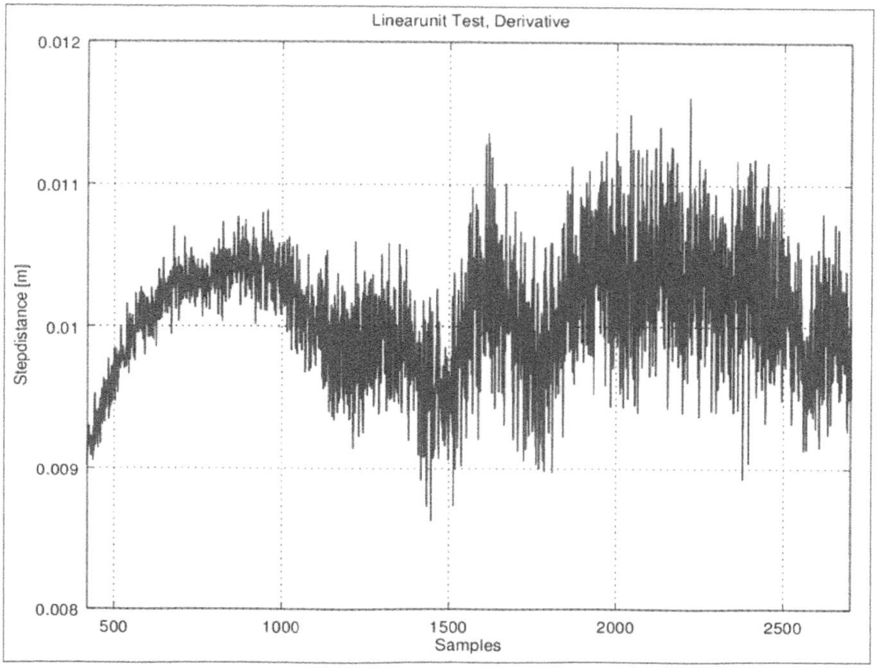

Figure 3.128. Zoom of sample 500 to 2000.

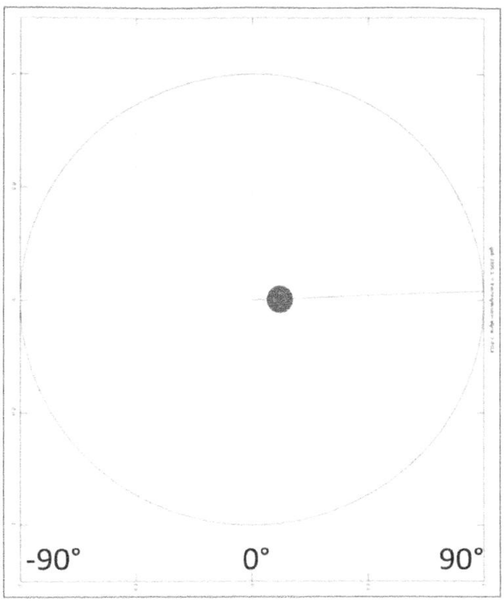

Figure 3.129. Orientation of the chip and deviation of the maximum peak.

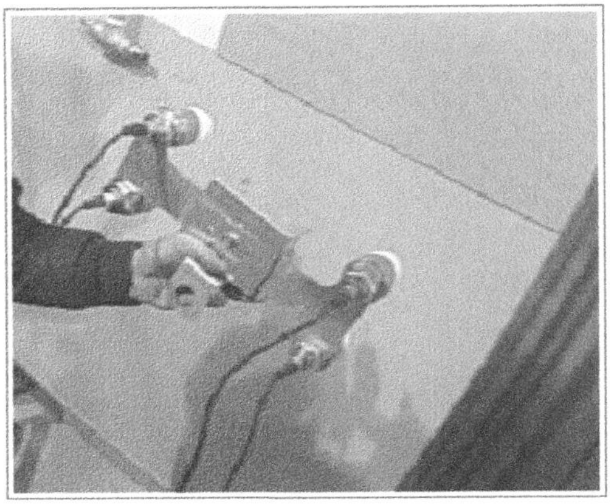

Figure 3.130. 1D radar mounting fixture.

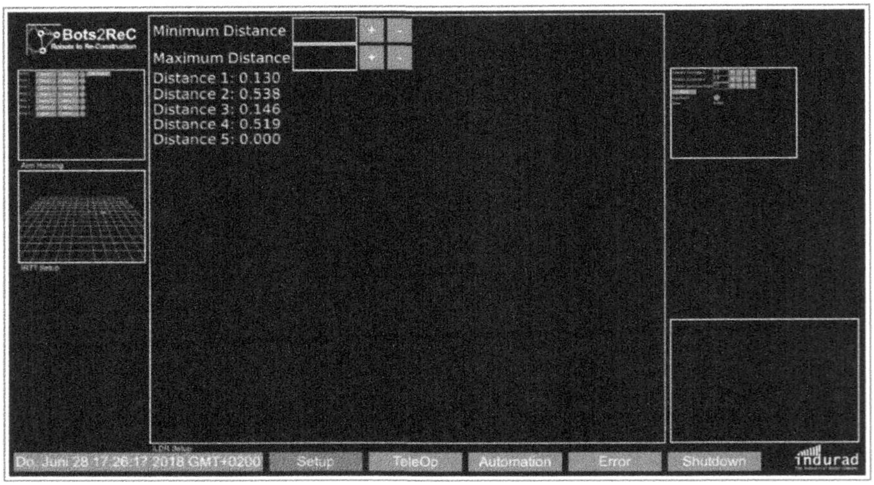

Figure 3.131. 1D radar setup screen.

It is behind the grinding disc. The antenna surfaces have later a distance of 5 to 10cm to the wall.

3.1.51.3 Setup screen and raw distance visualization

The software so far does not allow the configuration of the radars. The next step will allow to set the measurement rate and to define the number of strongest peaks that shall be passed to the robot control software. Also an offset correction will be implemented. A first screen on the iOAU that will be extended has been realized.

3.1.52 Outlook and Future Work to be Carried Out

First samples of the sensors have been developed. Version 1 was not functional, Version 2 is functional but needs optimization. Version 3 is by now available but has not been evaluated yet. Still the PLL loop filter as part of the high frequency generation path must be carefully analyzed and revised. All other hardware related issues are already adapted and corrected, including power management functionality that shall reduce the average power consumption to about 1,3 W. Peak power stays at about 2,5 W.

The interface to the Intel NUC computer has been partly implemented. The radar data stream is transferred. Configuration register are to be added and the motion data stream must be transferred as well. Extended features will allow a detailed setup and monitoring functions, e.g., like temperature and antenna pollution.

The accuracy has been checked in a first trial, the results are used to optimize the firmware that will operate coherent. In addition, mechanical adaptions have been made. This must be now verified with the hardware samples of version 3.

The arm movements must be performed in a direction that allows using the iLDR distance measurements in an optimal way. For example, symmetric situations are problematic, when the grinding direction is in parallel to two sensors next to each other. Then, two sensors are in front of the disc and two are at the rear. The first two sensors are measuring the distance to the surface before grinding, while the two rear sensors are measuring the distance to the processed surface. In this case it cannot be distinguished whether the surface is grinded to a certain depth or the alignment of the disc is wrong. Numerous variations of this example and further more a thinkable and will be evaluated during the ongoing development of the process control to allow a stable grinding process.

3.1.53 Summary and Conclusion

The hardware development for the main system, the 1D radar, are concluded in a preliminary version and will be improved and updated in a second revision, implementing several optimizations. The integration to the software and control is ongoing and continues together with the process development. The same is true for the second system, a force feedback sensor that is functional and integrated but currently not in place. It will be taken into service together with the implementation of an advanced control on the prototype and central process control system.

The third system to monitor the grinding result with cameras is technically available. However, the real camera usage and mount places must be analyzed after the first grinding experiences on the full operational process.

3.1.54 Further System Components

Deliverable Di3.3
Central aspiration and energy supply and safety architecture available

Authors: F. Becchi

3.1.55 Deliverable in the Context of the Work Program

Scope of this internal deliverable is to present proposed implementation of dust aspiration, system energy management and operational test safety. These central systems are used to supply the operational robotic unit during the tests performed and allow a safe testing environment. For the first V1 prototype, the systems are not necessarily designed according to final specifications. Emphasis here is on enabling the prototype rather than testing close to product solutions.

3.1.56 Central Aspiration

The mobile platform is integrated with a local dust aspirator device. This choice was preferred to a centralized aspiration unit to avoid reeled hose connection between the moving robot and the central unit.

After some preliminary discussions, the management of the connecting hoses was discarded not to increase too much the moving robot navigation constraints. In consideration also of a minimal aspiration pipe section of 40 mm the space occupation of a hose pipe reel would have a too strong impact on the overall moving platform dimensions and could not be integrated within the size restrictions. Fist tests will need to show the feasibility of the aspiration as an integrated device, which is now limited in filtering and storage capacity.

For this reason, an asbestos dust aspirator was customized and integrated directly in the platform.

A Pullman Ermator S13 was selected as commercial available starting unit for the following reasons:

- Dust aspiration capability is correctly dimensioned for the tool used in the robot;
- The device is already specific for asbestos dust management;
- The overall dimensions of the aspirator are compatible with the space availability of the platform.

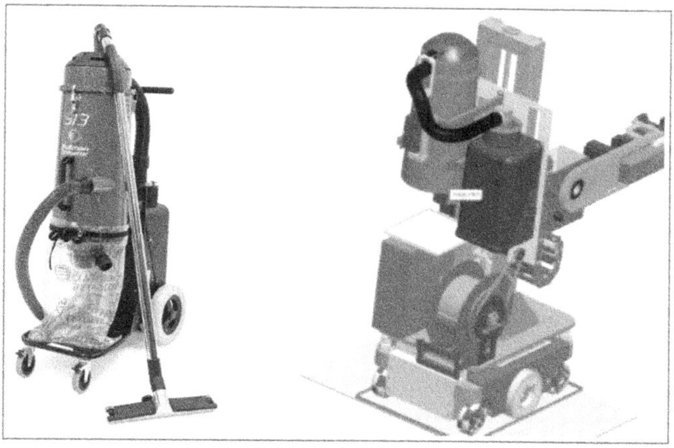

Figure 3.132. Standard asbestos dust extractor and integrated view on the platform.

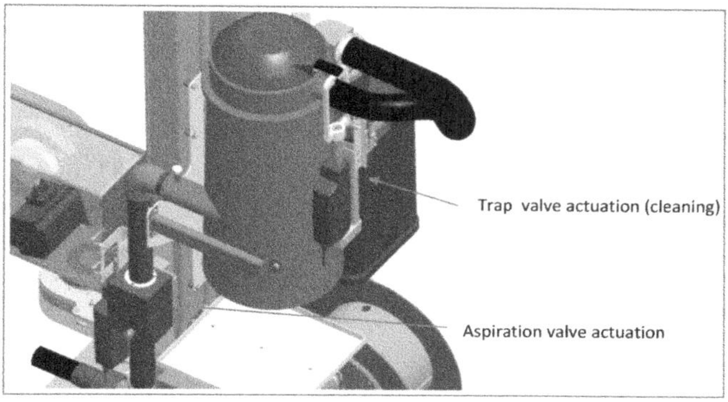

Figure 3.133. Filter cleaning automation.

The aspirator layout was optimized for the platform integration removing the supporting frame with wheels and repositioning the filter unit with respect to the motor unit.

A more in deep device customization was then request to implement the filter cleaning procedure:

In the standard usage the filter cleaning is performed replacing the aspiration pipe with a cap and actuating the cleaning trap valve several times manually.

This same sequence was implemented adding an electrical actuated valve on the aspiration port and adding an electrical actuator to move the cleaning trap valve.

A further system improvement that should be considered for prototype 2 is the automation of dust bag replacement (TBD).

In the reference system configuration the bag is held on the lower section of the filter box with a belt that is manually fastened on the upper bag section. A reliable automated bag replacement could be considered in the final implementation of the

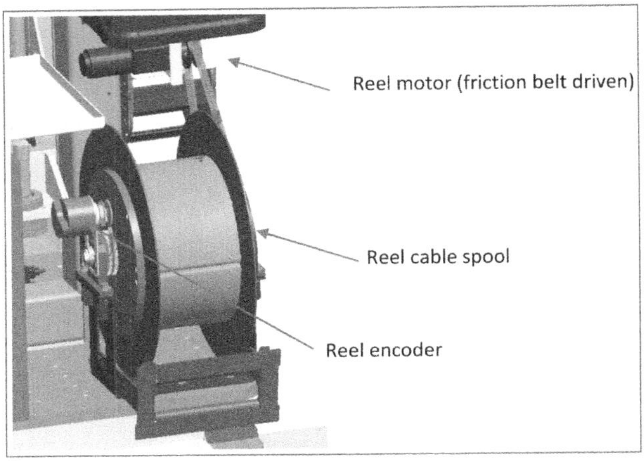

Figure 3.134. Cable reel.

system to allow the robot to change with no external help (cost vs convenience of this feature still to be assessed).

As visible in the above picture the integration of the actuated valve on the aspiration line was also used to reorient the suction pipe to better follow the robot arm in its displacement (on the other side of the platform a dedicate cable chain is used to connect arm actuators with motor driver located on the platform frame).

3.1.57 Energy Supply

The robotic platform is connected to site 230 VAC power lines with a 15 m cable rated for 5 KVA .

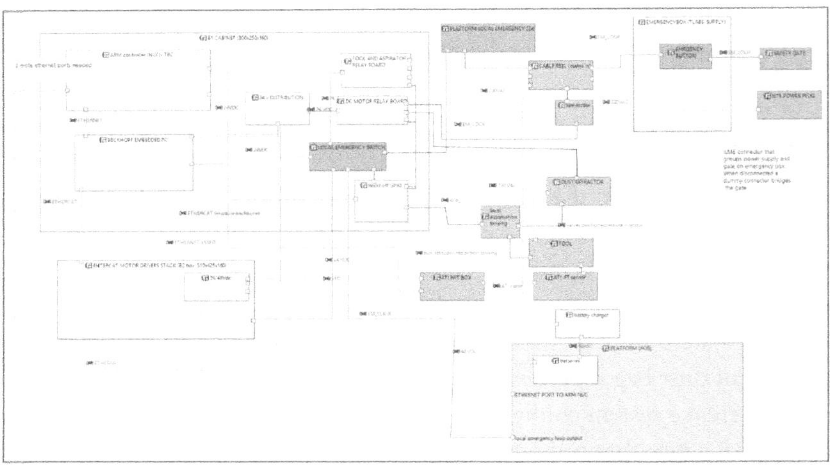

Figure 3.135. System physical block diagram.

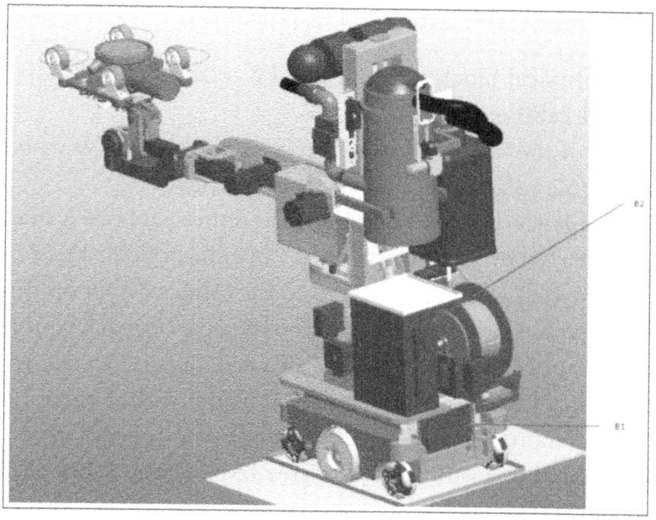

Figure 3.136. B1 and B2 localization in platform integration.

The power supply cable is winded on a custom made cable reel and connected to the platform main power line (on the same power cable some lines are dedicated to system safety management as below discussed).

The cable reel is actuated with a 24VDC motor integrated with a planetary gearhead connected to the cable reel with a friction belt transmission set so to allow cable release in case of overload. The cable reel is also provided with a multi-turn encoder used to measure the estimated amount of cable on the reel.

As shown in the system physical block diagram below reported the 230 VAC line is connected to the platform battery charger and to the asbestos removal tool and dust extractor.

The arm motor drivers board, supplied ad 48VDC, are then directly connected to the main platform battery used as accumulator so to assures when needed high peak power to arm motors. A 48VDC/24VDC converter is used to provide the service power to arm control units (including beckhoff embedded PC and arm NUC) and to drive the on/off electric actuators used for dust extractor cleaning and cable reel actuation.

In terms of electrical integration of the power system management of the platform the main system distribution box in integrated in an IP65 cabinet above identified as "B1" located inside the platform.

48V bus from batteries is connected directly inside the motor driver cabinet "B2" where also the DC/DC converter is integrated.

3.1.58 Safety Architecture

As visible in the physical block diagram an HW safety control loop in integrated in the platform B1 cabinet.

The main unit of the safety control loop is a redundant HW emergency switch that is connected to a redundant chain of emergencies.

The main emergency loop is linked via the tether cable to a main emergency switch that must be located out of the robot operation field. Physical fences provided with safety switch will be connected to the same safety loop through a junction box inside the main switch box (see following operational layout for reference).

To activate the robot platform, in case of need, for site to site transportation, a dummy emergency loop must be connected at the end of the robot tether. For this specific non operational case two emergency buttons positioned at the sides of the main robot structure ensure an acceptable level of user safety. The safety loop disengages the following devices:

- Robot arm actuation (via STO-safe torque off HW control embedded in the arm motor divers);
- Tool and dust extractor 230VAC power supply (via contactor controlled by the safety switch output);
- Platform translation motor (via two paired switch N.O. connected to platform wheels motor control board).

3.1.59 Outlook and Future Work to be Carried Out

The implementation of the works is in the final phase, all systems will be implemented for the V1 prototype. The safety architecture is however going to be finalized on the testing site after all initial robustness test after integration.

3.1.60 Summary and Conclusion

The central systems were implemented as planned, only the aspiration system was decentralized for design reasons and the efficiency of this solution is to be evaluated in the testing phase. In particular the aspiration and storage capacity / run time during operation will give valuable input to the next design phase for V2. The behavior and control of the cable reel will lead to first findings about the feasibility of this approach.

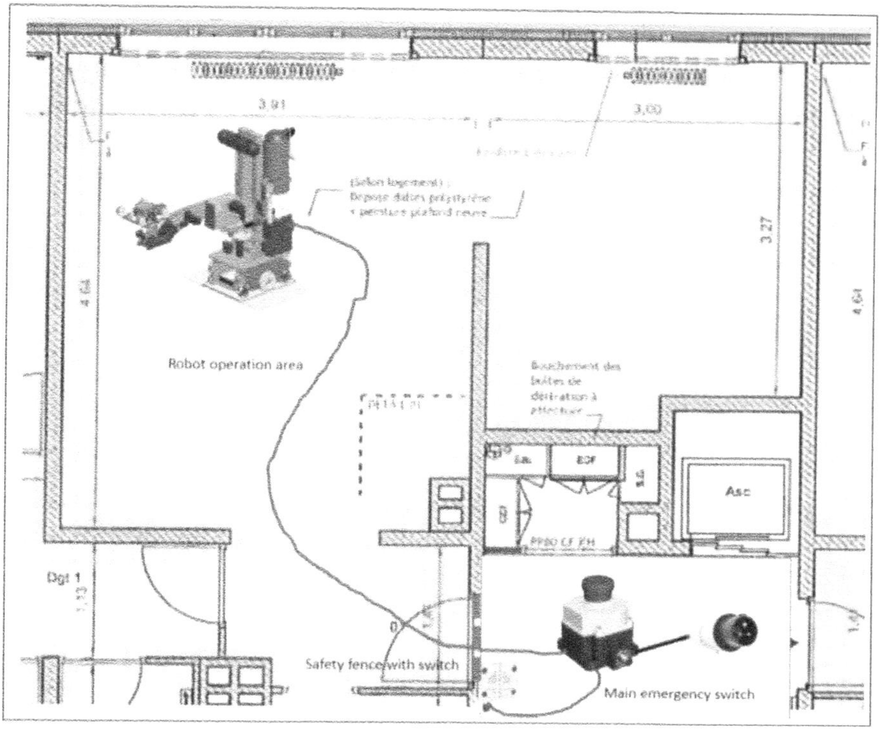

Figure 3.137. Robot operational field integration.

3.1.61 The Entire Systems

Deliverable 3.1
Robotic system V1, ready for the use in the test rehabilitation site

Authors: Daniel Martin (EUT), Ramón Carrasco(EUT), Oriol Orra(EUT)

3.1.62 Deliverable in the Context of the Work Program

The present deliverable report presents one main result of the first period of the project, where a first version (V1) of the Robotic Unit for asbestos removal has been developed. This phase ends with the integration of the different subsystems, designed and developed by the different partners involved in Bots2Rec project. In this document, a main overview of the Robotic Unit V1 is reported, which is a "demonstrator deliverable" as a result of the previous mentioned tasks.

3.1.63 Methods and Work Carried Out

The main components of the Robotic Unit V1 system have been shipped to the system integrators (Eurecat) facilities together with the system documentation (electrical and mechanical diagrams) and set up guidelines. Once the whole set of components has been received, Eurecat, supported by the different partners involved, has proceeded with the full integration, assembly and first test of the Robotic Unit V1. One example of this cooperation between partners on the integration has been a one week workshop where technicians and developers from all of the consortium partners attended for a detailed exchange on all technical levels.

3.1.64 Robotic System V1, Ready for the Use in the Test Rehabilitation Site

In this section the presentation of the Robotic Unit V1 prototype is made, describing the demonstrator deliverable stated in the project plan. This prototype is the result of the integration work and assembly of the different components that have been developed in this first period of the project.

The main components integrated on this first Bots2Rec project prototype are listed below, and shown in Figure 3Figure 3:

- Mobile Platform V1, described in D4.1
- Robotic Arm V1, described in D5.1
- Central Systems and auxiliary devices (aspiration, safety elements, cable reel, ...)
- A preliminary working version of the Sensor System for Environmental Perception, to be delivered on 01.08.2017 as Deliverable D4.2

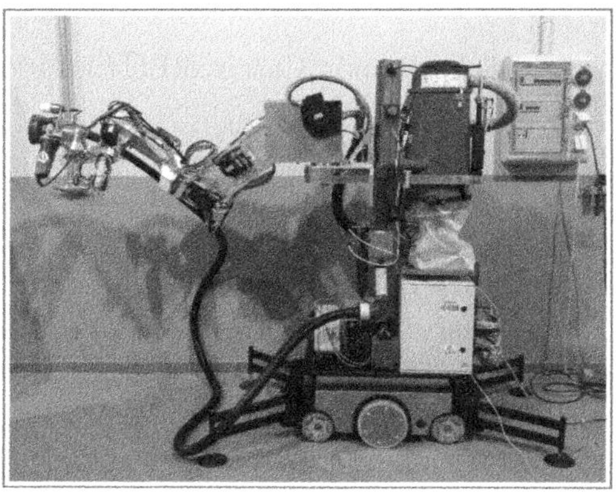

Figure 3.138. General Picture of the Robotic Unit V1.

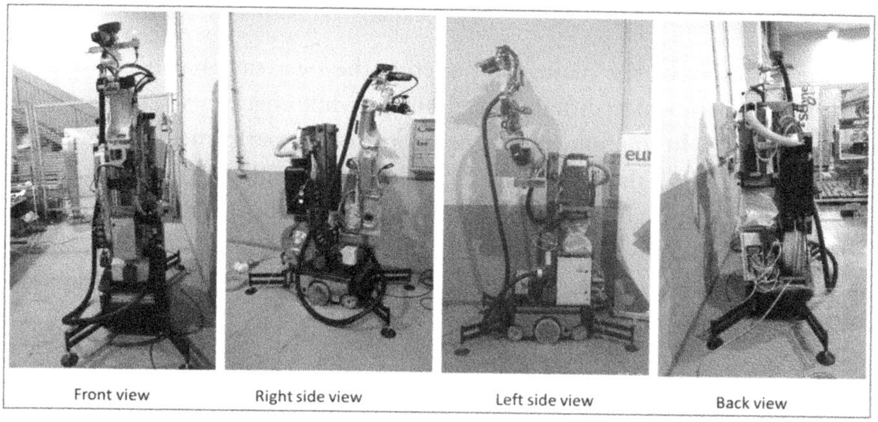

Figure 3.139. Different views of the Robotic Unit V1.

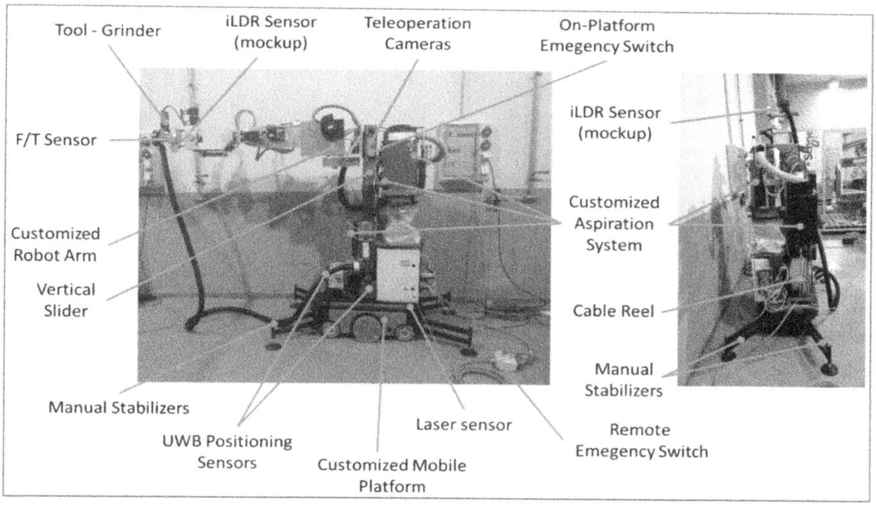

Figure 3.140. Robotic Unit V1 main subsystems.

- A mockup version of the Sensor System for Local Process Monitoring, to be delivered in 01.06.2018 as Deliverable D5.3
- A preliminary working version of the User Interface for Teleoperation, delivered as Deliverable D6.1 together with this Deliverable on 28.02.2017

3.1.64.1 System Assembly

The assembly of all of the components described in the previous section has been performed and all the mechanical and electrical interfaces have been implemented as defined during the design phase of the prototype.

3.1.64.2 Main Functionalities Tests

After the assembly of the main components, the next step has been to test the robotic system performing different motion tests while checking the system stability during these tests. In the following figures some of the performed motion tests are shown. Figure 6 shows the robot reaching its higher position, with the vertical linear axis on its upper limit and the arm completely extended in vertical position.

The most critical position regarding stability is shown in Figure 3.144. It corresponds to the arm completely extended to the front with the upper position of the vertical linear axis.

Figure 3.141. Mechanical assembly process of the Robotic Unit V1.

Figure 3.142. Prototype during electrical assembly process.

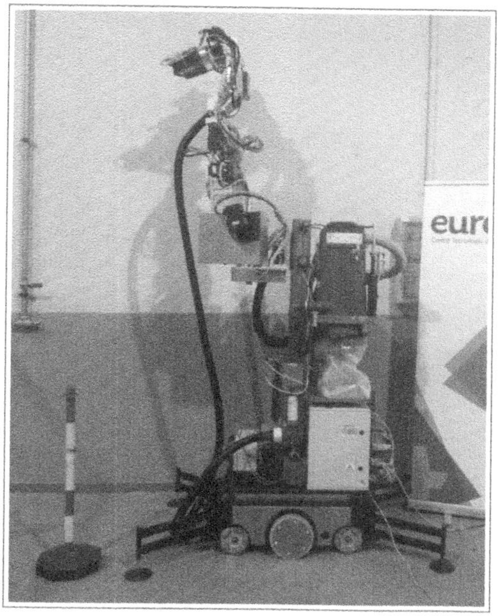

Figure 3.143. Robot positioned in the top of the vertical linear axis with the arm completely extended to the top.

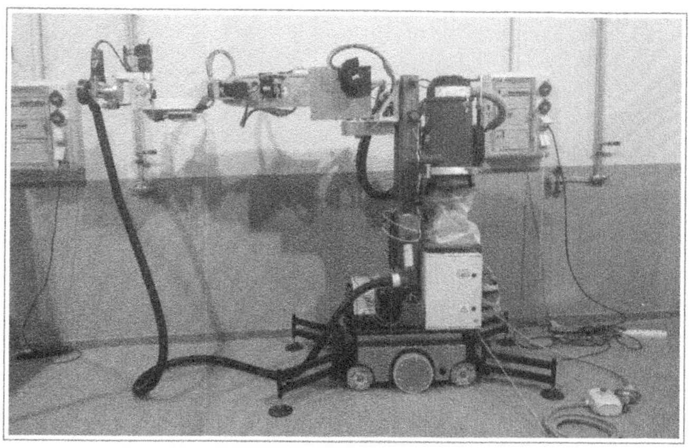

Figure 3.144. Robot positioned in the top of the vertical linear axis with the arm completely extended to the front.

Figure 3.145 shows the system reaching the floor, ready to perform grinding operations on ground surfaces.

Finally, Figure 3.146 shows the system operator interface for teleoperation in its current working status.

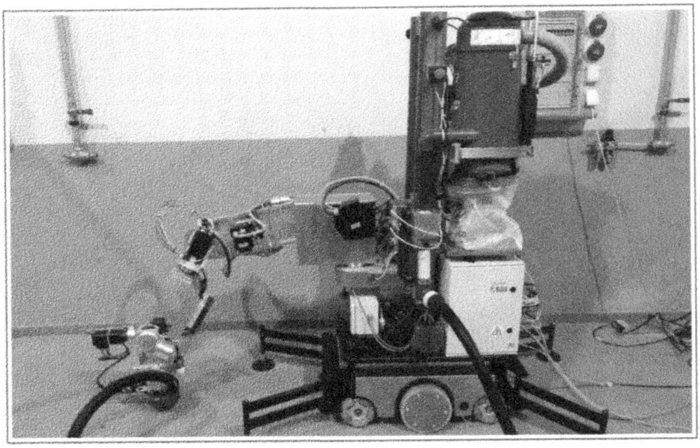

Figure 3.145. System reaching the floor.

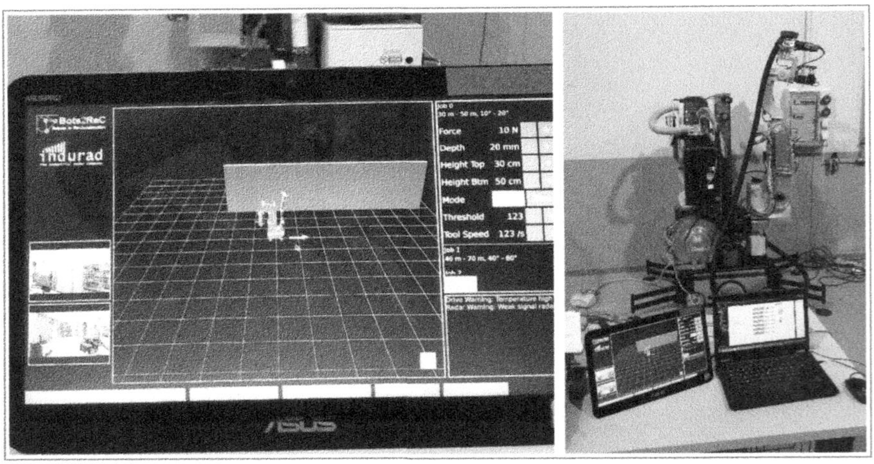

Figure 3.146. Robot Unit V1 operated by the preliminary user interface for teleoperation.

3.1.64.3 Interfaces test

At the time of writing the present deliverable report, a set of intensive tests for checking the proper operation of main control and software interfaces is in process. These tests are going to be performed in the system integrator facilities together with the first operation trials. This is in order to ensure that the major issues that could arise for the robot operation are detected and fixed before shipping the robot to the rehabilitation site, where fewer resources are available. We expect to finish the testing of these interfaces within the next month after the submission of this report.

3.1.65 Outlook and Future Work to be Carried Out

As shown in this report, the Robotic Unit V1 has been successfully integrated and its main functionalities are working. The mechanical and electrical assembly of the different subsystems has been done and tested. Nonetheless, more intensive validation tests (especially of control and software interfaces) need to be performed within the following weeks after this report. With this we can ensure a reliable function and operation of the system in different operational conditions.

As mentioned before, the next step once the prototype is ready, will be to perform the first set of operational test in system integrator facilities in order to detect any important issue in an early phase before shipping the prototype to the rehabilitation site, where fewer resources are available.

3.1.66 Summary and Conclusion

It can be concluded that the Robotic Unit V1 prototype is ready to be used in the rehabilitation side as planned. However, more intensive testing is recommended to be done before the shipment of the prototype, as well as performing a first set of operational tests in the system integrator facilities. This will ensure the systems reliability and avoid further problems on the rehabilitation site.

3.2 Second Robotic Prototype (V2)

3.2.1 Robotic Arm

Deliverable 3.3
Robotic system V2, ready for the use in the test rehabilitation site

Authors: Daniel Martin (EUT), Néstor García (EUT)

3.2.2 Deliverable in the Context of the Work Program

The present deliverable report presents one main result of the second period of the project, where the obtained feedback from the first prototype (V1) of the Robotic Unit for asbestos removal has been applied to an improved second version (V2) of the Robotic Unit. This phase ends with the integration and the complete validation of the different subsystems, designed and developed by the different partners involved in the Bots2Rec project. In this document, a main overview of the Robotic Unit V2 is reported, which is a "demonstrator deliverable" as a result of the previous mentioned tasks.

3.2.3 Methods and Work Carried Out

The main components of the Robotic Unit V2 system have been shipped to the system integrator (Eurecat) facilities together with the system documentation (electrical and mechanical diagrams) and set up guidelines. Once the whole set of components was received, Eurecat, supported by the different partners involved, proceeded with the full integration and assembly of the Robotic Unit V2. Learning from the past experiences and as an attempt to avoid the incurred delays and found problems with the Robot Unit V1, an extremely intense test is conducted at the system integrators facilities to validate the continuous operation of the integrated Robot Unit V2. This validation campaign prior to taking the robot to the testing site is very important to avoid future "debugging actions" that are very difficult and expensive to perform at the testing site (Valence).

3.2.4 Robotic System V2, Ready for the Use in the Test Rehabilitation Site

In this section the presentation of the Robotic Unit V2 prototype is made, describing the demonstrator deliverable stated in the project plan. The described prototype is the result of the integration work and assembly of the different components that have been developed in this second period of the project.

As described in the 2nd periodic report and indicated at the review meeting, the project is experiencing some delay which have affected this deliverable. The main reason that led to this delay have been:

- V1 prototype debugging required an extra effort due to unexpected failures.
- Implementation of prototype V1 not planned upgrades and the need for extra robot maintenance actuations including additional transport to integrator facilities and then back to the testing site.
- Delivery time of critical components (Harmonic Drives, ...)

Nevertheless, the integrated robot is available by M40, May 2018. The final version of this deliverable has been prepared after the system integration. It contains all the description of the V2 prototype fully integrated.

The robot (shown in Figures 3.147 and 3.148) is at integrator facilities (Eurecat). The main components integrated on second Bots2Rec project prototype are listed below, and shown in Figure 3.149:

- Two Mobile Platforms V2, described in Deliverable D4.3
- Robotic Arm V2, described in Deliverable D5.2
- Central Systems and auxiliary devices (cable reel, pipe reel ...)

Figure 3.147. General Picture of the Robotic Unit V2.

Figure 3.148. Other different views of the Robotic Unit V2.

- External aspiration system
- A working version of the Sensor System for Environmental Perception, described in Deliverable D4.2
- A functional version of the Sensor System for Local Process Monitoring, described in Deliverable D5.3
- A working version of the User Interface for Teleoperation, described in Deliverable D6.1
- A set of tools (see Figure 3):
  - Passive compliance grinding tool
  - F/T sensorized grinding tool
  - Scarifying tool

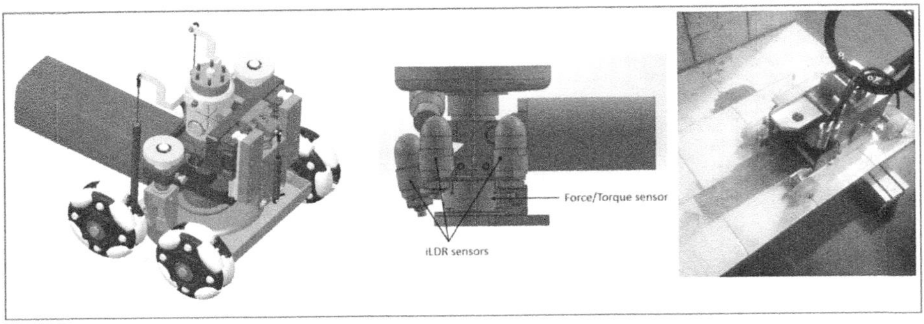

Figure 3.149. Tools of the Robotic Unit V2: Passive compliance grinder (left), F/T sensor grinder (middle), and Scarifier (right).

3.2.4.1 System Assembly

The V2 system assembly has been mostly conducted at the arm manufacturer site (TLABS) unlike the V1 prototype that was mainly assembled (arm to mobile platform) at the integrator facilities (EUT). This fact occurs due to the high level of integration of the arm and its cabinet, which acts as a structural element and need to be assembled directly to the mobile platform. Therefore, arm-to-platform assembly as well as main drive components of the arm have been assembled there.

3.2.4.2 Main Functionalities Tests

After the main assembly, the rest of elements as well as software and communications have been checked and validated at the integrator facilities. The main steps of the prepared validation plan are:

- 1.1 Mechanical integration
  - 1.1.1 All components are correctly assembled
  - 1.1.2 Mechanical interfaces between elements are correctly assembled
- 1.2 Energy Supply
  - 1.2.1 Cabling is correctly disposed and assembled
  - 1.2.2 Connectors and plugs are correctly assembled and connected
  - 1.2.3 Check voltage and correct supply of every component
  - 1.2.4 Measure
- 1.3 Safety elements
  - 1.3.1 Check emergency switch operation
  - 1.3.2 Check other emergency chain elements
- 1.4 Communications

- 1.4.1 Check communication cabling
- 1.4.2 Check configuration of communication elements (router, units IP, comm. speed, etc.)
- 1.5 Software
- 1.6 Check sensor data
- 1.7 Check software command of MP
- 1.8 Check software command of arm
- 1.9 Check software command of Aspiration
- 1.10 Check software of Cable/Pie reels
- 1.11 Integrated operation
  - 1.11.1 Conduct continuous operation test (similar to DiT2.4)
  - 1.11.2 Conduct continuous operation test6 (with load)
  - 1.11.3 Check of control loops

An intensive effort on the test of the integrated unit has been planned in order to ensure that the major issues that could arise for the robot operation are detected and fixed before shipping the robot to the rehabilitation site, where fewer resources are available.

The motions of the mobile bases have been tested (moving synchronously and independently). Besides, different arm motions have been tested while checking the system stability during these tests. In the following figures some of the performed motion tests are shown. Figure 3.150 shows the robot reaching its higher position and Figure 3.151 shows the robot reaching the lower part of the wall.

The most critical position regarding stability is shown in Figure 3.152. It corresponds to the arm completely extended to the front.

Figure 3.153 shows the system reaching the floor, ready to perform grinding operations on ground surfaces.

3.2.4.3 Interfaces Test

At the time of writing the present deliverable report, a set of intensive tests for checking the proper operation of main control and software interfaces is in process. These tests are going to be performed in the system integrator facilities together with the first operation trials. This is in order to ensure that the major issues that could arise for the robot operation are detected and fixed before shipping the robot to the rehabilitation site, where fewer resources are available. We expect to finish the testing of these interfaces within the next month after the submission of this report.

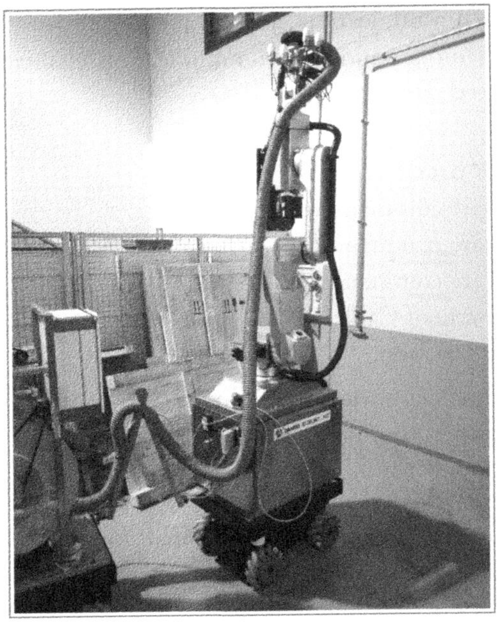

Figure 3.150. Robot with the arm completely extended to the top.

Figure 3.151. Robot reaching the lower part of the wall.

3.2.5 Outlook and Future Work to be Carried Out

As shown in this report, the Robotic Unit V2 has been successfully integrated and its main functionalities are working. The mechanical and electrical assembly of the different subsystems has been done and tested. Nonetheless, more intensive validation tests (especially of control and software interfaces) need to be performed within the following weeks after this report. With this we can ensure a reliable function and operation of the system in different operational conditions.

Figure 3.152. Robot positioned in the top of the vertical linear axis with the arm completely extended to the front.

Figure 3.153. System reaching the floor.

As mentioned before, the next step once the prototype is ready, will be to perform the first set of operational test in system integrator facilities in order to detect any important issue in an early phase before shipping the prototype to the rehabilitation site, where fewer resources are available.

3.2.6 Summary and Conclusion

It can be concluded that the Robotic Unit V2 prototype is fully integrated and working. However, more intensive testing is recommended to be done before the shipment of the prototype, as well as performing a first set of operational tests in the system integrator facilities. This will ensure the systems reliability and avoid further problems on the rehabilitation site.

Deliverable 5.2
D5.2 Robot Arm V2 tested and ready for integration to the robotic system

Authors: F. Becchi

3.2.7 Deliverable in the Context of the Work Program

This deliverable is an output from WP 5: Customization of the robotic arm: Design, control and implementation and in particular from Task 5.5 Testing of the sub system.

The deliverable resumes the outcomes of the test procedure implemented on the V2 robotic arm before its shipment to the integration site.

3.2.8 Methods and Work Carried Out

With reference to the Robotic System V2 architecture, this report focus to the set of subsystem integrated with the robotic arm.

More specifically the modules that will be covered in this test report are

- o Robotic Arm

In line with TLABS QC system FAT (final acceptance test) is managed through a test procedure aimed at the incremental verification and validation of all relevant subsystems.

The main structure of the test sequence covers the different topics resumed in the list below reported:

1.0 arm integration

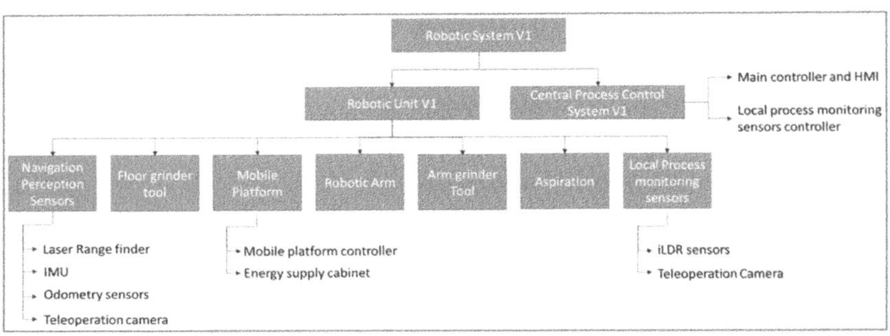

Figure 3.154. General system architecture.

Figure 3.155. Arm assembly.

1.1 robot arm power supply
1.2 robot arm electrical interfaces
1.3 emergency loop ok
1.4 motor calibration
1.5 ethercat network configuration ok- motors
1.6 ethercat network configuration ok- IOS
1.7 tool
1.8 arm calibration check
1.9 arm test sequence

The following section will details the results of the test performed.

3.2.9 Robot Arm Test Report

3.2.9.1 Arm Integration

Scope of this first test is to check that robot arm and main subsystems are correctly integrated according to the reference CAD model.

The reference CAD model considered for this purpose is the T000120A100.asm shown below for reference. The following table summarizes tests done and results achieved.

As visible in the above table all robot arm subsystems are correctly installed. Tool is currently not included in the integration. A review on the "reference" tool was performed and a new tool is currently in construction.

The following pictures show some views of the system being integrated.

Table 3.3. Tests and results.

§	Short	Description	ok	nok	NA	Ref.	Remarks
1.0	arm integration	robot arm and main subsystems are correctly integrated according to reference CAD model	x				
1.0.1	robot arm is correctly assembled		x				
1.0.2	robot arm is integrated on robot cabinet		x				
1.0.3	tool is integrated on the robot			x			tool integration is postponed with respect to arm integration
1.0.4	cabinet is integrated with the motor drivers, Beckhoff CPU and other auxiliary components		x				
1.0.5	robot arm is electrically connected to robot cabinet		x				

3.2.10 Outlook and Future Work to be Carried Out

Within the next weeks, the following test will be performed in order to ensure, that the robotic arm is ready for the final integration at Eurecat facilities.

Figure 3.156. Arm assembly.

Figure 3.157. Arm integration.

Test schedule:

1. Robot arm power supply
2. Robot arm power electrical interfaces
3. Emergency loop ok
4. Motor calibration
5. Ethercat network configuration ok- motors network
6. Ethercat network configuration ok- I/O network
7. Beckhoff arm calibration test
8. Tool
9. Arm test sequence

This deliverable will be updated accordingly as soon as the tests are finalized.

3.2.11 Summary and Conclusion

The robotic arm for the V2 prototype was manufactured and assembled. First tests have been performed with satisfying results. Before the robots will be shipped to the Eurecat facilities for the final integration, further tests will be done according to the above listed schedule. After completion, this deliverable will be updated and finalized.

ANNEX: Full Table To Be Completed

§	Short	Description	ok	nok	NA	Ref.	Remarks
1.0	arm integration	robot arm and main subsystems are correctly integrated according to reference CAD model					
1.0.1	robot arm is correctly assembled		x				
1.0.2	robot arm is integrated on robot cabinet		x				

§	Short	Description	ok	nok	NA	Ref.	Remarks
1.0.3	tool is integrated on the robot		x				tool integration is postponed with respect to arm integration
1.0.4	cabinet is integrated with the motor drivers, bechoff CPU and other auxiliary components						
1.0.5	robot arm is electrically connected to robot cabinet						
1.1	robot arm power supply	the robot arm power supply is correctly managed and is ready for system integration					
1.2	robot arm electrical interfaces	connection btw robot arm and platform are compliant with the defined interface					

§	Short	Description	ok	nok	NA	Ref.	Remarks
1.3	emergency loop ok	the emergency loop is correctly implemented and working					
1.4	motor calibration						
1.5	ethercat network configuration ok- motors	all motors are correctly mapped of the beckhoff PC via ethercat protocol					
1.6	ethercat network configuration ok- IOS	all IOS are correctly mapped of the beckhoff					
1.7	tool	tool can be activated b dedicated beckhoff output signal					
1.8	arm calibration check	beckhoff sw manage robo calibration.					
1.9	arm test sequence	simple test sequence is implemented o beckhoff. Moto are moved on b on and the low level configuration contro stops motion in case of need.					

Deliverable 3.3
Robotic system V2, ready for the use in the test rehabilitation site

Authors: Daniel Martin (EUT), Néstor García (EUT)

3.2.12 Deliverable in the Context of the Work Program

The present deliverable report presents one main result of the second period of the project, where the obtained feedback from the first prototype (V1) of the Robotic Unit for asbestos removal has been applied to an improved second version (V2) of the Robotic Unit. This phase ends with the integration and the complete validation of the different subsystems, designed and developed by the different partners involved in the Bots2Rec project. In this document, a main overview of the Robotic Unit V2 is reported, which is a "demonstrator deliverable" as a result of the previous mentioned tasks.

3.2.13 Methods and Work Carried Out

The main components of the Robotic Unit V2 system are being shipped to the system integrator (Eurecat) facilities together with the system documentation (electrical and mechanical diagrams) and set up guidelines. Once the whole set of components is received, Eurecat, supported by the different partners involved, will proceed with the full integration and assembly of the Robotic Unit V2. Learning from the past experiences and as an attempt to avoid the incurred delays and found problems with the Robot Unit V1, an extremely intense test will be conducted at the system integrators facilities to validate the continuous operation of the integrated Robot Unit V2. This validation campaign prior to taking the robot to the testing site is very important to avoid future "debugging actions" that are very difficult and expensive to perform at the testing site (Valence).

3.2.14 Robotic System V2, Ready for the Use in the Test Rehabilitation Site

In this section the presentation of the Robotic Unit V2 prototype is made, describing the demonstrator deliverable stated in the project plan. The described prototype is the result of the integration work and assembly of the different components that have been developed in this second period of the project.

By the moment of presenting this deliverable, the complete integration of the robotic system V2 still in progress because of different delays. As described in the 2nd periodic report and indicated at the review meeting, the project is experiencing

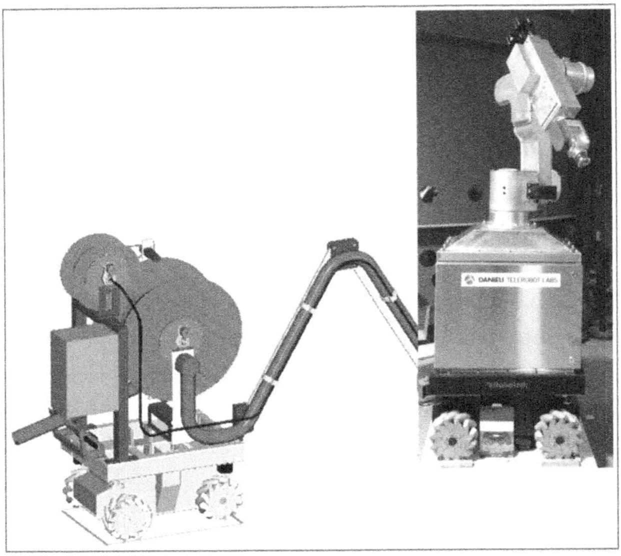

Figure 3.158. General Picture of the Robotic Unit V1.

some delay which affects this deliverable. The main reason that led to this delay have been:

- V1 prototype debugging required an extra effort due to unexpected failures.
- Implementation of prototype V1 not planed upgrades and the need for extra robot maintenance actuations including additional transport to integrator facilities and then back to the testing site.
- Delivery time of critical components (Harmonic Drives, ...)

The availability of the integrated robot is expected by M36, January 2018. The final version of this deliverable will be prepared after this system integration. It will contain all the description of the V2 prototype fully integrated, tested and ready to be used at the testing site.

The current status (shown in Figure 3.158) is:

- Mobile Platform of FRONT unit is available and ready for integration and available at arm integrator facilities (TLABS)
- 7 DoF arm is mechanically integrated as well as the electrical cabinets, which act at the same time as structural element of the FRONT unit. It is missing the integration of most of the electrical equipment of the arm. This works are in progress at arm integrator facilities (TLABS)
- Communication elements and sensors will be installed at system integrator facilities (Eurecat)

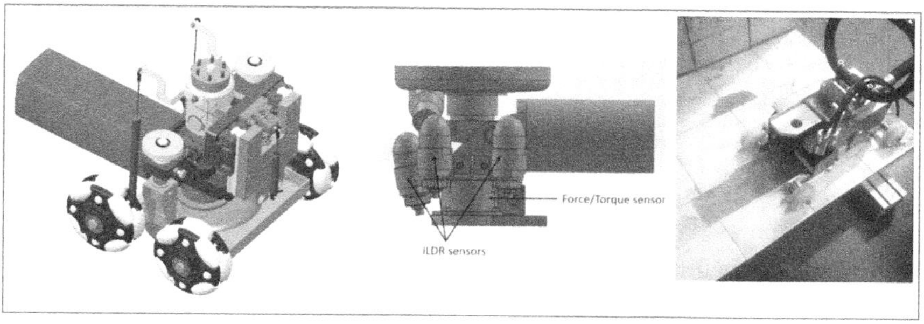

Figure 3.159. Tools of the Robotic Unit V2: Passive compliance grinder (left), F/T sensor grinder (middle), and Scarifier (right).

- Software integration at system integrator facilities (Eurecat)
- Connection between FRONT and REAR unit still in progress.
- New set of Local process monitoring still in progress.
- All components of REAR unit still in progress

The main components integrated on second Bots2Rec project prototype are listed below, and shown in Figure 3:

- Two Mobile Platforms V2, described in Deliverable D4.3
- Robotic Arm V2, to be described in Deliverable D5.2
- Central Systems and auxiliary devices (cable reel, pipe reel …)
- External aspiration system
- A working version of the Sensor System for Environmental Perception, described in Deliverable D4.2
- A functional version of the Sensor System for Local Process Monitoring, described in Deliverable D5.3
- A working version of the User Interface for Teleoperation, described in Deliverable D6.1
- A set of tools (see Figure 3.159):
  - Passive compliance grinding tool
  - F/T sensorized grinding tool
  - Scarifying tool

3.2.14.1 System Assembly

The V2 system assembly will be mostly conducted at the arm manufacturer site (TLABS) unlike the V1 prototype that was mainly assembled (arm to mobile platform) at the integrator facilities (EUT). This fact occurs due to the high level of integration of the arm and its cabinet, which acts as a structural element and need to be assembled directly to the mobile platform. Therefore, arm to platform assembly as well as main drive components of the arm will be assembled there.

3.2.14.2 Main Functionalities Tests

After this initial assembly, the rest of elements as well as software and communications will be assembled, checked and validated at the integrator facilities. A validation plan is being prepared accordingly, which main steps will be:

- 1.1 Mechanical integration
  - 1.1.1 All components are correctly assembled
  - 1.1.2 Mechanical interfaces between elements are correctly assembled
- 1.2 Energy Supply
  - 1.2.1 Cabling is correctly disposed and assembled
  - 1.2.2 Connectors and plugs are correctly assembled and connected
  - 1.2.3 Check voltage and correct supply of every component
  - 1.2.4 Measure
- 1.3 Safety elements
  - 1.3.1 Check emergency switch operation
  - 1.3.2 Check other emergency chain elements
- 1.4 Communications
  - 1.4.1 Check communication cabling
  - 1.4.2 Check configuration of communication elements (router, units IP, comm. speed, etc.)
- 1.5 Software
- 1.6 Check sensor data
- 1.7 Check software command of MP
- 1.8 Check software command of arm
- 1.9 Check software command of Aspiration
- 1.10 Check software command of Cable/Pipe reels
- 1.11 Integrated operation
  - 1.11.1 Conduct Integrated Robot Unit V2 continuous operation test (similar to DiT24 test)
  - 1.11.2 Conduct Integrated Robot Unit V2 continuous operation test (with load)
  - 1.11.3 Check of control loops

An intensive effort on the test of the integrated unit is planned in order to ensure that the major issues that could arise for the robot operation are detected and fixed before shipping the robot to the rehabilitation site, where fewer resources are available.

3.2.15 Outlook and Future Work to be Carried Out

As shown in this report, the integration of the Robotic Unit V2 has been initiated. By the time of writing this report, some tasks regarding the arm assembly and its integration to the mobile platform are in progress at the arm developer facilities.

After that, all the V2 system components will be received at the system integrators facilities in order to finish the integration process, including hardware and software.

Next step will be an intensive validation test campaign at the system integrator facilities will be conducted in order to ensure a reliable function and operation of the system in different operational conditions.

3.2.16 Summary and Conclusion

It can be concluded that the Robotic Unit V2 prototype is delayed and, therefore, not ready yet to be used in the rehabilitation side as planned. Some integration task still in progress and some of them will be initiated within next months.

One of the lessons learned with the previous prototype is the fact that before the Robotic Unit V2 prototype can be considered as ready to be used in the rehabilitation site as planned, extremely intensive tests must to be conducted in the system integrator facilities. This will ensure the systems reliability and avoid further problems on the rehabilitation site and related delays.

3.2.17 Radar Sensors

Task 4.2 Sensor Systems for environmental perception

The autonomous mobile front and rear units use 2D-SLAM (simultaneous localization and mapping) algorithms to navigate inside buildings through rooms and around obstacles. Each unit is equipped with two iSDR-C (indurad Scanning Dynamic Radar – Compact, Figure 3.160) sensors mounted on the front-right and rear-left side of the mobile base of the each unit as depicted in Figure 3.160.

The iSDR-C was still a prototype at the early stages of the project. Its development has been advanced in the last two years, and it reached the point of product maturity. The radar sensor performs a 360° round scan with reliable measurement of distances to any surface capable of reflecting the radiated microwave in a frequency band centred around 77 GHz. The iSDR-C offers an extremely high update rate of up to 1600 measurements per second and an accuracy of down to < 10 mm depending on the measured range and the radar-configuration.

The raw data consisting of sampled received radar waves and the associated angles are sent via Ethernet to an embedded computer (Intel-NUC) for digital signal processing. The results of the processing are arrays of frequency-spectrum power intensities corresponding to the scanning radial line from the sensor. These frequencies correspond to distances from reflecting objects around the sensor. The peaks in the spectrum caused by these targets and the incident angle are converted to cloud-points. These points are sent to the main ROS-computer and used by the 2D-SLAM algorithms developed at IGMR-RWTH. Figure 3.161 through Figure 3.162 show typical iSDR-C radar images and the corresponding ROS cloud-points.

Task 5.2 Sensor system for local process monitoring

The control and placement of the grinding tool by the robot arm requires accurate measurements of the distance between the tool and the wall. The arm of the V2 robot is equipped with five iLDR-C (indurad linear dynamic radar – compact, Figure 3.163) sensors as depicted in Figure 3.164.

iLDR-C V1 sensors were tested on the V1 robot in the early stages of the project. The current iLDR-C sensors mounted on the V2 front unit robot are of the latest version, V4, at this time. None linearity, misplacement of the radar chips and detection of very close objects are critical issues that were faced during workshops with the V1 and V2 robots, which were addressed in deliverable D5.3. Since then a lot of work has been done to resolve these issues and mitigate their impact on

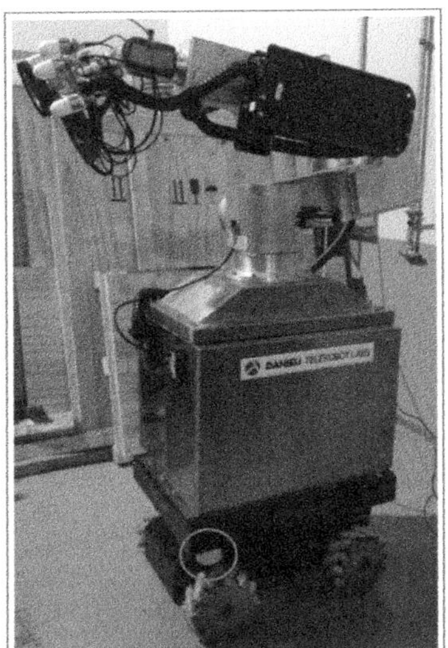

Figure 3.160. iSDR-C mounted on the front unit.

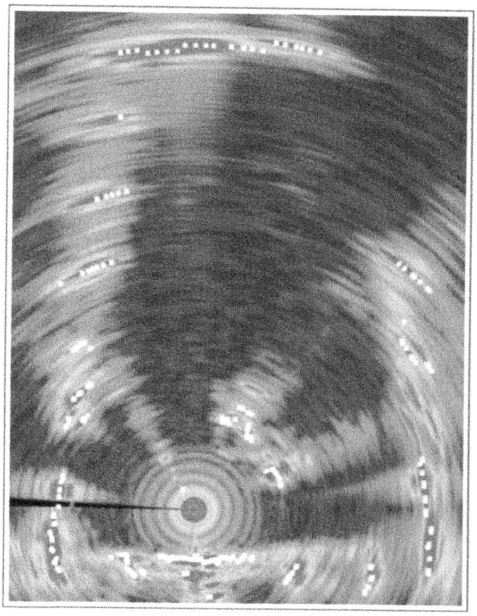

Figure 3.161. Radar image showing cloud-points.

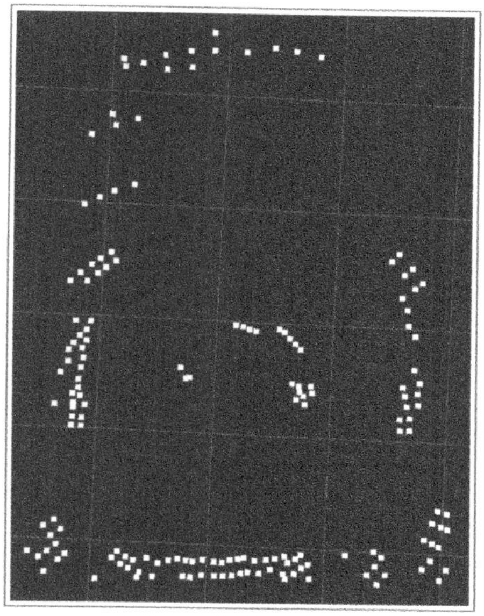

Figure 3.162. Cloud-points in ROS visualization (RViz).

Figure 3.163. iLDR-C front view showing the PTFE lens.

the accuracy of the measurement. The following paragraphs will briefly discuss the work done.

The linearity of the radar measurements was investigated in a series of experiments using a linear axis with a stepping motor. The following diagrams show the relationship between the real distances to a target compared with radar measurement. The plot at the top of Figure 3.165 shows a linear relationship at first glance. Zooming in to region around zero in the third plot of the figure shows a none-linear offset. Figure 3.166 shows the same measurements taken varying speeds (1 mm/s–10 mm/s).

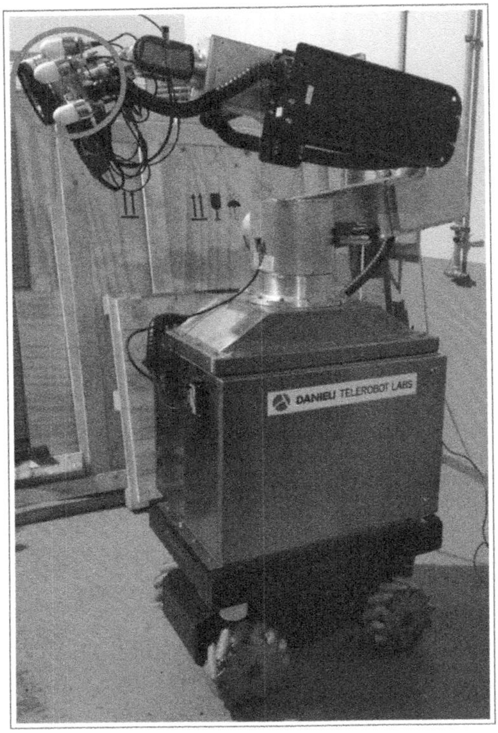

Figure 3.164. Five iLDR-C mounted on the tool.

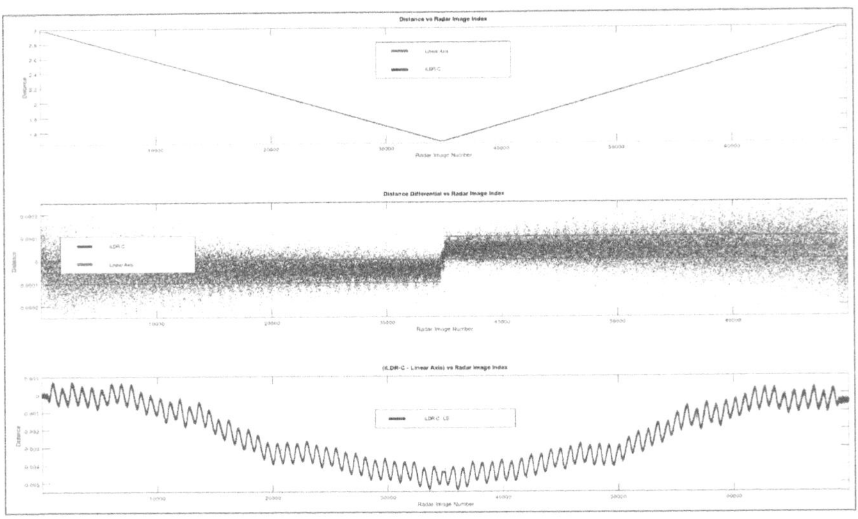

Figure 3.165. Linear axis moving with constant speed of 1 mm/s.

Further investigation of these and several other experiments led to corrections applied to the setup and configuration of the PLL (phased locked loop) and radar parameters (bandwidth, start and end frequencies, etc.).

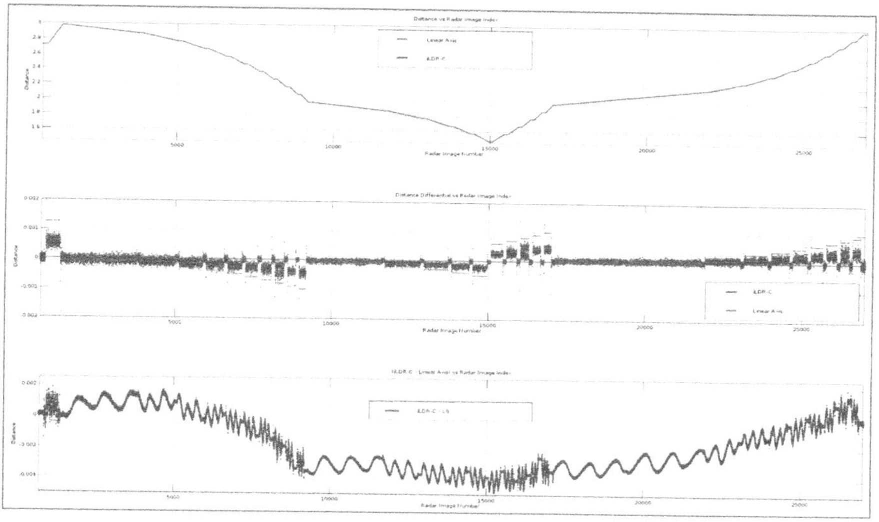

Figure 3.166. Linear axis moving at different speeds (1 mm/s–10 mm/s).

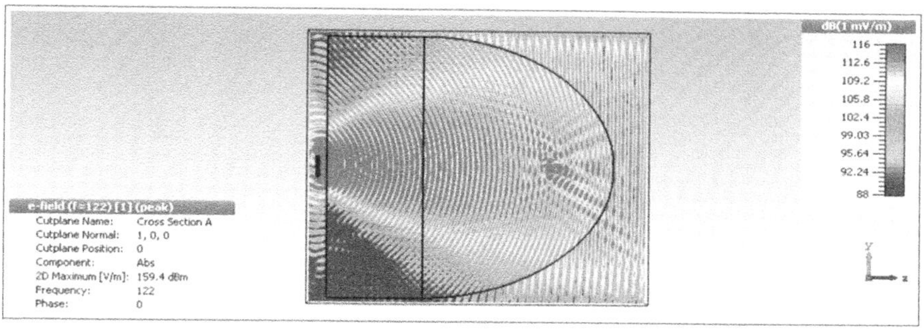

Figure 3.167. Simulating the propagation of the electromagnetic waves thought the lens.

The issue concerning the misplacement of the lens had a bigger impact on the accuracy of the radar measurements. Figure 3.167 shows a simulation of electromagnetic field propagation in a PTFE lens, which converts the spherical field to a plane wave and achieving 32 dB.

In reality though, a bi-static radar, i.e. a radar with separate transmitting and receiving antennas cannot be optimally a placed under a radome. The first iLDR-C versions suffered of this misalignment. The following figures shows simulation results of the placement of the radar chip in iLDR-C V4.

Figures 3.168 and 3.169 show simulation results of the receive and transmitting antennas optimally placed to mitigate the misalignment due to the nature of bi-static radars.

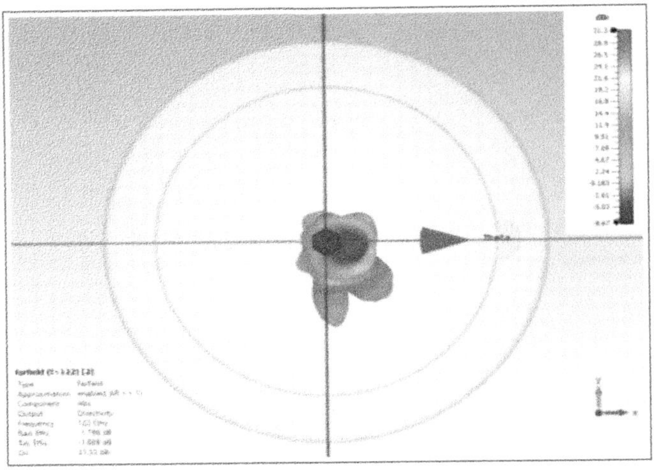

Figure 3.168. Receive antenna.

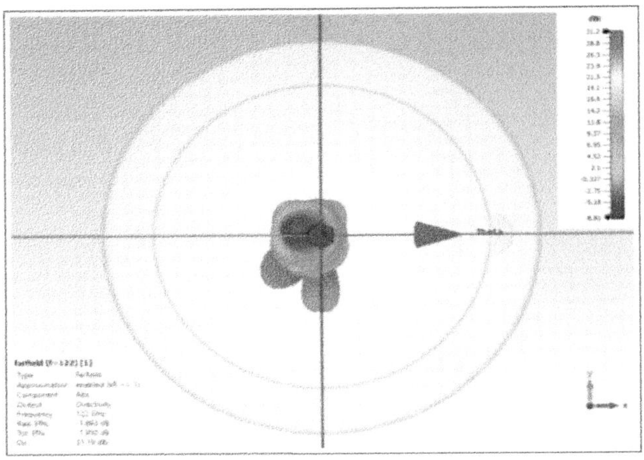

Figure 3.169. Transmitting antenna.

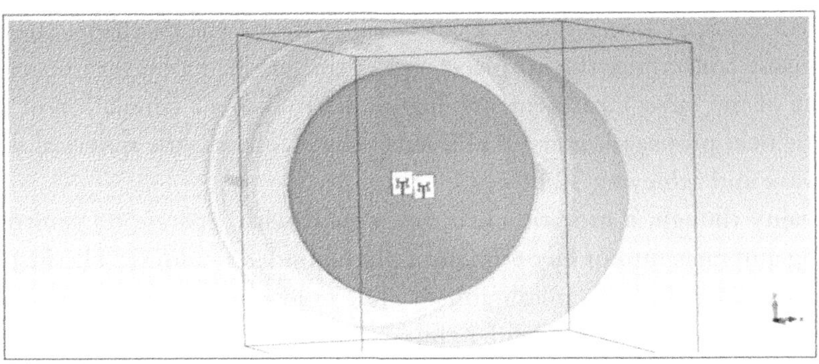

Figure 3.170. Alignment of transmitting and receiving antennas under the lens.

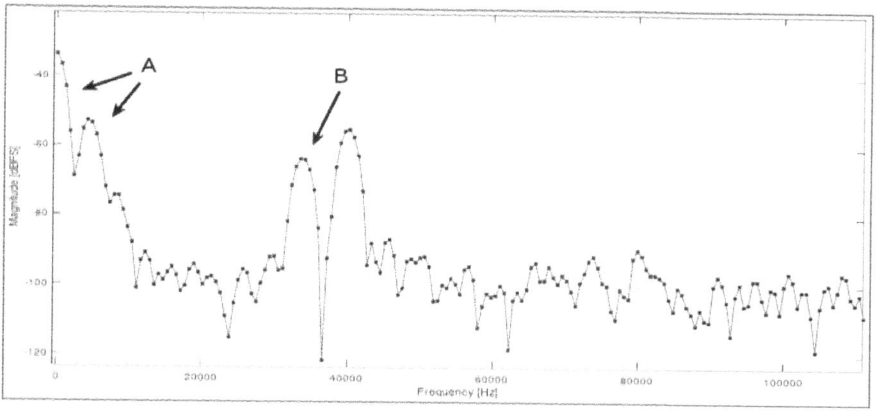

Figure 3.171. Frequency spectrum showing the effect of the lens. A: Static clutter due to the lens. B: Targets 80 cm away from the sensor.

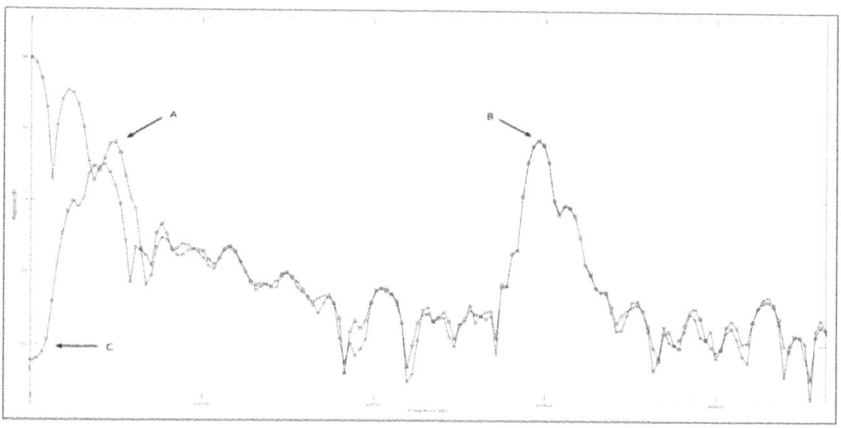

Figure 3.172. Effect of applying a high-pass filter to remove the reflections of the lens. A: Close target. B: Far target. C: DC-leakage and lens attenuated by 80 dB.

The PTFE lens affects the detection of targets close to sensor. Reflections of target in the range of 10 cm in front of the lens are difficult to separate from reflections of the lens. The reflections of the lens are always present and hence are considered static clutter. The following frequency spectrum plot shows the effect of thelens.

Several methods for the mitigation of the impact of the lens on close targets were investigated. The first and simplest method is applying a high-pass filter with a cutoff frequency corresponding to the reflection of the lens. Several filter designs and configurations were tested.

Figure 3.172 shows the effect of a 9th order Chebychev 2 IIR high-pass filter implemented in five second- order sections. This approach was not satisfying

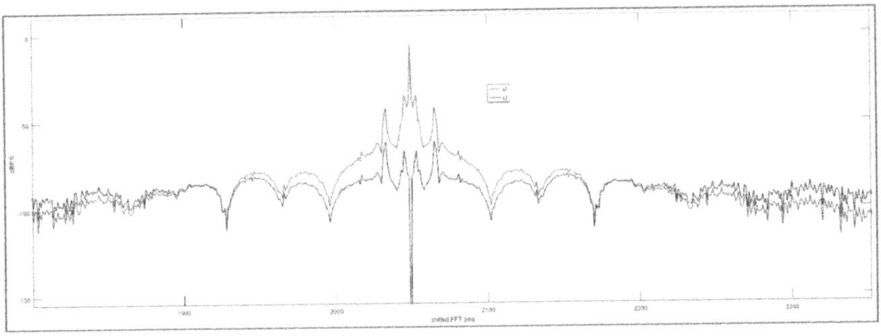

Figure 3.173. SVD method. S1: Original signal showing high peak in the middle of the plot corresponding to static clutter. S2: Result of removing the lens, as it was the strongest signal in the measurement.

enough to be adopted. The high-pass filter damped the signals of the targets as they moved closer to the lens.

The second approach was the method of SVD (singular value decomposition). Using this method, the radar signals could be separated according to the spectral law of linear algebra in several matrices representing signal and noise spaces.

Figure 3.173 shows the effect of applying the SVD method to remove static clutter of the measurement. In this case, the signal of the lens was the strongest. Hence, it corresponded to the largest singular value. Subtracting the matrix of the signal space corresponding to this singular value removed the static clutter of the lens from the total signal.

The SVD method was considered from a theoretical point of view and was further not implemented in the project. The third method investigated was the simplest and most effective one. A pre-recorded signal corresponding to the lens is subtracted from each measurement. This pre-recorded signal was measured with no targets in 20 meters in front of the sensor, leading to the lens being the only or at least the strongest target. This method is currently implemented in the signal processing software of the iLDR-C system on the V2 robot.

Figures 3.174 and 3.175 show screenshots from the signal processing software running on the V2 robot. The effect of subtracting the pre-recorded lens signal is clearly demonstrated.

3.2.18 Further System Components

Task 3.4 Implementation of central systems

In this task the central systems, that are not part of the robotic units or the central process control system were layout, implemented and tested. For the V2 robot the suction system is external to the robot to provide more suction power and to be

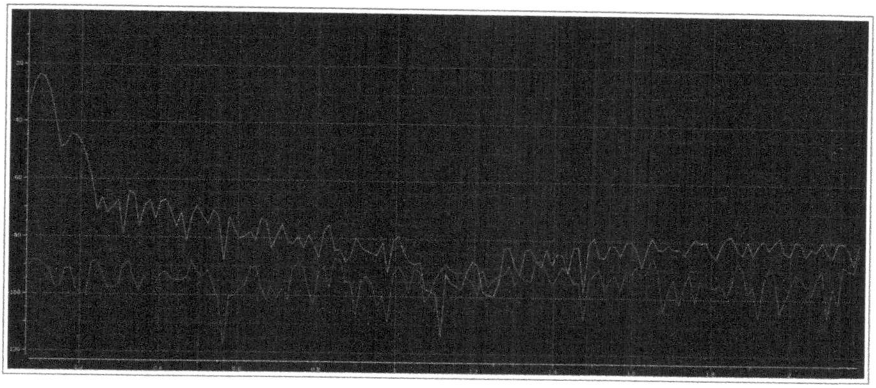

Figure 3.174. White plot shows the original signal. Red plot shows the signal after subtracting the lens siganl.

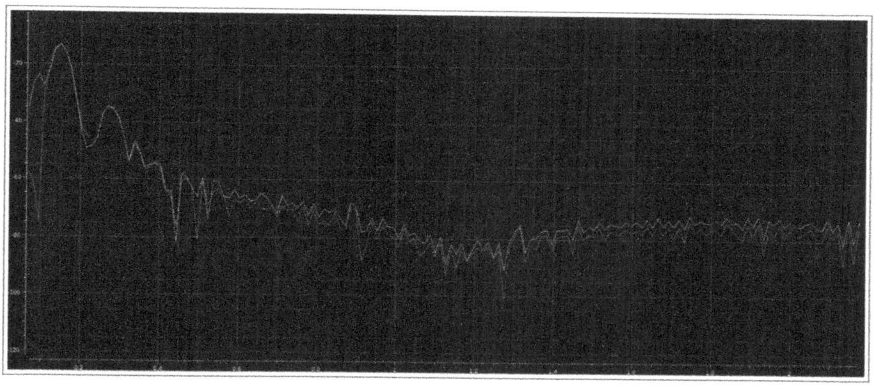

Figure 3.175. The red plot shows a strong peak by a wall 4 cm in front of the sensor. The wall is easily separated from the strongly attenuated lens reflection.

able to centralize the collection of contaminated material from all units in operation. In this way, the robot was provided with an aspiration hose connected to the suction system, which is picked up and released when the robot moves. These pipe-reeling system was designed, implemented and, then, tested in manual and automatic mode. The robot was also provided with two safety switches, one for each mobile platform, so both buttons can safely stop the robot, during the testing and benchmarking phase operation. When pressing the emergency stop, the wheels and joints of the arm are blocked in such a way that they prevent the movement of the robot as well as putting pressure on any object by the effect of gravity, increasing the safety of the system. The emergency stop was tested, with both units working in tandem configuration and working independently (in this case, each button stops the corresponding mobile platform).

A reel unit was integrated on the robot following platform as discussed in the second periodic report. To extend the reach of the robotic system without affecting

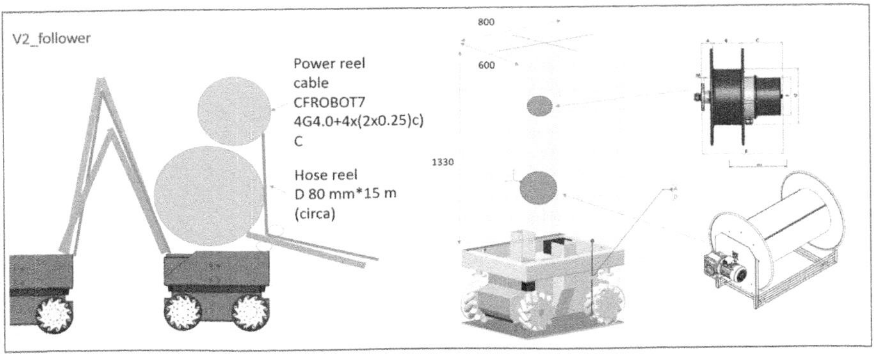

Figure 3.176. Reel unit components.

dust removal performance the original concept of a single fully integrated robot evolved in two paired robotic platform with the flowing one supporting the operation of the main robotic arm with aspiration hose reel and power supply cable reel. The following figure summarizes first approximation detailed requirement for the two reels to be integrated and a rough dimensional reference for the development of the reel units.

As for the first prototype after a first discussion with industrial reels supplier we opted for a fully custom design with the aim of optimizing dimensional integration. As visible in the integration pictures below reported the overall concept of the two reels is based on the same design implemented and tested for the cable reel for the V1 prototype.

As in the first design, a sliding belt transmission is implemented between the reel and the motor where, also in this case, drum position is directly read with an independent belt transmission, read on the Ethercat interface with a dedicated multiturn encoder.

The novel feature introduced in this case is the fully programmable reeling unit on the two reels:

Cable/hose position is guided along the reel thanks to a motor driven guide. Cable/hose position can be as consequence controlled to follow the reel rotation and to guide the correct cable/hose winding. This is useful for cable reel but is of paramount importance for the hose reel where large hose diameter aspiration pipe fills nearly completely the reel (see picture above). The reel unit is completed by a local control cabinet (see picture below) that integrates local Ethercat module, connected to the main controller, and the needed power distribution.

Task 4.3 Control of the mobile platforms

The two reels unit integrated in the auxiliary robot platform are integrated in terms of control in the same Ethercat network of the main robot arm drivers. The local

Figure 3.177. Mechanical integration.

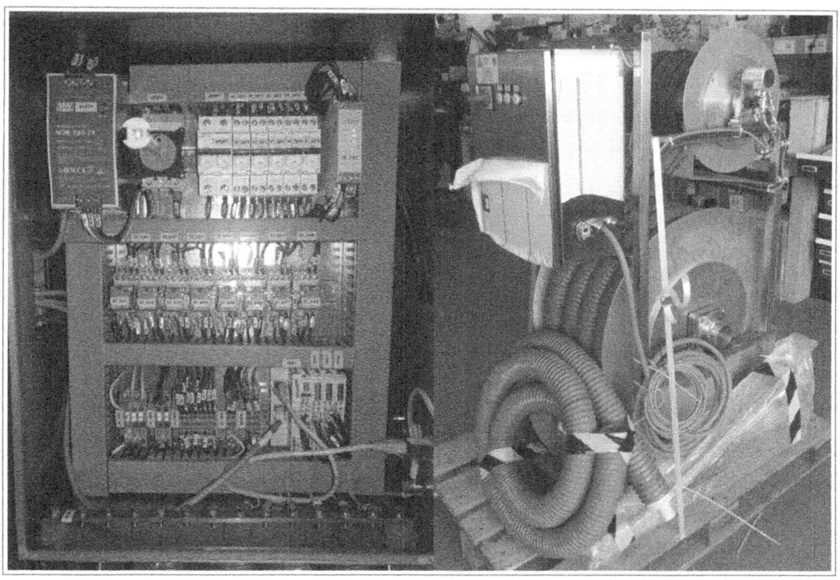

Figure 3.178. Final integration, system ready for shipment.

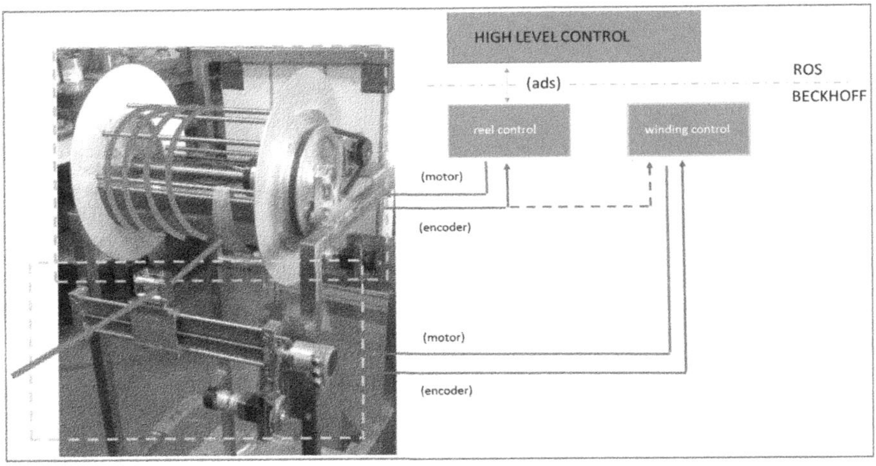

Figure 3.179. Reel control.

network implemented on the reel unit manage four multi-turn absolute encoders and four electric motor.

The processes managing the reels are integrated in the Beckhoff CPU installed in the main robot cabinet as ,at least in the actual implementation, there is no need to use the reel unit decoupled from the robot arm.

The reel/unreel control can be controlled directly from the ADS interface via ROS, to manage cable/hose length together with platform motion control. Alternatively, reels can be left "idle" so that the cable/hose is unreeled passively (reeling is always actuated).

Winding guide control is a parallel low-level process running independently by the reel control: the reel rotation angle is fed to the winding guide control that as consequence adapt cable position so the cable always follow (with actuated or idle motion) the reel.

3.2.19 Entire System

Task 3.5 Assembly and validation tests of the robotic system

Robot arm integration and testing was performed right at the end of the 2nd reporting period, as anticipated in the previous technical report. The full robot arm was integrated and tested in DTLAB facilities in Genova before shipment to integration site. Detailed test report can be found on D5.2.

Some pictures of the robot arm integrated under test extracted from this deliverable are here reported as reference. As visible the main goal of a cleaner integration of robot arm different subsystem can be considered met. The amount of exposed components is limited and the complete electrical integration is performed in the

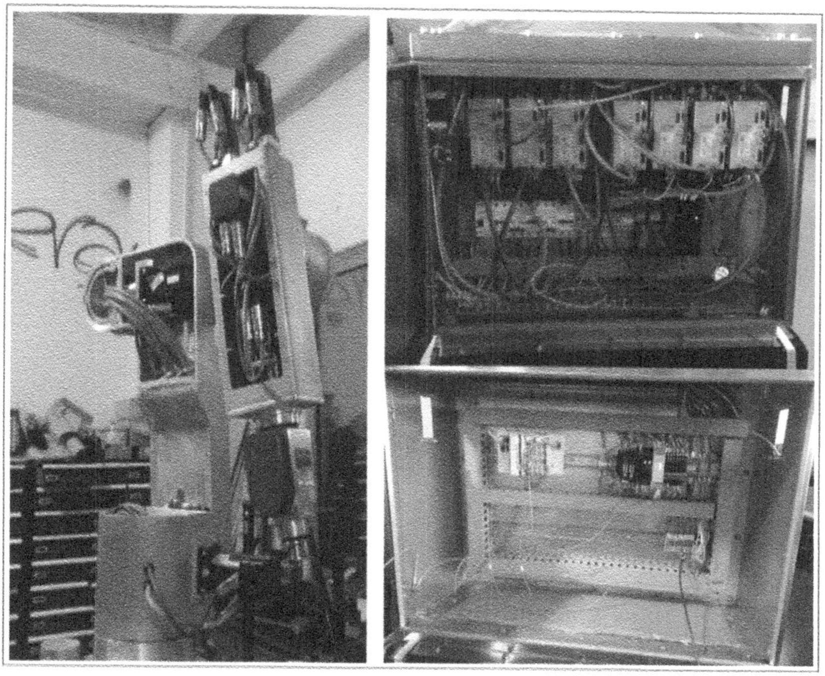

Figure 3.180. Robot arm integration (details).

robot supporting structure, leaving a significant free space for further modules integration. On the other hand, even if the low level joint control is completely different than the V1 prototype (different motors and different drivers) thanks to the adopted Ethercat protocol the new robot arm was seamless integrated on the upper control layer.

The main components of the Robotic Unit V2 system were shipped to the system integrator (Eurecat) facilities together with the system documentation (electrical and mechanical diagrams) and set up guidelines. Eurecat, supported by the different partners involved, proceeded with the full integration and assembly of the Robotic Unit V2. Learning from the past experiences and as an attempt to avoid the incurred delays and found problems with the Robot Unit V1, intense tests were conducted at the system integrators facilities to validate the continuous operation of the integrated Robot Unit V2. In this task, it was tested the electric, mechanical, communication and software interfaces between the arm, the two mobile platforms, the cable-reel unit, the external user-interface unit and all the sensors. Tests were also carried out of autonomous and manual navigation of the platforms in tandem formation and operating independently. The arm control was also tested in manual and autonomous mode, being able to move the tool from one position to

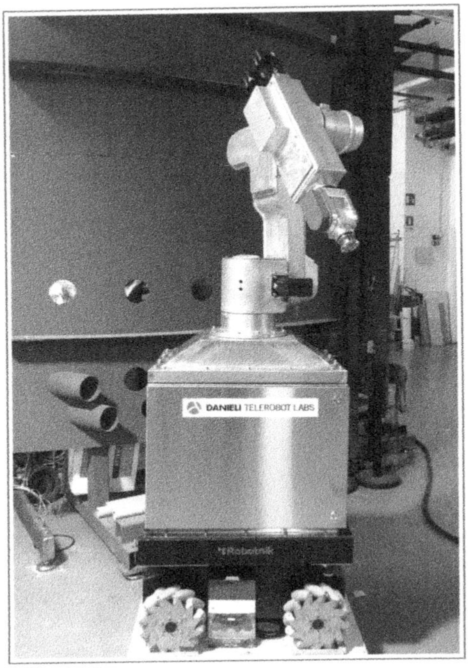

Figure 3.181. Robot arm integrated on the robotic platform.

another and to land softly on the wall. Besides, cable and hose pick-up and release tests were made, in manual and automatic mode (i.e. synchronized with the mobile platform movements). Once the whole robot was fully functional, it was sent to the RWTH facilities to perform grinding tests repeatedly and in different operational conditions, among other tests, to finally validate the robotic system.

DOI: 10.1561/9781680837155.ch4

Chapter 4

Algorithms for Removing Asbestos

By Tobias Haschke, Sami Charaf Eddine, Kamal Mohy El Dine, Siddharth Maraje, Fabian Herb, Oriol Orra, Diego Fogliacco, Tim Detert, Claudia Cornely, Daniel Martin, Albert Fullana Puig, Ramón Carrasco Jimenez, Matthias Rabel and Jean-Christophe Fauroux

4.1 User Interface

Deliverable 6.1
User Interface for Tele-Operation

Authors: Tobias Haschke, Matthias Rabel

4.1.1 Deliverable in the Context of the Work Program

The User Interface for Tele-operation (UITO) (Task 6.3) is the point of contact for the end user of the robotic system that is developed within Bots2ReC. It is the instance for reducing complex robotics behavior into intuitive, safe and comfortable ways of interaction. Therefore it is embedded into work package 6 where the high-level task planning and the necessary algorithms are developed. The UITO is communicating with the CPCS of the robotic system and creates the first solution for interacting with the robotic system via the self-written control system. It is the first task within WP6 and requires a first system architecture for the basic tele operated movements. Along the development of the UITO it is also necessary to define the first levels of the CPCS for avoiding time consuming adapting operations in the ongoing project. Beside this, it is also necessary to have supervision for the whole project because the user interface is going to be extended in Task 6.4. According to this schedule it is important to create a UI that is easy to maintain and to extend during the ongoing project.

4.1.2 Methods and Work Carried Out

For developing the UITO a systematical approach of designing the software architecture is chosen. The documentation of this process is done with the Unified Modeling Language (UML) which is state of the art for displaying the content of software solutions vivid and comprehensive. The most important advantage of using UML is that the results are understandable for every consortium member even if they do not have further expertise in computer science. Out of this a basic way for discussions and feedback loop is found. Referring to the common workflow for using UML for software development, in a first step the identification of the stakeholders for the project was carried out. Afterwards the targets for the UITO are defined and corresponding Use-Cases are written. Finally the necessary activity diagrams are confirmed for solving the Use-Cases with a well-defined flow of data. During this process multiple feedback loops are used for informing the affected consortium members and for finding a consensus of the right solution. Especially the feedback of the end user Bouygues is essential as they have the expertise in removing asbestos and can provide important details about the conditions on a construction site. The feedback loops are organized via a Microsoft Word Document where the functionality of comments is used for asking questions, setting remarks and sending responses. Beside the documentation of the software architecture there are also discussions about the hardware for displaying and interacting with the robotic unit. These discussions are also part of Task 6.3. For the development of the UITO the hardware definitions of Task 3.1 and the technical requirements of Task 3.2 are taken into account. As the partner Indurad is developing the UITO they also use the input from Task 4.2 for the environmental perception and the first results from Task 6.1.

4.1.3 Identification of Requirements

The identification of requirements is a classical approach to define the content of the UITO on a textual basis. It is a solution to define the use-cases with words so that every participant of the discussion can follow it easily even without deeper knowledge of computer science. At first the stakeholders of the software solution are defined. Out of this the main targets of the UITO and the interaction with the different roles of the stakeholders are outlined. The next steps are models for the systems behaviour with use-case diagrams and corresponding activity diagrams. The Unified Modelling Language (UML) is chosen to document the software product. The typical workflow for defining these aspects can be seen in Figure 4.1.

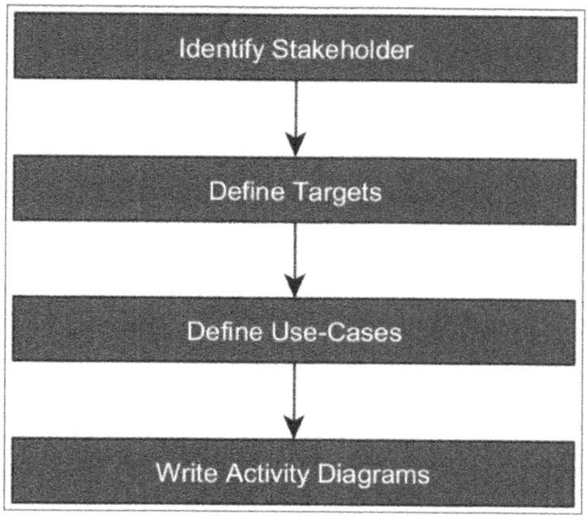

Figure 4.1. Steps for the development process.

4.1.3.1 Stakeholders

The different stakeholders represent the groups of people or systems that are affected by the user interface for tele-operation. For this system there are the following stakeholders:

- Construction worker (User)
- Developer (Engineers of the Bots2ReC Consortium)
- Construction Company (Company that uses the product)
- Product Owner (Company that sells the product)

4.1.3.2 Targets

These targets are meant to be top-level-requirements that need to be fulfilled by the UITO. They are solution-neutral, consistent and relevant for the project.

4.1.3.2.1 Template for Targets

For the definition of targets the following template was used. It is common in the description of tasks in the understanding for UML and is enough to define abstract targets without mentioning the solution.

ID	###
Target	What should be achieved?
Stakeholder	Which stakeholders are linked to this target?
Effect on Stakeholders	Which effects can be expected for the stakeholders?
Boundary conditions	Are there fix boundary conditions for this target?

It is essential to define the stakeholder who is linked to this target and how he is affected by it. Out of these ideas the targets can be used to understand the influence of the software product more precise.

4.1.3.2.2 List of Targets for UITO

ID	T-1
Target	Tele-operate the robotic unit inside of construction site
Stakeholder	• User
Effect on Stakeholders	The User does not need to move the robotic unit manually
Boundary conditions	The CPCS can only convert the user input according to the abilities of the mobile platform. Active collision avoidance is also necessary.

ID	T-2
Target	Tele-operate the robotic arm for asbestos removal tasks
Stakeholder	User
Effect on Stakeholders	• The User does not need a protective suit any more • The User is not working in a hazardous environment
Boundary conditions	• The Robotic Unit needs to be stable during the whole process • The asbestos removal task can be fulfilled completely (process parameters as necessary)

ID	T-3
Target	Get feedback about the robotic system that enables the tele-operation
Stakeholder	• User
Effect on Stakeholders	• The user has a vivid and intuitive feedback of the actual state • The robotic unit can be tele-operated more precisely
Boundary conditions	• The connections between the systems (Wi-Fi, Ethernet, etc.) have data limits that are limiting the amount of data for the interface

ID	T-4
Target	Ensure a pure safe tele-operational process for human and machines
Stakeholder	• User
Effect on Stakeholders	• The User is not able to crash the robotic unit • The Robotic Unit cannot be harmed by wrong input commands of the user
Boundary conditions	–

ID	T-5
Target	Give feedback about the removal result of the robotic unit
Stakeholder	• User

ID	T-5
Effect on Stakeholders	• The User needs to get feedback about the removal success
Boundary conditions	• The connections between the systems (Wi-Fi, Ethernet, etc.) have data limits that are limiting the amount of data for the interface

ID	T-6
Target	Enable the user to manage errors and issues
Stakeholder	• User
Effect on Stakeholders	• The User do not need to enter the hazardous environment for solving expected issues like sensor restarts
Boundary conditions	• The performed actions are limited to the low level access to every machine (e.g. for restarting devices)

ID	T-7
Target	Create a user interface that is easy to use and easy to learn
Stakeholder	• Construction Company
Effect on Stakeholders	• The Construction Company does not need much time in teaching employees (User) for the automated asbestos removal
Boundary conditions	• As easy as possible but as complex as necessary! A common understanding of robotics behaviour is mandatory

ID	T-8
Target	Create a user interface that is easy to maintain and to extend
Stakeholder	• Product Owner
Effect on Stakeholders	• The Product Owner can extend the user interface with new functionalities and maintain it without huge effort
Boundary conditions	• Properties of the used programming language is the limiting factor

4.1.3.3 Use-Cases

Use-Cases describe with normal words (no technical terms) the interactions of actors with the system. Actors are often members of the identified stakeholders as e.g. the User itself. Every Use-Case can be defined with nouns and a single verb. It is solution-neutral and only describes **what** needs to be done by the system and **not how** it is done.

4.1.3.3.1 Template for Use-Cases

For a clean and readable structure of the Use-Cases the following template is used. It contains all necessary information about the Use-Case and can be read quite easy.

ID	###
Use-Case	<Verb> <Noun>
Brief Description	Brief description of the Use-Case
Prerequisites	What needs to be done before this Use-Case?
Results	What is the outcome of this Use-Case?
Typical Process	Steps towards the results

Every project member involved used this template to edit the list of Use-Cases during the documentation of possible Use-Cases.

4.1.3.3.2 List of Use-Cases for UITO

ID	UC_001
Use-Case	Give input for tele-operation
Brief Description	Use an input device for giving commands to the UITO
Prerequisites	The system is ready to receive tele-operational commands
Results	The commands are used for the robotic movement
Typical Process	• The commands are read into the system • They are converted to robotic commands
Graph	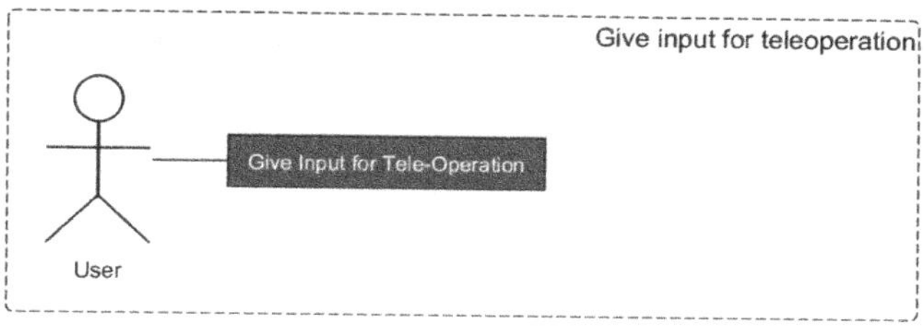

ID	UC_002
Use-Case	Communicate with Hardware
Brief Description	Enables the CPCS to communicate with every hardware of the robotic system (sensors, actuators, tools)
Prerequisites	The Robotic Unit is powered on
Results	The CPCS can send specific commands to the robotic unit and gets feedback about it
Typical Process	• Establish connection with the hardware • Check if everything is running as wanted • Give status about the actual situation (everything is fine, errors or warnings)
Graph	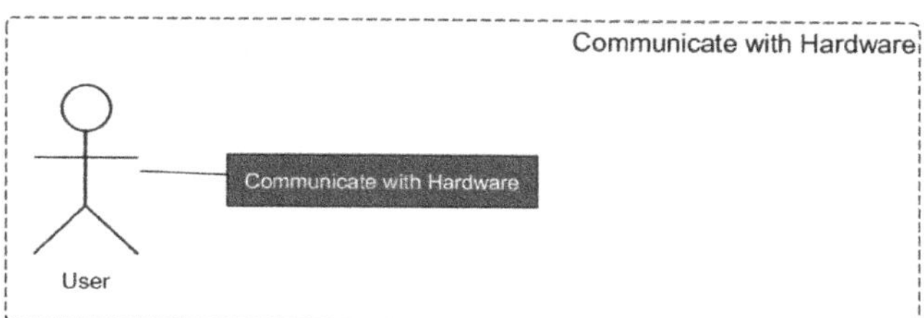

ID	UC_003
Use-Case	Visualize system data
Brief Description	The User is able to see system output on the screen of the UITO
Prerequisites	The System is powered on and connected to the UITO
Results	The User can easily see the actual State of the Robot
Typical Process	• Convert robotic data into visual information like 3D view of scenery
Graph	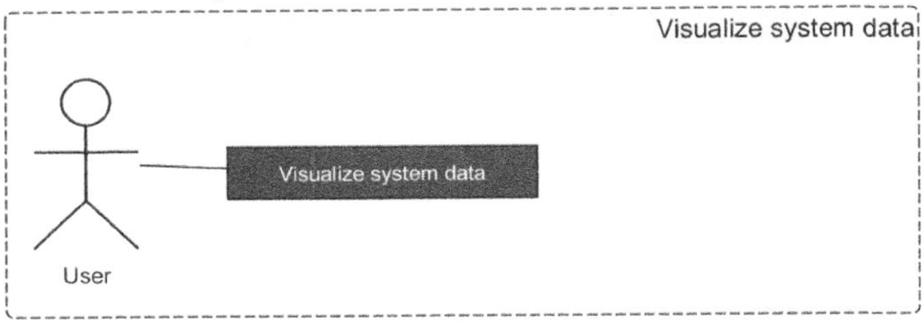

ID	UC_004
Use-Case	Move Robotic Unit
Brief Description	Move the robotic unit in x, y and theta inside of the construction site
Prerequisites	The Robotic Unit is powered on and calibrated.
Results	The Robotic Unit reaches the new position as the user expected
Typical Process	• Recognize User input • Process movement commands • Execute Commands Give Feedback
Graph	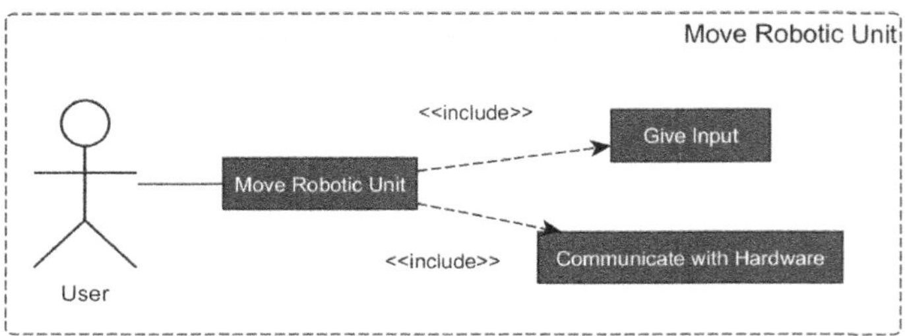

ID	UC_005
Use-Case	Move Robotic Arm
Brief Description	Move the robotic arm in space to reach walls or to look around with the camera
Prerequisites	The system is powered on and the robotic unit is calibrated
Results	The arm reaches the new position as the user expected
Typical Process	• Read Joystick Input • Convert it to arm movement • Execute motion
Graph	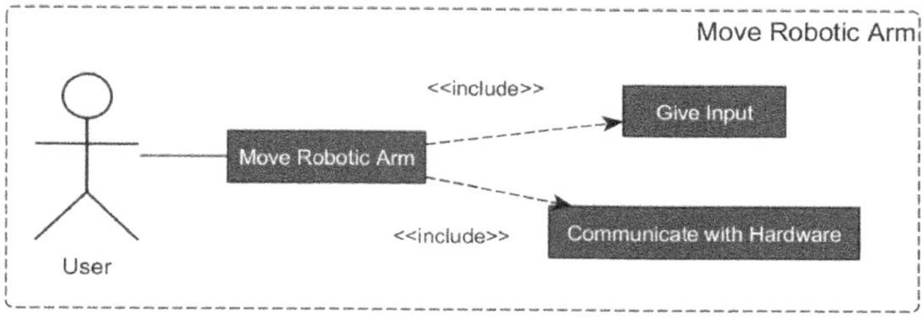

ID	UC_006
Use-Case	Configure removal process
Brief Description	Setup the specific process parameters for removing asbestos (disk speed, pressure, max. speed, etc.)
Prerequisites	The system is turned on and ready to be configured properly
Results	Save the configuration for the removal process (dynamically?)
Typical Process	• Open actual configuration • Allow changes Save new configuration
Graph	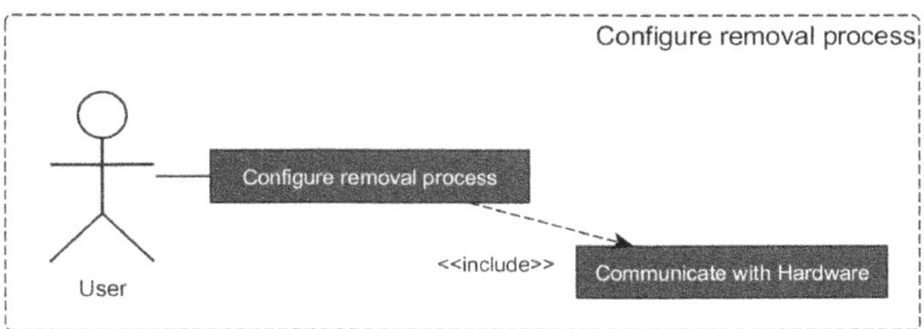

ID	UC_007
Use-Case	Perform grinding movements
Brief Description	Enable the tool and the aspiration for removing asbestos
Prerequisites	The robotic Unit is in 'Ready to Remove' status
Results	The robot allows the execution of tele-operational commands with running tool and aspiration
Typical Process	• Give command for turning on tool and aspiration • Check if command is allowed • Run tool and aspiration and allow tele-operation of the arm
Graph	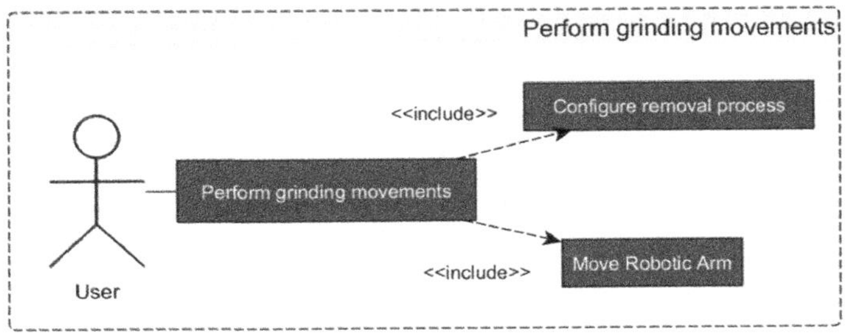

ID	UC_008
Use-Case	Power Up Components
Brief Description	Run the boot process of the robotic system after turning on the power
Prerequisites	The robot has a charged battery
Results	The robotic system is ready to use in tele-operation mode
Typical Process	• Wait for upcoming hardware • Solve errors if necessary
Graph	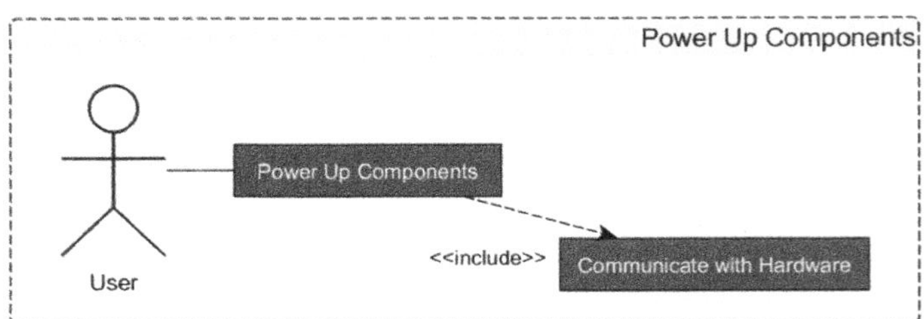

ID	UC_009
Use-Case	Shutdown Components
Brief Description	Shutdown the robotic system without harming it
Prerequisites	The system is turned on
Results	The system is shut down without being harmed
Typical Process	• Wait for shutdown of every component • Allow to switch power off afterwards
Graph	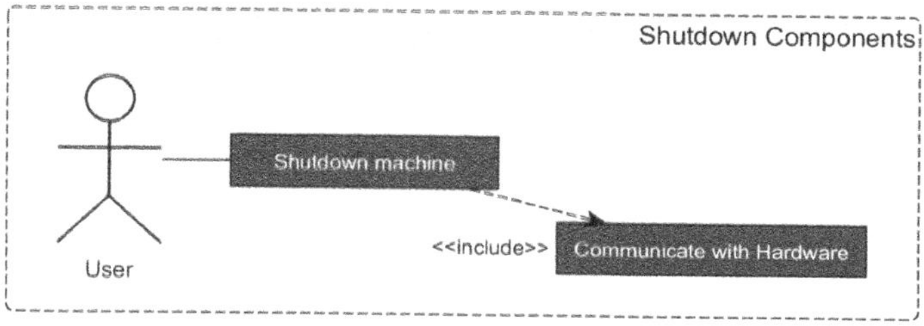

4.1.4 Activity Diagrams

Every Use-Case can be divided into smaller activities that enable the system to give the desired outcome. These activities and all the corresponding decisions and alternative ways can be shown in an activity diagram. There is only **one** activity diagram per Use-Case. It is a documentation of all allowed processes during the Use-Case. In the case of Bots2ReC it is more convenient to focus on the main tasks as tele-operating the robot and not on the small Use-Cases like communicating with the hardware that are necessary anyway.

Flow Charts can be used for the more vivid description of interaction between different subsystems and their network. It is also a good way to define different stages within the software solution and to manage the behaviour. As seen in Figure 4.2 there are multiple outcomes possible for the boot process. On the one hand it is possible to run into the error handling process to solve upcomming issues during the boot process or it can end up in a successful boot process.

The error handling flow is divided into two parts. The system will try to solve a detected error on its own first and asks the user for help afterwards if there was no solution. The last option of doing the recovery behaviour is a restart of the whole system that might be solved during a new boot process.

Getting the system moving is most essential for fulfilling the task of removing asbestos. The combination of tele-operating the platform and the arm are not used for now in combination. Out of this the tele-operation of the mobile platform and the arm are split into two different workflows. As the movement of the arm needs more stability than the movement of the platform itself there are different bounding conditions for each tele-operational task. Bots workflows are taken into account for the UITO.

If the user enables the system to remove asbestos there are quite a lot constraints that need to be fulfilled before being able to start the tool and the aspiration. Related to this the first step for the removal process is a well-defined question and answer system for the actual state of the robot and if the user shall be granted for starting the removal process. It is necessary that the robot is stabilized firmly for running the removal process. This is represented within the robot mode that is handled by the CPCS. After getting access to the command of running the tele-operational removal task the user is just using the workflow of the robotic arm tele-operating process that was given before.

4.1.5 Basic Information

4.1.5.1 Software Development Tools

In relation to the defined Use Cases the consortium members defined the operational tools for fulfilling the tasks. The programming language C++ is used for the programming of the logical behaviour of the robot. The user interface itself needs to provide more visual effects and complexity that can be dealt with using by mark-up languages. For the software architecture of the high control system the open source framework ROS (Robot Operating System) is used. It is based on the publish-subscribe pattern which makes it highly modular and adjustable for the use

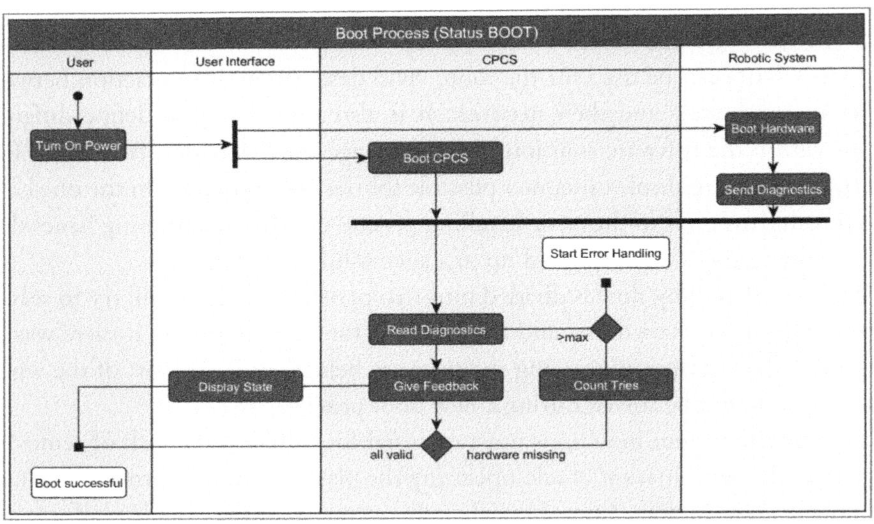

Figure 4.2. Flow Chart of the boot process.

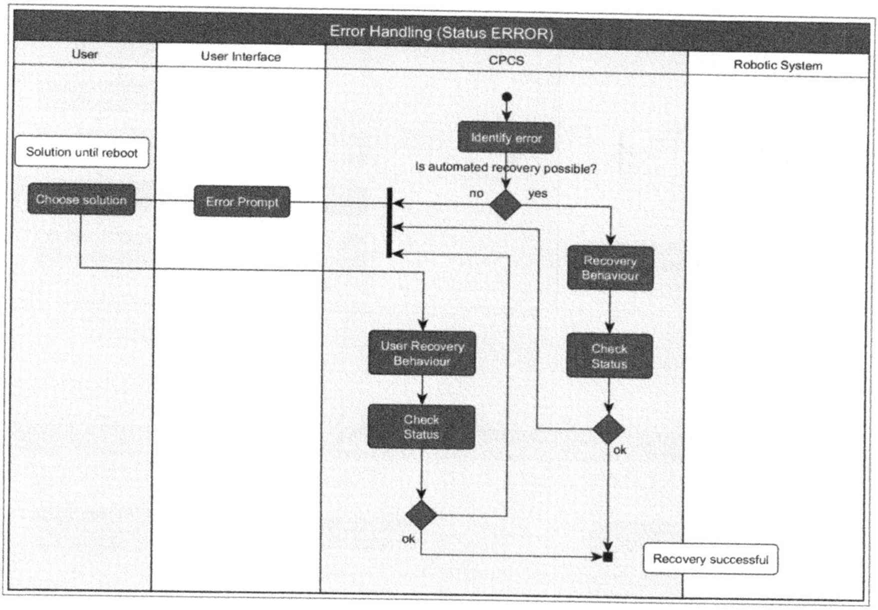

Figure 4.3. Flow Chart of the error handling.

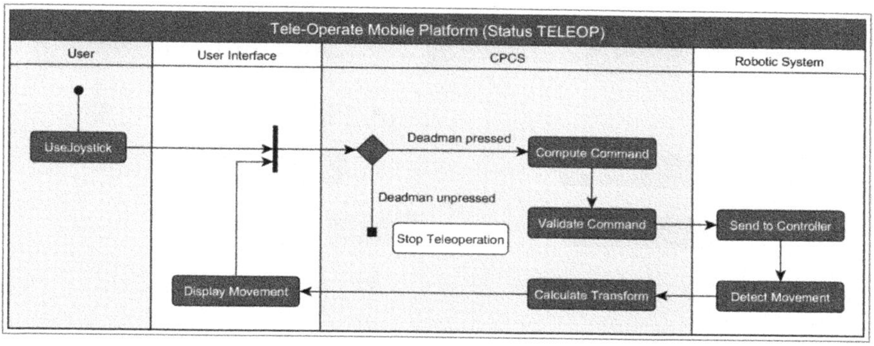

Figure 4.4. Flow Chart of the tele-operation of the mobile platform.

of different kind of robots. The common language of ROS is C++ which was one of the main arguments for choosing this framework because of its hardware-related programming language.

Because the UI is implemented by Indurad the existing framework of Indurad for user interfaces is used. It is a C++ framework exposing a QML Qt Meta Language or Qt Modelling Language interface and it is written with Qt Creator. The output is a standalone C++ Project that can be launched by an executable of ROS later on. Out of this the first integration of the framework into ROS is already manageable.

Out of this the main structure of the UITO (User Interface for Tele-Operation) is already predefined. There is the pure user interface that cares about the

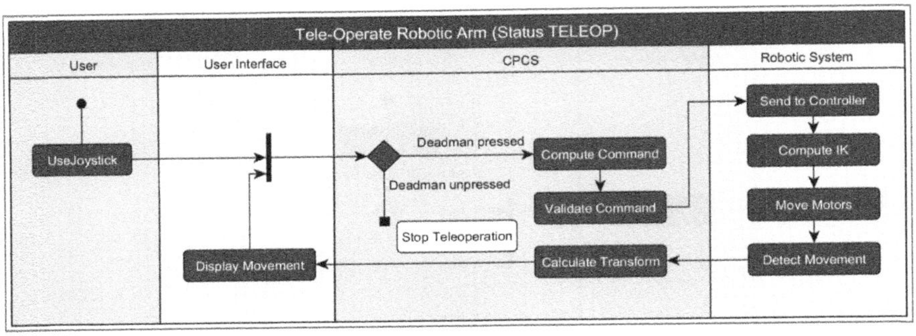

Figure 4.5. Flow Chart of the tele-operation of the robotic arm.

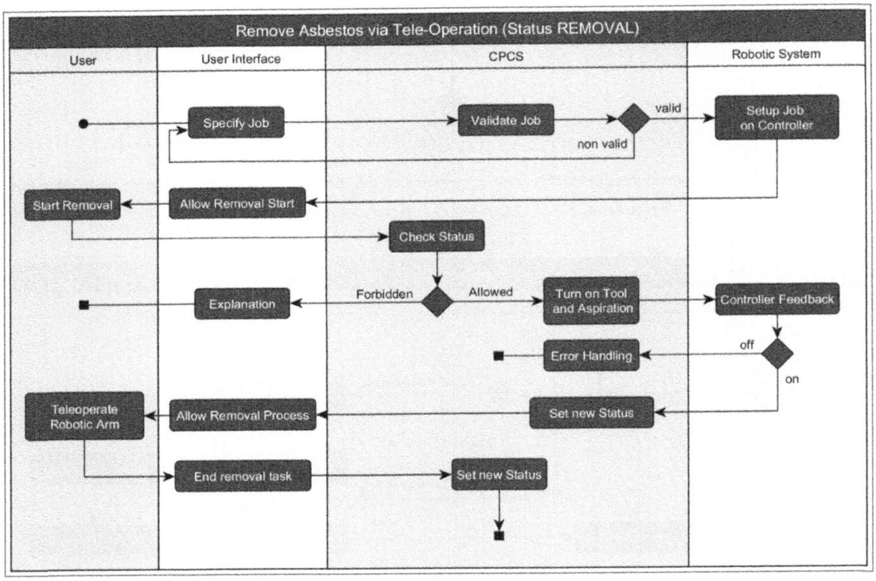

Figure 4.6. Flow Chart of the asbestos removal process.

representation of data and the handling of user interactions (QML). On the ROS site there is a software solution that translates the commands from the user site into technical signals for the robotic system. As an example: The user is moving a joystick to tele-operate the mobile platform and this mentioned ROS code is translating it into velocity and position commands. The interface in between is just to make sure that the QML of Qt Creator and the typical ROS commands can interact with each other. All in all, the very abstract system layout can be defined like this:

Table 4.1. List of the software units their IDEs and the used programming language.

#	Logical Parts of the User Interface	IDE	Language
1	User Interface	Qt Creator 5.2 and higher	C++, QML
2	Logic behind the interface (on the ROS side)	Whatever you prefer e.g. Qt, Eclipse, etc.	C++
3	Interface between ROS and QML	Whatever you prefer E.g. Qt, Eclipse, etc.	C++

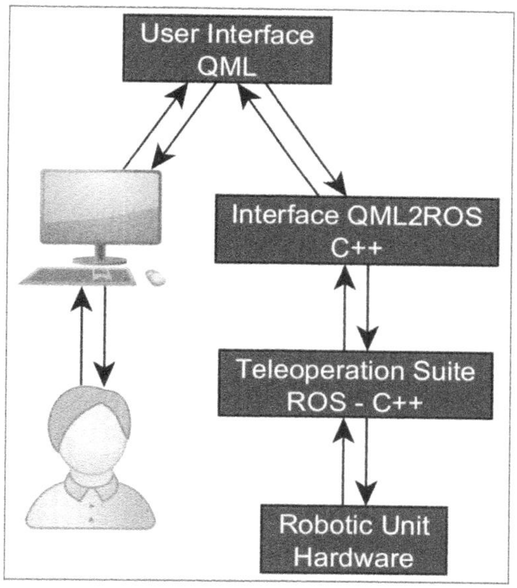

Figure 4.7. Global Layout of the UITO.

4.1.5.2 Interface between Indurad Framework and ROS

The interface "QML2ROS" is defined by typical ROS messages that are used for transferring the data. In the following picture the necessary messages are displayed.

The String messages are necessary for sending text-based information and the Bool messages are most likely to send states or commands between the two software instances. Sending the Int16MultiArray allows the software solution to define error codes and status messages for the user interface. The Pose message is used for showing the 3D environment of the robot. It is necessary to give positions and orientation to the STL-files of the robot. The usage of the different information is explained later on in detail.

As this is the first state of the user interface there need to be extensions during the overall project when the global CPCS is designed. Out of this the shared information between the UI and the CPCS can be extended easily by adding further message types that are used for interaction. A good example can be the usage of environment descriptive objects like boxes, planes or meshes that could be taken from the shape_msgs of ROS. The implementation of these objects types is straight forward and underlines the freedom of development within this software solution.

4.1.6 The UITO

Out of the requirements list and the different Use-Cases a user interface could be realised. In relation to the mentioned requirements the UITO is built out of modular windows, menus and task bars. Here is a list of all the objects that can be handled by the user.

#	Module	Content
1	3D-Window	Show the virtual representation of the robotic unit and the construction site
2	Status messages	Show warnings, Errors or Task Results
3	Menu bar	Enable the user to choose between different views, actions and tasks. Perhaps to open task specific windows or higher setting options
4	Task specific windows	Specific information about the ongoing action (e.g. the tele-operation of the mobile platform)
5	Robot Camera View	Actual video from the different cameras mounted on the robots
6	Step by Step window	The user is guided through different steps of the removal process. This bar can be found on the ground/bottom of the UI

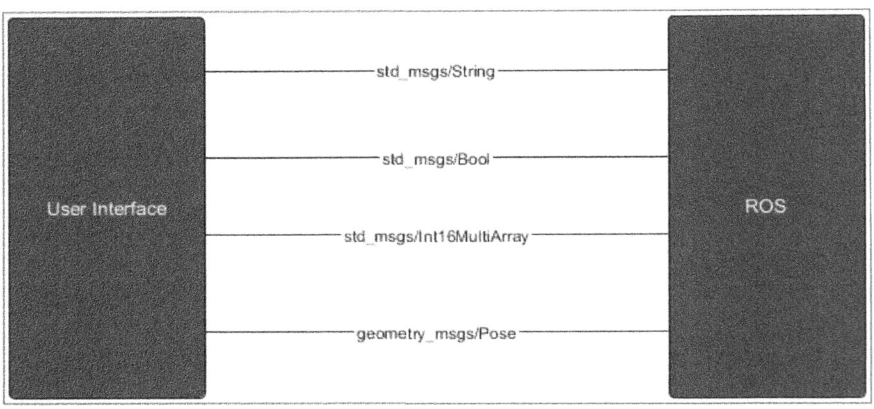

Figure 4.8. QML2ROS.

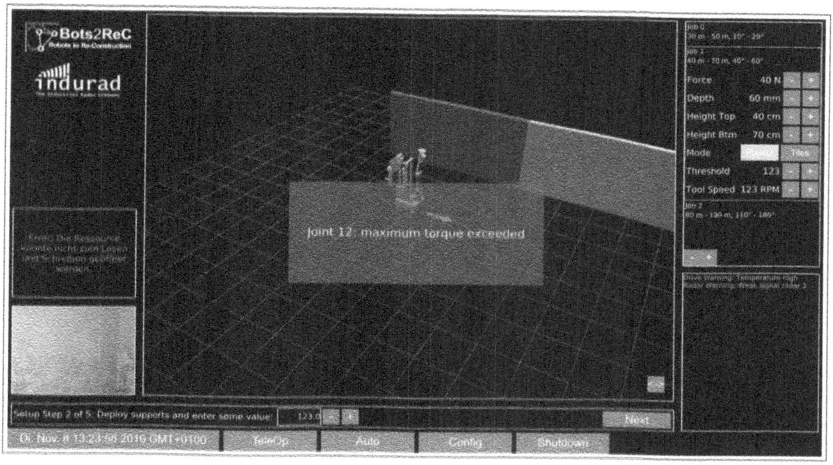

Figure 4.9. User interface without comments.

The screenshot in Figure 4.9 gives a first impression about the implemented UI. The centered view is a 3D View of the robotic system and its current state. It is located within a 3D map that can be build by 3D Slamming algorithms or imported by 3D Scans made by professional construction site rebuilders.

This task of autonomous map generation is part of the tasks 6.1 and 6.2 and is going to be clarified in the future. The final data structure for the environmental description is going to be defined during the testing phase of the first prototype. For now the user interface is implemented in a generic way so that it can be adapted to every solution that might be defined for this mentioned data structure in the future. Within the 3D View it is possible to see the current robot state, its position inside of the flat and the properties of the flat itself. The robot positions are updated with the ROS data at runtime and allow the tele-operation in most of the cases without any further details. If the sensor signals are not clear enough and the 3D view might get unprecise it is possible to enlarge the actual view of the mounted cameras to clarify the situation. Beside this two different ways of representing the robot state and its surrounding it is possible to specify the job that is going to be run during the removal task. While only the typical grinding task will be executed and represented during first tests, the user interface can be extended in the future for different tools, processes or surface circumstances.

Within the UITO it is also possible to have the feedback about warnings and errors of the robotic system. In the right lower corner of the screen the warnings are displayed. Warnings are issues that are not perfect but that can be handled by the robotic system. Only when they get worse they can turn into an error. Error messages are written in the center of the screen as shown in Figure 4.10. Errors will force the robot to stop its task execution. Out of this the centered banner in red was chosen for the user interface.

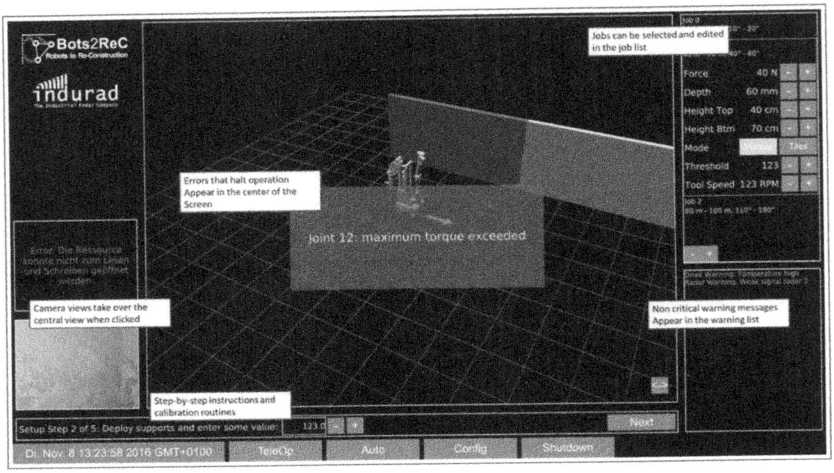

Figure 4.10. User interface with comments and explanations.

Beside this mentioned functionality the user interface of tele-operation is also able to show the different states of the robot. For now there are states as "TeleOp", "Auto", "Config" and "Shutdown" implemented. The best way how to organize the system states for the user will be clarified during the tests of the prototype V1. The aim is to create an easy and robust way to deal with the robotic system. The feedback of Bouygues and its experienced employees will be taken into account for the design of these tools.

4.1.6.1 Hardware for tele-operation

Beside the normal input of words and commands on the screen the system need a device for the tele-operation movement. Of course this is also possible with a common keyboard but probably it is also a good idea to look for a more intuitive and flexible device. During the investigations concerning this device the following possibilities have been checked by the consortium members.

#	Device	Properties	Example
1	Normal Joystick	Typical Desktop Solution **Advantages:** • Cheap • Already implemented in ROS **Disadvantages:** • Only for clean rooms without dust and dirt	

#	Device	Properties	Example
2	Rugged Joystick	Very rugged and safe Joystick boards with additional functions as emergency stop. **Advantages:** • Made for tough conditions • Additional functionality over buttons and switches **Disadvantages:** • Expensive • Not implemented in ROS **Product:** • Control Chief	 
3	Game Controller	Typical gaming solution for PS3 or Xbox **Advantages:** • Easy to use • Rugged and cheap • Already implemented in ROS **Disadvantages:** • Perhaps not rugged enough?	 
4	Rugged Gaming Controller	Rugged version of the gaming controller **Advantages:** • Rugged setup extra for construction sites **Disadvantages:** • Not implemented in ROS **Product:** • Humanistic Robotics	 

#	Device	Properties	Example
5	Keyboard and Mouse	Normal desktop solution **Advantages:** • Already necessary for the normal input of text and klicks • Cheap • Everybody knows it **Disadvantages:** • Not as intuitive as other devices	 

After several investigations the PS3 Controller was chosen as the input device for tele-operation. In correspondence with Bouygues it was pointed out, that the area for tele-operation is not as rough as mentioned. Furthermore, the PS3 controller can be covered with plastic foil as well and is cheap in comparison to other devices. Related to other devices it gives a very intuitive solution for tele-operating the arm and the mobile base at the same time. Another positive aspect about the PS3 Controller was the fully integrated functionality in ROS that made the integration straight forward. During the first tests the usefulness could be pointed out.

4.1.7 Outlook and Future Work to be Carried Out

Within the Task 6.3 the UITO is created successfully on software site. The desired functionality is shown and proven in simulation. The manipulability of the mobile platform with the PS3 Controller was shown on hardware site as well. Accordingly to the hardware delay of the robotic arm as pointed out in the reports of WP5 it was not possible to debug the UITO with the complete robotic system in the loop. Therefore these tests are planned for the beginning of March when the whole robotic unit is assembled at Eurecat in Barcelona. The estimated time for debugging this software suite is less than a week and should be straight forward so that there will be a working interface between the UITO and the robotic unit in the mid of March. Finally the task will be reached with a delay of two weeks and there are no needs to adjust the goals of WP6.

4.1.8 Summary and Conclusion

All in the entire user interface for tele-operation was implemented successfully. It is able to represent the current robotic state and to allow the end user to manipulate essential parts of the robotic system in an intuitive way. For reaching this deliverable the methods of the Unified Modeling Language where used for documenting the software architecture and for sharing ideas and thoughts on a vivid basis. A consensus about the software solution was found via sharing the Word Documentation and attached comments. As pointed out in chapter 7 the success of the UITO was shown in simulation for the whole prototype and for the mobile platform on the hardware site as well. The test of the whole unit will be carried out as soon as possible at the beginning of March. In according to the user interface the display methods for environmental objects are developed and improved in Task 6.1. The resulting data sets will be integrated into the UITO continuously. The user interface may be also affected by the development of the CPCS later on but the estimated influences are quite low.

Deliverable 6.4
Interface for job specification layout, implemented and documented

Authors: Tobias Haschke, Fabian Herb

4.1.9 Deliverable in the Context of the Work Program

The User Interface for the Bots2ReC Project is the abstraction layer between the complex robotic data elements and the future operator. According to the development of the CPCS and the Use Case automation the User Interface is a key factor for the acceptance of the product within the market. Related to Task 6.3 it is the consequent progression of the tele-operated version implemented before. Beside the past UI version it is also connected to the development processes of the other tasks within WP 6 as it needs to enable the operator for using the entire robotic system with the desired purpose. Beside the relation to the CPCS development it is also represented to BYCN for the confirmation of usability.

4.1.10 Methods and Work Carried Out

As this version of the user interface is a progression of the tele-operated version the development strategies and the chosen framework remains the same. The whole system is tailored for an All-In-One (AIO) PC from Asus for enabling a tablet-like behavior within a robust case. The AIO is chosen as it allows typical PC hardware interfaces as USB and Ethernet for a robust solution within construction environments.

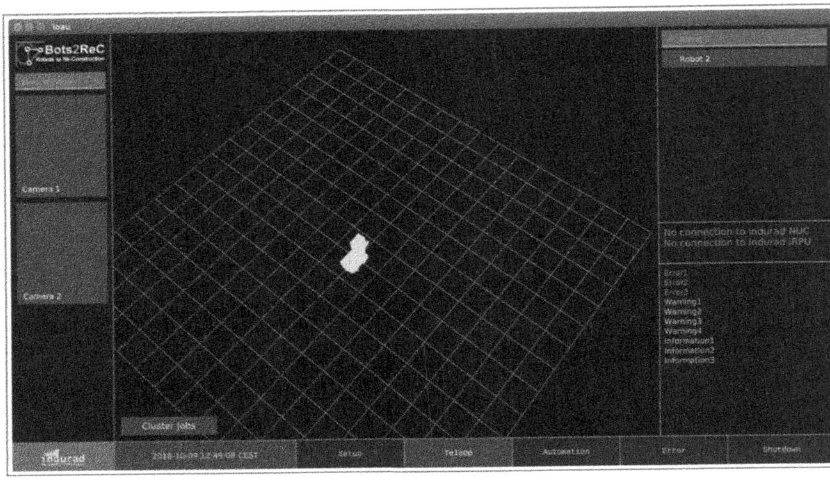

Figure 4.11. New straightened design of the User Interface.

On the software side the AIO is running a minimal linux image with the UI application as startup sequence. Out of this the operator does not recognize anything of the operating system in the background as the only available screens are the ones from the UI. This solution is used for simplifying the use of the system without enabling the operator to mistune the system manually. For the implementation the UI Framework QML by Qt is still in use as it is easily linked to plain C++ code. Out of this the interface between the CPCS programs and the UI can be developed quite flexible as they are based on the same programming languages.

As an addition the QML framework natively supports touch-enabled solutions and well rendered output screens for a vivid, fresh and intuitive usage of the UI. Out of the combination of high range of feature and the high performance of C++ it is often used for custom displays as the one at hand for the Bots2ReC use case.

4.1.11 User Interface for Job Specification (UIJS)

During this development phase the whole design of the UI for tele-operation was cleaned and well-balanced for the new added features. This means that the first solution for tele-operation was still fully functional but the overall appearance had to be straightened out for intuitive workflows for the operator. The main idea of the new design is to keep the usage of the UI as simple as a typical and well-known app on nowadays smartphones. This lead to compact button and panel positions with clean understandable rules.

In comparison to the former solution the already developed panels remained the same. The 3D view for the robots model status is still cantered by default and can

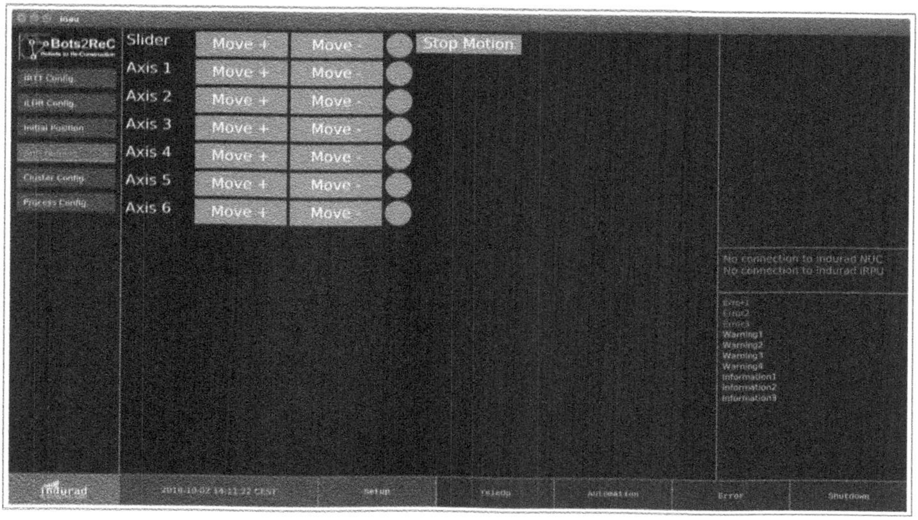

Figure 4.12. Feature Arm Homing.

be swapped by the camera input panels from the left side. Also still the same are the boards for the network status and the message board on the lower right side.

A new feature that was necessary is the robot selection menu in the upper right side as the development of V2 forced the development team to control both robots from within the UI. During the CPCS extension in the meantime five discrete stages for the asbestos removal robot where defined that are changeable by the operator:

- Setup – Calibration and Configuration processes
- TeleOp – Teleoperation of the robot in form of the mobile base and the tool
- Automation – Autonomous task execution based on the global environment data model
- Error – Robot is unable to solve a special situation alone and is waiting for external advices
- Shutdown – Status for shutting down the robotic system

Especially the Setup Stage changed a lot for this deliverable, as the setup of the environment model and the initial state of the robots are essential for the task execution process.

Since this release, it is possible to home the arm of V1 manually at the beginning without being an expert in robot engineering. This simplification is one huge step to the intuitive usage of the robotic system through this simple interface panel with direct robotic feedback.

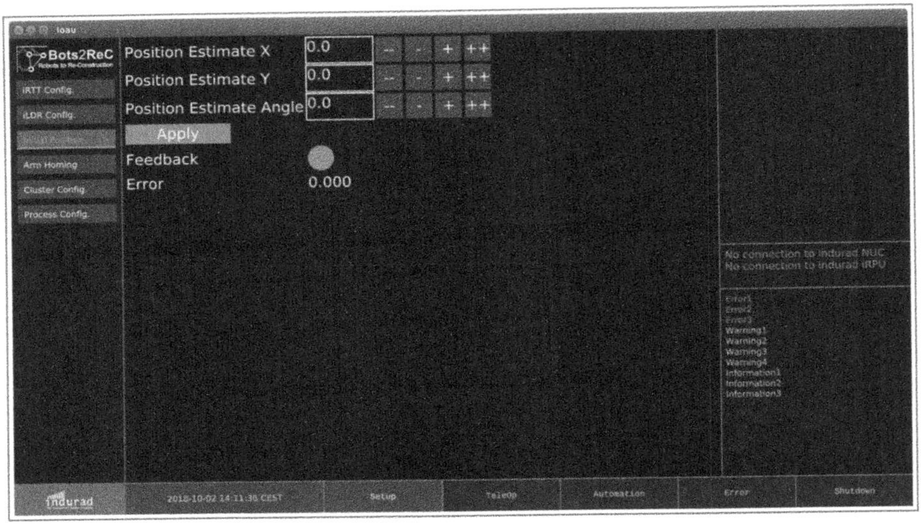

Figure 4.13. Feature Initial Pose.

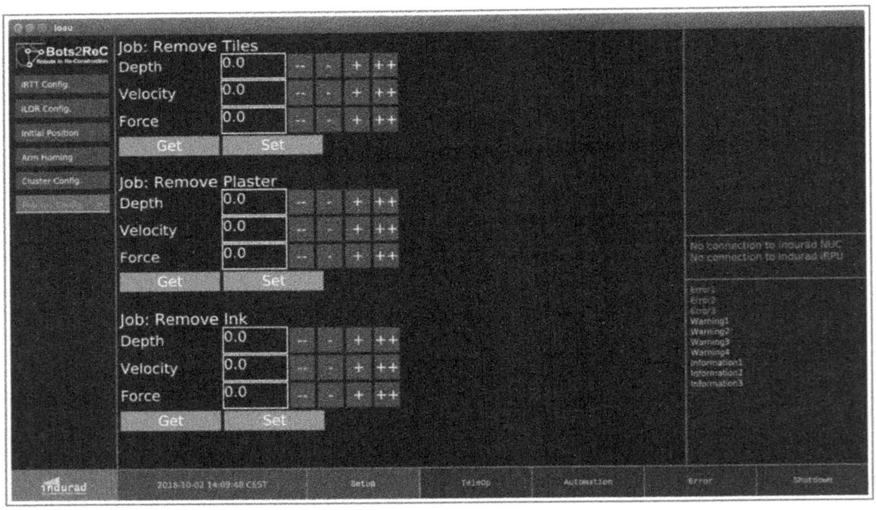

Figure 4.14. Feature Job List.

During the Setup Stage each robot need an initial position within the known map for being able to localize itself. As the mapping and removal execution processes are separated and most likely also run on different robotic systems this step is necessary for every boot sequence on a construction site. To reduce the complexity the current error and a feedback signal are used to give the operator a visual and numeric idea about the current initial pose accuracy that was configured.

As the semantic map is the semantic and geometric data representation after the mapping process this is also used to add the process specification. Each element can be setup with a specific parameter set that can be specified during the Setup Stage. In this case Figure 4.14 shows the possible setup of three processes which can be adapted to the current use case.

4.1.12 Delays and Aspired Compensation

The entire UI is implemented and deployed on the target AIO for the Bots2ReC project. Its functionality is entirely tested within the ROS framework but due to the hardware challenges on V1 a real-scenario test is still outstanding.

4.1.13 Outlook and Future Work to be Carried Out

The tests with the real robot will start immediately after solving the mentioned hardware challenges. Due to the modular software layout, the integration of these features into the real robot should be remarkably small.

4.1.14 Summary and Conclusion

Overall, the user interface has reached a well-tailored solution for the use cases at hand. It is simple, intuitive and represents the workflow of the robotic system. After the integration into the real robot system and the real-scenario validation this should be the final release of this user interface for the Bots2ReC project.

Task 2.4 Cooperative robotic control strategies

The payload (equipment) has been distributed in two bases as the robot footprint could not be increased due to application limitations (size of the apartments to be decontaminated). Both platforms have an umbilical cable between them and need to be moved as a single system. The robot should be easy to operate in manual and automatic modes. The system should keep the distance between the platforms avoiding tension in the cable.

To avoid the cable damage, a mechanical solution was developed by TLABS, a kinematic chain with 3DOF and two steel links that protect the cable and that act as a physical limit if the maximal extension is reached, in this case the structure is strong enough to withstand the motor torques. This structure imposes a limitation in the minimum distance, and also in the relative orientations of both platforms. The minimum distance is about 20 cm between platforms, while the structure allows angle differences of $+/-90^\circ$ of each end joint. The mentioned structure can be seen in the next figures:

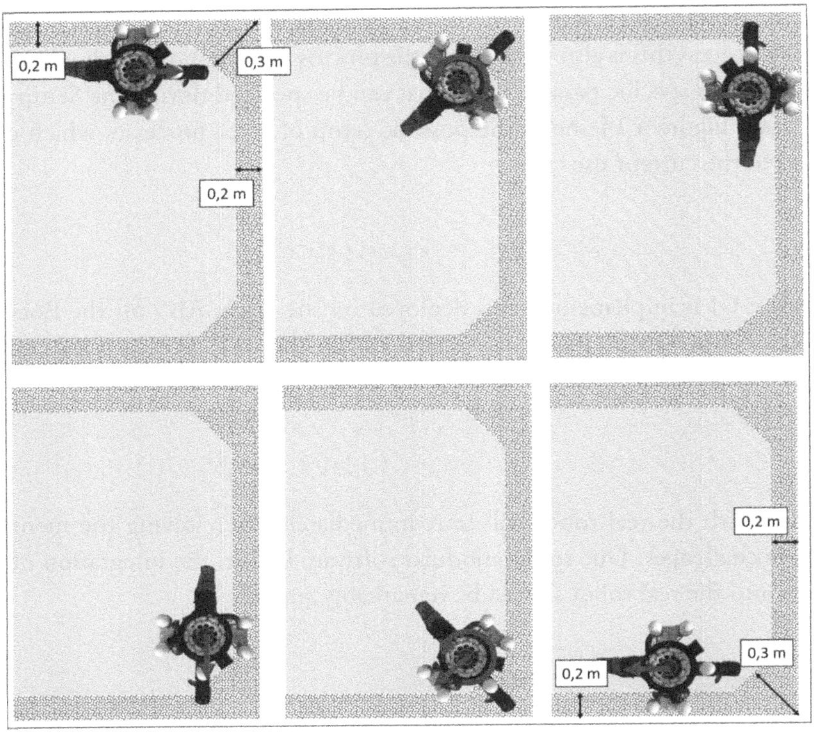

Figure 4.15. Wall coverage floor.

Figure 4.16. Robotic Tandem with cable and aspiration bridge.

With the described tandem system, the objective was the developmet of a control strategy that permits the coordinated control of two platforms. With this objective, the following three concepts have been analysed:

- Concept 1 – **Following reactive**: the rear platform just follows the front one in a reactive way. The rear platform "sees" the front one either by beacons or QR codes, etc. Speed references are commanded in a way that the objectives of following/keeping distance to obstacles are weighted and a resulting speed is computed. This concept can be implemented with or without localization, the differences are explained next.

1.1 Both platforms are localized: the rear platform tries to keep **a fixed relative pose w.r. to the front platform based on the estimated pose.**

1.2 With no localization of the rear platform: in this case the rear platform tries to keep **a fixed relative pose w.r. to the front platform based on the measured relative pose (e.g. beacons)**. The main advantage of this approach that the requirements in terms of localization are relaxed as only the front platform requires localization, while the rear platform is just acting in a reactive mode. A disadvantage is that the rear platform cannot benefit from the local costmap obstacle avoidance of the move_base stack.

- Concept 2 – **Following planning**: both platforms are correctly localized. The rear platform uses a trajectory planner and computes poses relative to the estimated pose of the front platform and updates the navigation goal continuously.

  2.1 The rear platform plans path (with preferred planner) to the updated front platform robot pose or fixed transform (tf) relative to front platform. The standard planners do not introduce restrictions in distance nor orientation. The controller may stop the front platform if the distance between platforms is not kept or will adjust speed.

  2.2 **The rear platform plans to a fixed pose that is relative to the front platform.**

  2.3 **The rear platform follows the historic frame sequence of the front platform**. This should allow e.g. turning a corner.

- Concept 3 – **Pure teleoperation**: both platforms receive the joystick reference and operate synchronized in omnidirectional mode. When the tandem system moves linearly, both platforms receive the same command. This concept has been implemented in a client-server configuration where the front platform connects to a topic that is sent via a rostful server to the rear platform that reads this reference and inputs to its own joystick multiplexor.

  3.1 The angular speed is sent according to a fixed distance and this distance is kept manually by the operator. This leads to some adjustments while the robots move, but turned out to be a simple and effective solution.

  3.2 The angular speed is sent according to a measured distance. This should be easy, as the mentioned markers should permit an update of this value in real time.

The operation of the tandem system is in open loop, speeds are sent and the control and odometry errors can lead to some misalignments of the platforms. The joystick allows to control each platform independently thus permitting

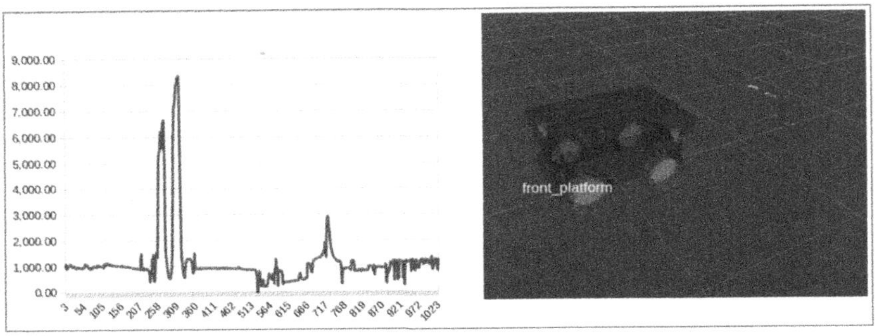

Figure 4.17. Single laser intensity and detected markers on front platform.

Figure 4.18. Reflectors installed in the front platform.

single motions to align the tandem in orientation, to keep the required distance or to perform complex operations as passing through narrow spaces where the operation in more autonomous modes becomes too complex.

The three described concepts have been implemented and tested in simulation (1.1, 1.2 ,2.1, 2.2) and the concepts 1.2, 2.2 and 3.1 tested with the real robots.

For the implementation of the reactive mode QR codes were initially considered, but finally discarded due to the dust. The final solution selected was based on the detection of markers with a higher reflectivity. This method has proven to operate even in dust conditions. The laser range finders are providing not only the distance but also the intensity of the reflected beams. In a short range, the intensity perceived on the markers is much higher than non reflective objects, even in dust conditions. The next figure shows the intensity of a single laser scan. Two peaks can

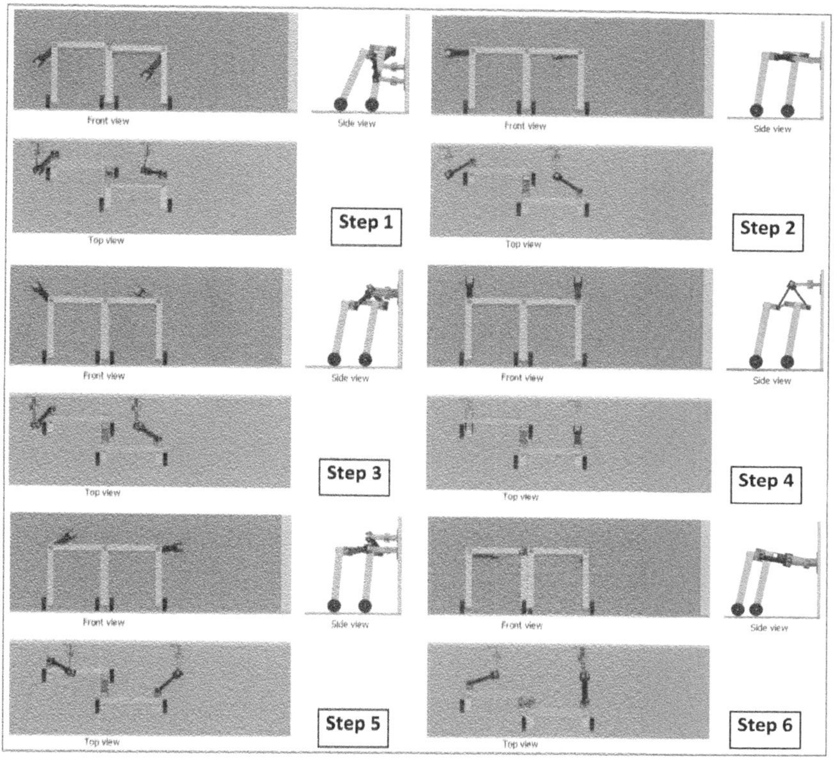

Figure 4.19. Grinding process using a p-bot in side by side connection.

be clearly identified corresponding to the markers. The position of the two markers (x1,y1,x2,y2) permits to compute a transform relative to the laser link and finally to the robot base link. This information is finally used by the reactive system to estimate the front platform pose and to command motion commands.

From the localization point of view: in autonomous mode, each platform is tracking its own estimated pose. The distances of the reactive system can be compared with the distances of the localizations of platform one and two and a divergence should make the system stop. This is the considered software checking routine that is not integrated yet as long-term tests weren´t able to be performed.

Figure 4.18 gives an impression of the real scenario two platform following each other.

4.1.14.1 Work package 4: Customization of the mobile platform: Design, control and implementation

The entire work package 4 was closed within RP2 by the Deliverable 4.3 already submitted and approved by the last review meeting. Nevertheless, the integration of hardware showed that some iterative implemented modifications were necessary.

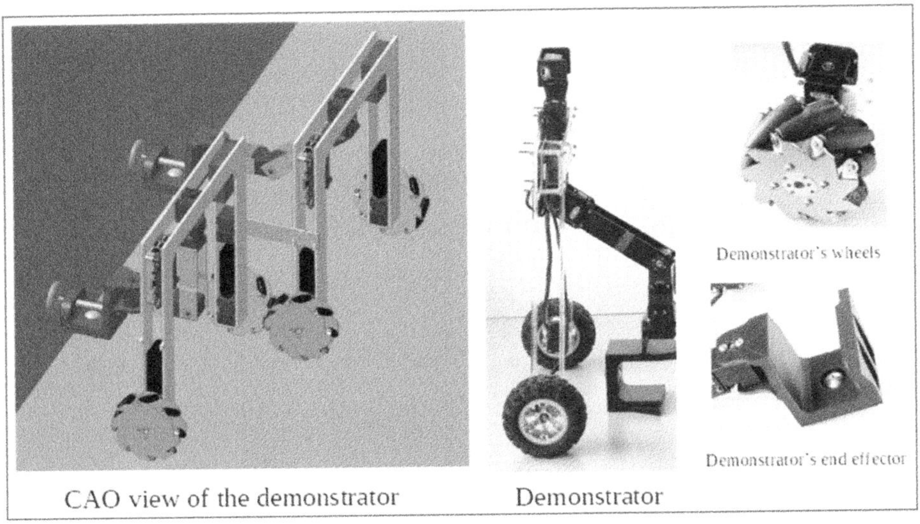

Figure 4.20. Demonstrator of the grinding process.

Figure 4.21. iSDR-C.

To keep the form they are documented within the related tasks even though they are officially closed already. In principle, all the modifications and optimizations were caused by necessary improvements during the integration and commissioning phases in WP3 and WP6.

4.2 Semantic Mapping

Deliverable 6.2
D6.2 Sensor fusion and data evaluation implemented

Authors and contributors:
CHARAF EDDINE Sami, RWTH Aachen University

4.2.1 Deliverable in the Context of the Work Program

Deliverable **D6.2** (Sensor fusion and data evaluation implemented and tested) is part of task 6.1 (Environmental perception and mapping). The main objectives of task 6.1 are:

- to **combine data coming from various sensors** via the defined interfaces from task 3.1 (e.g. vision, RGBD cameras, Lidar, radar) used in task 4.2 **to build a map** of the environment
- to **identify processing** areas such as
  - ground surface,
  - walls and
  - ceiling
- to **detect areas** that should be **avoided** during automated asbestos removal:
  - doors,
  - windows and
  - areas of too complex geometry (e.g. pipes, lighting, electric devices)
- to continuously **update the map** by several robots working simultaneously and share the map between these robots

The achievement of these objectives will allow the creation of an intuitive map based user interface for the task allocation by the operator.

4.2.2 Suggested Approach

Solving task 6.1 requires two kinds of maps, one of them for navigation in possibly low-visibility environments, the other one for **task allocation, planning and fulfillment**. The map for navigation will be used solely internally for computational purposes and is investigated in task 6.2, based on the the sensor developments for the mobile platform (iRTT, iSDR, iSDRcompact described in D4.2 and LIDAR described in D4.1). The map for task fulfillment is visible to the user and thus has

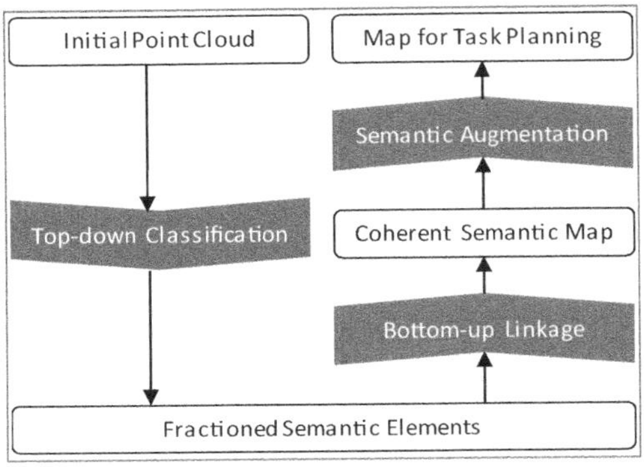

Figure 4.22. Main processing steps of the proposed algorithm.

to be displayed in the user interface. This map is the central element of interaction with the user, as it will provide information about processing and non-processing areas within the contaminated rooms.

The mapping approach presented in this section does not need any sensor fusion as solely a 3D laser scanner, provided by Bouygues, captures the recorded data.

4.2.3 Mapping for Task Fulfilment

We aim to enable task planning and assignment based on a lightweight visualization of the robot's environment within a user interface. To do so, we propose a computationally efficient approach that consists of three consecutive main steps, as shown in Figure 4.22. First, we apply a **top-down classification** (sec. 3.1) to divide the initial point cloud into several clusters and classify them as wall, ceiling or floor, solely based on geometrical considerations. Second, we perform a **bottom-up linkage** (sec. 3.2) and integrate the classified objects, grouped as single room entities, into a hierarchical tree data structure to build a coherent semantic map. Third, we **augment this map** (sec. 3.3) by detecting doors, windows and other elements that might impede asbestos removal. Finally, we infer a lightweight coherent polygon model from the created semantic map in order to visualize it within the user interface. The user is now able to make decisions for further task planning and can assign tasks to certain semantic objects.

4.2.3.1 Top-down classification

The input data of our algorithm is a single point cloud as shown in Figure 4.23. We segment this point cloud into separate clusters as shown in Figure 4.24, using an implementation of the region-growing algorithm provided in PCL.[1] It is worth

1. http://pointclouds.org/

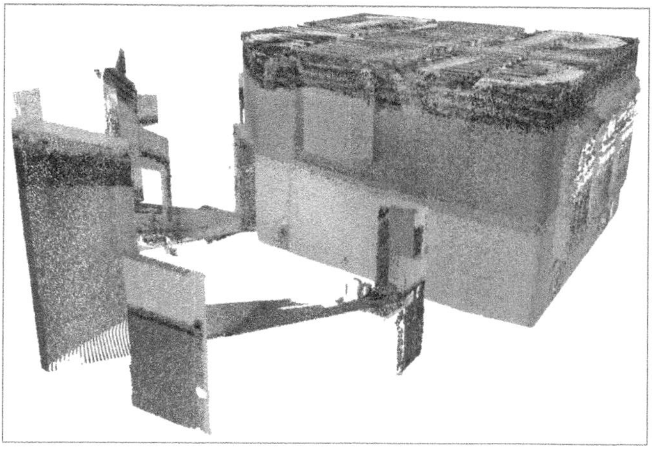

Figure 4.23. Initial point cloud.

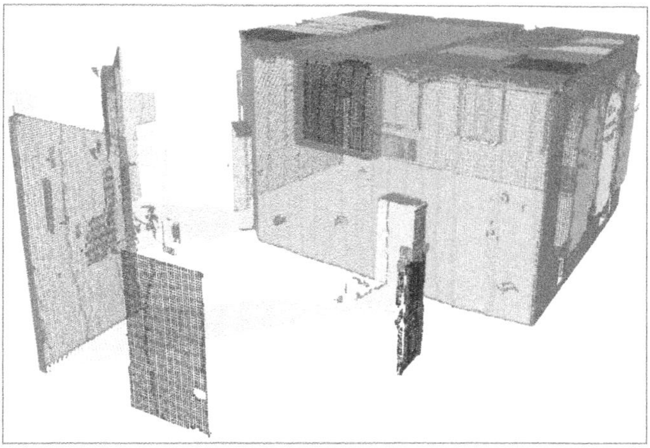

Figure 4.24. Results of the region growing algorithm. Each color represents a certain cluster, the red color indicates dismissed clusters due to too small cluster size.

mentioning, that the walls in Figure 4.24 are clustered correctly, but – and this is crucial to our application – the single walls are not yet grouped as one group feature walls. This is shown by the different colors of the single walls. The same effect happens to the ceiling of the room: Instead of being clustered as one single element, it is clustered into multiple non-coherent elements. We then use the principal component analysis to calculate the respective bounding boxes for each region. Afterwards we classify the estimated bounding boxes according to their geometry and obtain walls, floors and ceilings, as shown in Figure 4.25.

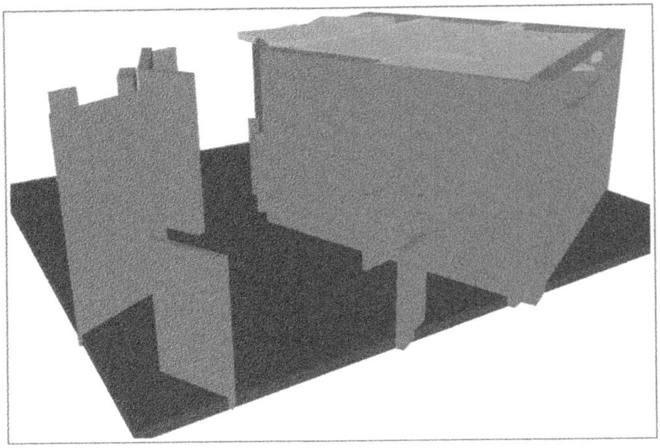

Figure 4.25. Classified bounding boxes of the clusters. Red boxes characterizes walls, green boxes are referred to as ceiling clusters and blue boxes represents floor clusters.

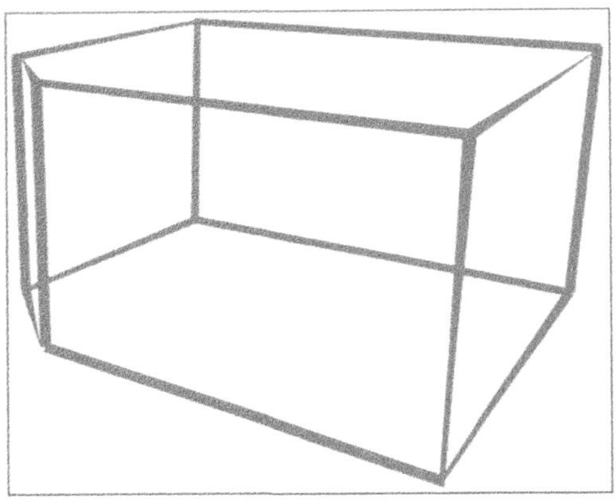

Figure 4.26. Linked and approximated polygon model inferred from the scene.

4.2.3.2 Bottom-up linkage

Our aim is to concatenate the detected, yet fragmented, clusters in order to obtain a coherent semantic map. First, we search for the adjacent bounding boxes of every wall and obtain a list of all wall elements and their neighbored walls, so called **linkage**. Given this description of the overall environment, we then determine single room subsets using a graph cycle search on the wall edges and store this information in a semantic objects tree structure, so called **grouping**. In the last step we **approximate** all regions by their best fitting planes and intersect any neighboring planes with each other. The resulting intersection points represent the semantic map as a coherent polygon model of the robots environment (Figure 4.26), enabling a lightweight visualization in the user interface.

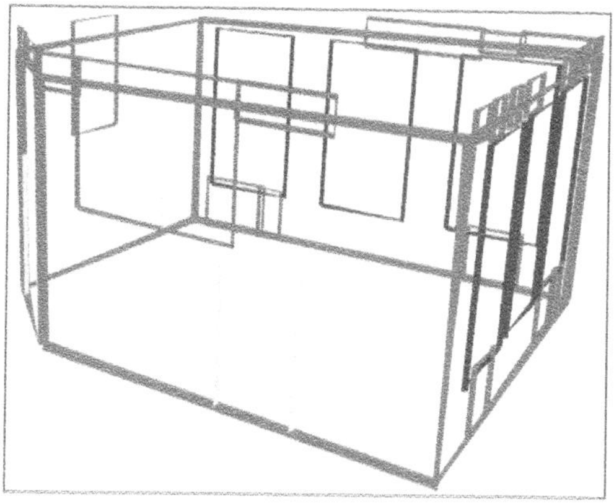

Figure 4.27. Polygon model with extracted doors, windows and other structures visualized. Blue polygons indicates windows, yellow polygons doors and pink polygons shows not further classified noticeable features which should be concerned during task planning.

4.2.3.3 Semantic augmentation

After the creation and linkage of the semantic map we further examine the classified clusters in order to identify subfeatures, e.g doors and windows within wall clusters or lamps within ceiling clusters. The final map is shown in Figure 4.27.

4.2.4 Validation

To demonstrate the performance and accuracy of our deduced model, we finally compare the polygon model from Figure 4.27 to the initial point cloud from Figure 4.23. As shown in Figure 4.28, the polygon model matches the initial point cloud very well.

4.2.5 Outlook and Future Work to be Carried Out

As mentioned in Sec. 2, the current approach relies on the point cloud of a 3D laser scanner from Bouygues that is a) very expensive and b) requires the point cloud to be captured before the removal process. This also implies that the Bouygues 3D laser scanner is a separate unit that is not mounted on the robotic system and needs additional manual operations. To circumvent this drawback, we will try to develop a low-cost solution, integrated to the robotic system, that is able to capture the 3D point cloud required for our semantic mapping approach. As our previous trials using a single Kinect mounted on a Turtlebot were not promising, we plan to develop a Multi-Kinect-sensor and build the overall point cloud representing the environment in the scope of task 6.2.

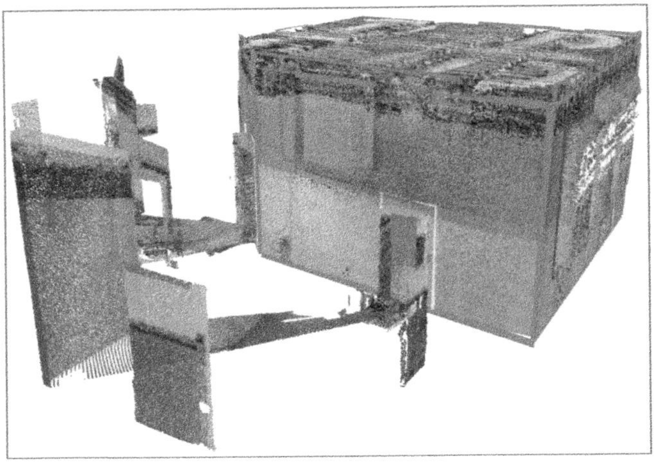

Figure 4.28. Visualization of the initial point cloud and the resulting polygon model.

4.2.6 Summary and Conclusion

The algorithm for semantic mapping based on a 3D point cloud has been developed successfully. The main purpose of the algorithm is to translate the huge initial point cloud to a lightweight and coherent substitute made of polygons and to enable task planning of the asbestos removal. In order to pass the inferred information to the user interface, we transform the detected and classified group features from the semantic map into a coherent polygon model. We obtain:

- appealing coherent 3D representation of the semantic objects
- class labels of every semantic object
- editability of the objects for defining process related properties (e.g. surface condition: plaster, paint or tiles)

4.3 Artificial Task Planning

Deliverable D2.3
Algorithm for precise, dynamic and centralized control strategies

Authors and contributors:
Tobias Haschke, RWTH Aachen
Albert Fullana Puig, EURECAT

4.3.1 Deliverable in the Context of the Work Program

For controlling multiple tethered mobile robots with the aim of removing asbestos from normal flats, a multi-layer control strategy is implemented. Due to the requirements set by the construction industry (Bouygues Construction), the main objective of the task execution is time efficiency. This objective is explained by the enormous costs for an unlet apartment house during the refurbishment time, as the building cannot generate any income during this period. Out of this, the whole control implementation is aiming for a fast and robust task execution.

The dominant layers are *task planning* and *task execution*. Within the *task planning* layer an optimal strategy for solving the whole removal problem is generated. Due to the uncertain environment of construction sites, this optimum may differ during the execution time, which leads to a continuous planning process throughout the whole execution time. This already reconciled to the *task execution* layer, which aims on the one hand for the execution of the pre-planned tasks and needs to deliver valid process data for updating the systems state. This data is used for re-planning the whole execution problem during runtime again to adapt to any changes or mislead assumptions.

In the following chapters, the individual mechanisms of actions are explained for each layer individually and how they interact. Within the first chapter, the underlying model of the environment and the robot is explained. Build on this the next chapter will explain some optimization problems and their solution to fulfill the task planning layer. Finally, yet importantly, the task execution chapter will point out how the currently optimal tasks are executed in a robust and efficient way.

4.3.2 Methods and work carried out

Modeling of the problem

Before an optimization problem can be formulated an adequately exact representation of the problem in case of a mathematical model needs to be developed first. This chapter will explain the different model elements and its context for the following steps. During the completely modelling process, a fully parameterized description was put into focus.

Also the problem formulation can be split in two different layers. The first subproblem deals with the optimization of operational points in front of a wall that needs to be cleaned from asbestos. This can be defined as an optimization problem as different robots can have different properties as reach lengths of the robotic arm, tool size or a different force distribution. Especially within the Bots2ReC Project this property of the problem formulation is essential, as the project will have multiple different prototypes (V1 and V2) that still need to work together.

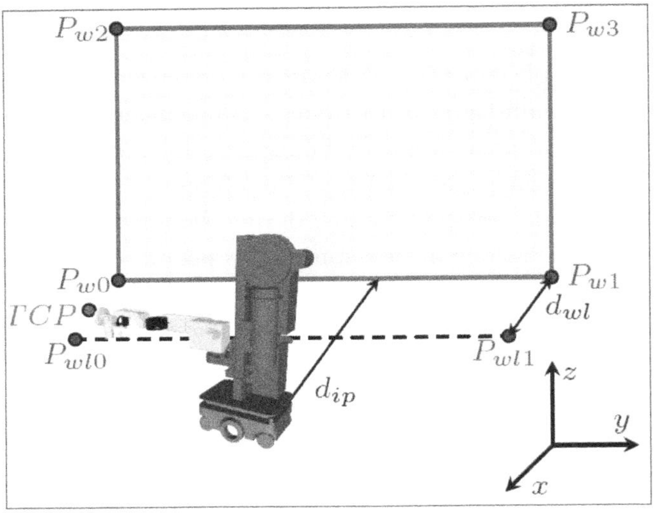

Figure 4.29. Geometric modelling of a wall task.

These different properties will be used for formulating the cost of each robot to fulfill a certain task. The individual costs need to be modelled to enable an optimal task planning. In Figure 4.29 the geometric modelling of a wall task is shown that contains the entire wall as a polygon representation. This data format is equal to the semantic mapping approach developed in WP6 in Tasks 6.1 and 6.2.

The additional information for each robot is the working line distance *dwl* that is represented by *Pwl0* and *Pwl1*. This is an individual distance, which is best for each robot and its reach size. Additionally, each robot is defined with a maximum working width that can be processed. Due to the non-linear force transmission dependent on the distance to the operational point it is also necessary to formulate cost functions for these characteristics.

Those cost functions are shown in Figure 4.30 and are fully parametrized by 3 parameters a, b and c. Out of this it is possible to adapt this quadratic cost function to any new prototype easily without affecting the model or the planning strategy. As each task need to cover a complete area of a wall, ceiling or floor element the vertical and horizontal movements need to be combined to trajectories. As the cost functions for horizontal and vertical movements can be different (e.g. tool is working better in direction of the gravitational vector), two main trajectory options are defined. They are displayed in Figure 4.31.

The trajectories are parametrized over the tool size of each individual robot. Which means that the trajectories will differ for different grinding disc diameters and their maximum speeds. Summing up the first sub-problem the removal task splits each wall element into the desired number of tasks which are optimal for each robot. Afterwards the optimal removal trajectory is selected for each element.

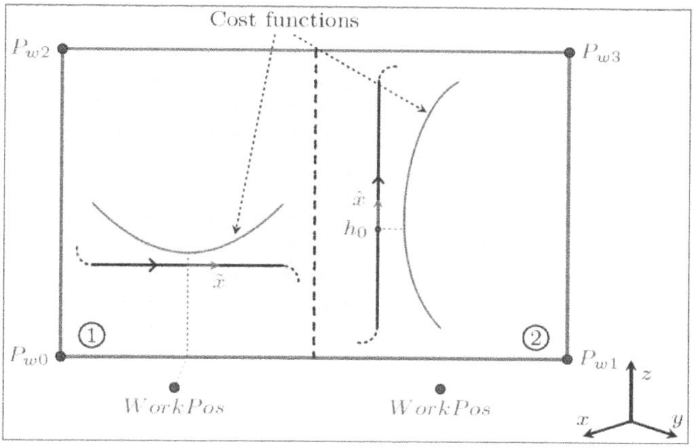

Figure 4.30. Cost functions for coping with higher distances.

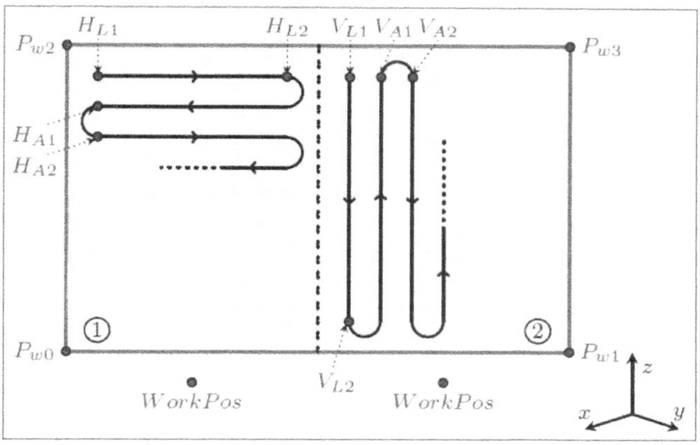

Figure 4.31. Removal trajectories in horizontal and vertical orientation.

The second sub-problem is anchored on a higher abstraction and managing level. It is the formulation of the entire plans that need to be synchronized throughout the whole team of robots. This sub-problem addresses typical questions as "Which robot is executing the task?" and "When is the task executed?". Before the answers can be given for these questions the model needs to be developed.

The data format for storing the entire plan is represented in Figure 4.32 and can also show the different abstraction levels. The *RobotsPlan* contains the complete global plan which defines each individual movement of each robot. This plan is stored in the CPCS. As it is not really efficient to send the complete tasks to every robot individual sub-levels are introduced. They are called *RobotJob* and contain a list of task for one individual robot. The last two sub-levels are lists of so called *ScheduleActivities* which can contain four different shapings: *Travel*, *WaitBeforeTask*,

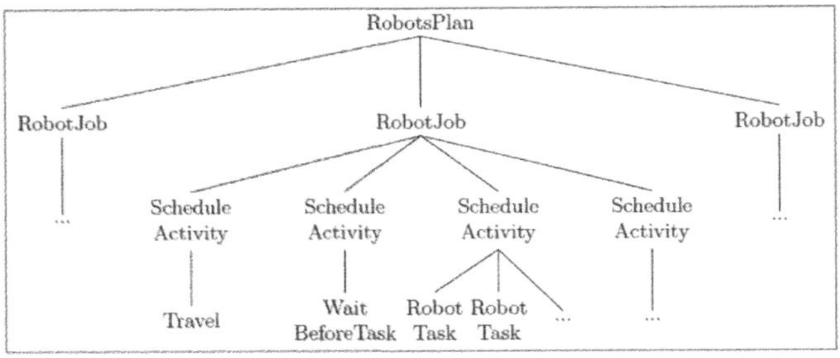

Figure 4.32. Data Format of tasks.

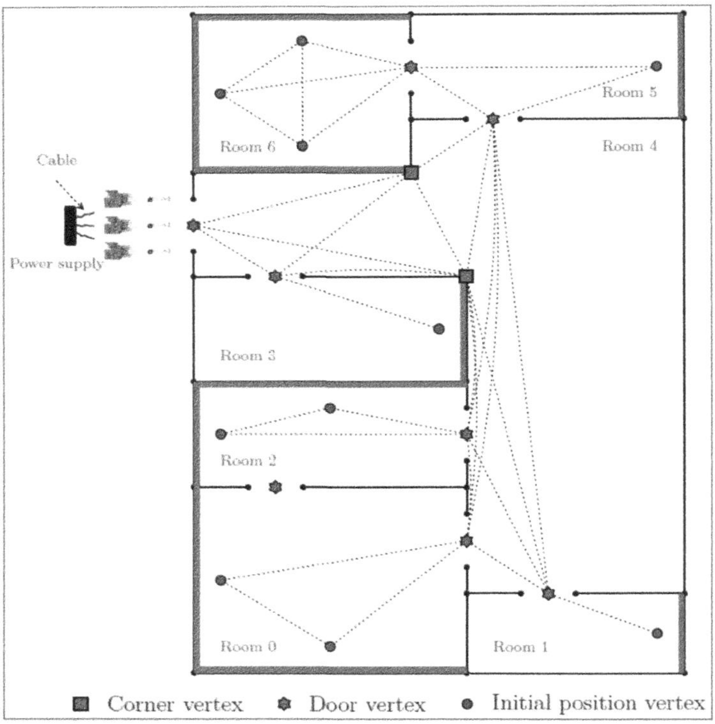

Figure 4.33. Graph generation.

WorkOnTask and *WaitAfterTask*. The robot itself without causing high computational effort can process each individual ScheduleActivity easily.

Beside the data format a graph is generated as a fundamental data model for the task planning of the second sub-problem. This graph represents the entire floor plan with the user-defined tasks, the corresponding initial position for each task and typical characteristics of the flat as doors and corners. The connection of the mentioned vertices generated the entire graph as it is shown in Figure 4.33 that is the

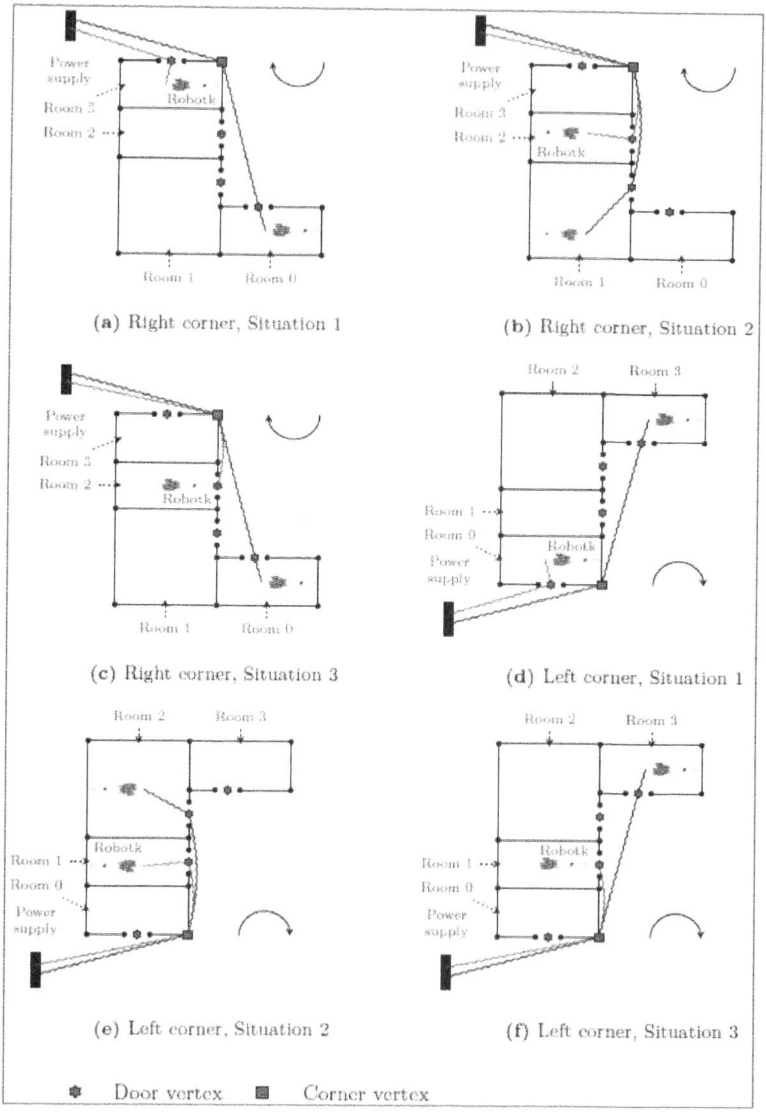

Figure 4.34. Typical cable crossing problems with corners in hallways.

basis for the optimal solution of the second sub-problem. As the mobile manipulators of the Bots2ReC project are tethered to a fixed power supply during the removal process this cable handling becomes very important for the entire planning process. As an example there are six different situations shown in Figure 4.34 that can occur with embossed corners in hallways. These cable-crossing problems are formulated in logical rules and can be tested for each individual task and removal situation.

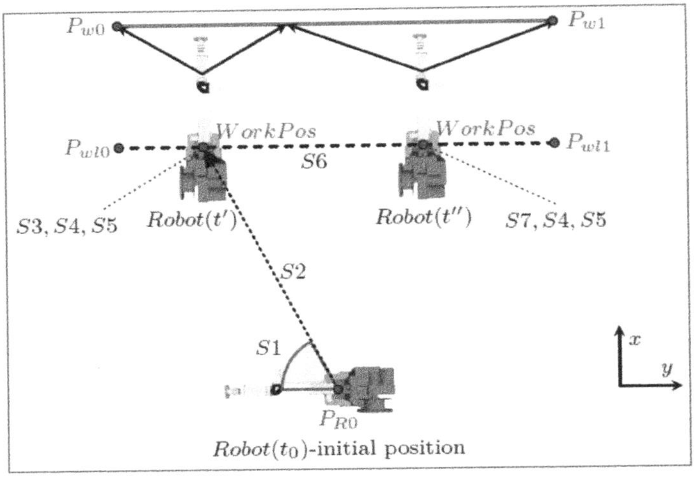

Figure 4.35. Solution for the first sub-problem.

Task Planning

The task planning algorithm solves the different problems that are defined in the previous chapter. For solving the first sub-problem for one individual wall the entire costs for the removal process are added and minimized. This means that the travelling costs and the costs for the trajectory execution are added up to one complete cost value. In this case it is the total amount of time that a robot needs to fulfil the whole task.

In Figure 4.35 a possible solution for one wall is given for a single robot. The wall is split into two areas that are processed after each other. For each area the optimal trajectory (either vertical or horizontal movements) is selected and the correct point of operation in front of the area is defined. Beside this also every translational movement of the mobile base is computed as well. As this is a simple minimization of an objective function the solution converges towards the global minimum and can be solved fast.

The solution of the second sub-problem is more complex due to the time-depending cable situation and the interdependent characteristics of the whole robot fleet. Mathematically it would be possible to set up an entire tree for every possible solution of the problem but this would be very time consuming. Having in mind that the environment on construction sites is unstable and sometimes even unpredictable another method is chosen for solving this problem: an auction between the different robots. This method is much faster compared to the absolute formulation of the problem and can be executed repeatedly with the current process information for adjusting the current plan if planned processes are getting out of scope.

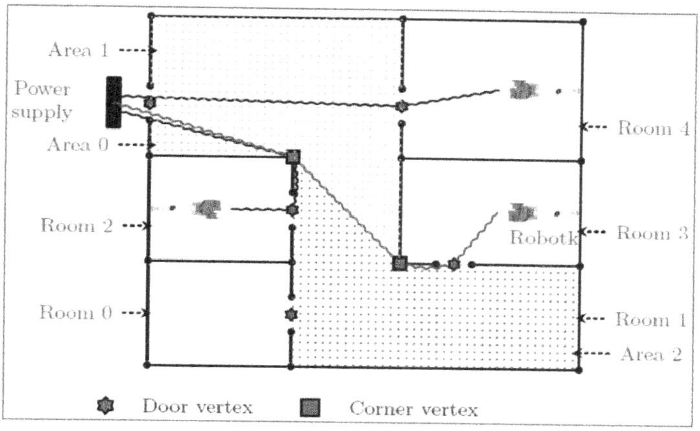

Figure 4.36. Removal situation within a normal flat.

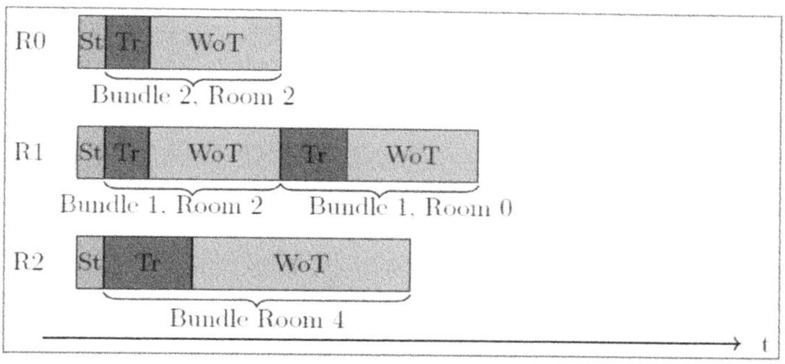

Figure 4.37. RobotsPlan.

This approach is leaned into the direction of reactive behavior that is often used for mobile robots and their task execution.

Out of this the auctioneer on the CPCS knows all the robotic properties and computes their individual costs for fulfilling each known task in the tasks list. Afterwards different heuristics can be used to bring tasks to auction (e.g. the closest task to the entrance, the room with the higher number of tasks, etc.) and each robot will set its cost as an individual bit. The robot with the lowest cost will win the task and can add the necessary *ScheduleActivities* to its schedule. This is done until the auctioneer sells every task. During every bit the cable-crossing conditions are checked and when a forbidden cable situation occurs, a high penalty value is thrown as a bit to ensure that this robot do not win this task. In Figure 4.36 a typical removal situation with three robots in a normal flat is given.

It also displays the different areas of the hallway that are stretched out by the robot cables when they start to touch corner or door vertices. A typical *RobotsPlan* and the individual schedules are shown in Figure 4.37.

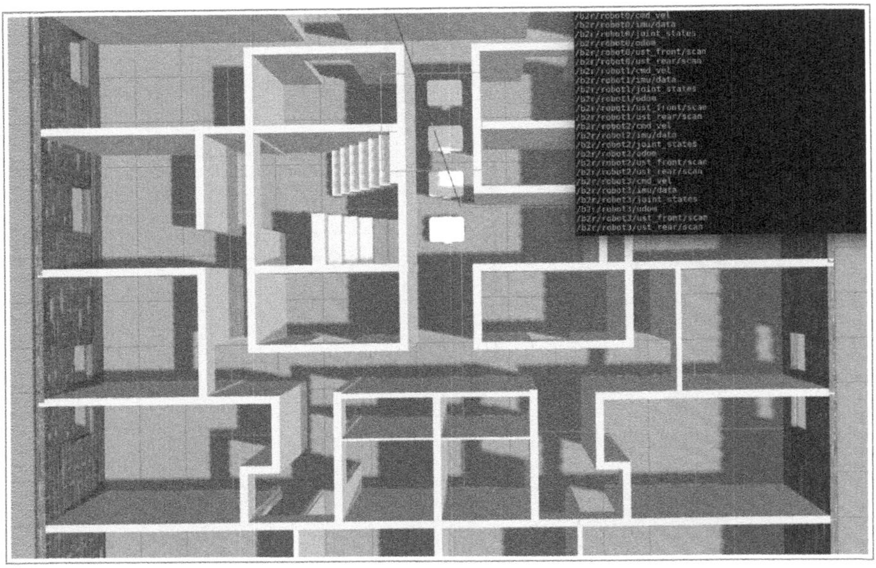

Figure 4.38. Simulation environment for multi-robot strategy.

Task Execution

The last element within the task handling chain is the *Task Execution* during run-time. In addition, this chapter can be split into two sub-elements that are in control of the execution: the environment and the task managers. Following with the previous description of the multi-robot strategy problems, the behavior of each robot within the environment has a repercussion to other robots. For this reason, an environment manager is executed in the cpcs, which is in charge of updating the information available regarding all robots. Each robot will receive positions of other robots and more importantly, its costmap for autonomous navigation will be updated with the cables position from the power supply to all other robots. This measure will prevent any cable crossing, thus, trapped or fallen robots. In Figure 4.39, the costmap update of one robot is shown in a simulation environment representing an entire flat created for the Bots2ReC project (Figure 4.38).

Once the previously explained algorithm is executed, each robot will have a planned task, and the cpcs will update each robot costmap taking into account the tasks assigned. Note that a robot costmap will now include all robot cables as obstacles except its own cable. When a new task is assigned or the position changed, the robot will notify the cpcs which will update and publish all other costmaps. This communication is implemented using ROS services where the environment manager acts as a server and the robot as a client and when the service is called each costmap is published on each robot costmap topic.

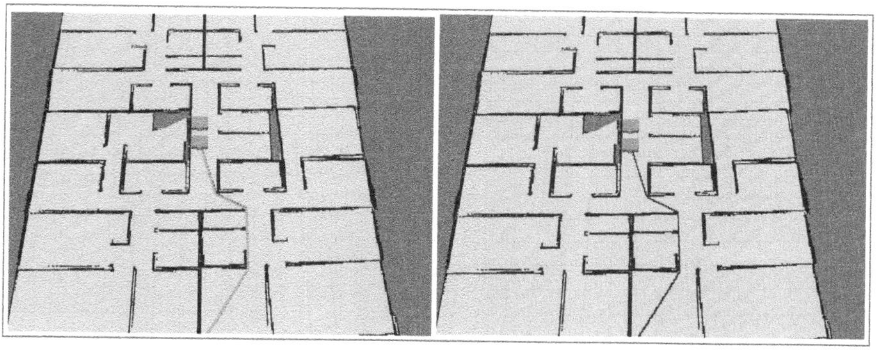

Figure 4.39. Costmap update with cable as obstacle.

Concerning the task manager, different *Modes* are used to establish robust robotic behavior that are managed by a typical *StateMachine* that will survey the global system status. For the Bots2ReC robot the following *Modes* are created:

- Boot
- Calibration
- Teleoperation Base
- Automation Base
- Teleoperation Arm
- Automation Arm
- Error
- Shutdown

Each Mode contains so-called *hierarchies*, which can affect the robotic behavior for each mode directly. The *AutomationBase* mode can be used to explain this strategy. During the autonomous movement of the mobile base to a new goal the whole robotic arm needs to be in the *drivingposition* to ensure a stable movement throughout the whole movement. If the arm is not in this position it is moved towards it before the base starts moving. Furthermore, every arm movement is prevented during the autonomous movements of the base. When a button on the user interface or the remote control is pressed the mobile base can flip directly into the *TeleoperationBase* mode. All those rules and behaviors are stored within the hierarchies that are strictly linked to each mode.

For simplification of the requirements for changing between different modes a central state machine on each robot takes care of the entire robot data. This means that, for example, the state machine will survey the hardware status of the robot and will generate simple Boolean flags that can be read directly by the execution modes. This reduces the logical handling of the hierarchies and can make the logic

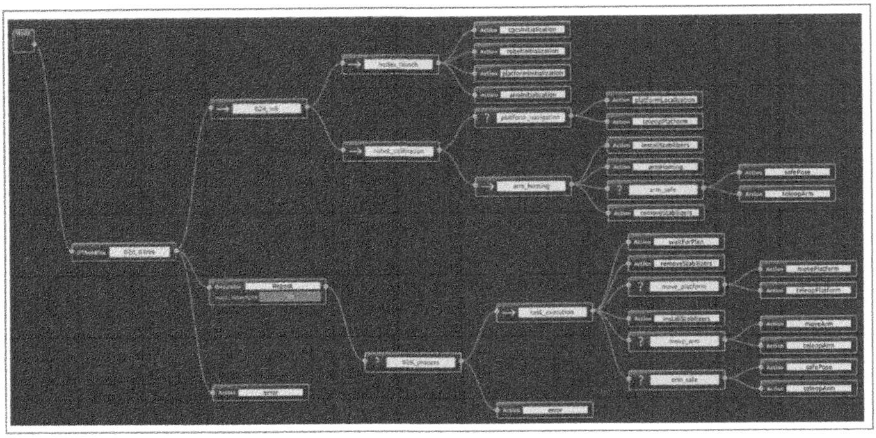

Figure 4.40. Behavior Tree for Bots2Rec robot task execution.

much more understandable. This state machine is implemented using a structure called Behavior Trees. They are extensively used in gaming and AI for representing an agent behavior and are far more readable than classical state machines as a lot of connections between states are implicit within the behavior tree definition. The execution order is from left to right and up to down. Basically there are three types of nodes: Actions, Controls and Decorators.

- *Action*: it will define a specific execution *Mode* or a simple action to develop.
- *Control*: these nodes define a tree behavior, how its child nodes will be executed. For example, we use *Sequence* () (fail if one of its children fails) or *Selectors* (?) (success if one of its children succeeds).
- *Decorator*: with one only child, typically used for invert its child result or repeat it until fail or a number of attempts.

The previous explanation was necessary to understand the Bots2ReC state machine, defining the behavior of a single robot, represented in Figure 4.40.

The first node (B2R_Tree) is a *control node* which manages the higher level of the behavior algorithm. It simply tells: "**If** the initialization (boot and calibration) of the robot succeeds, **then** repeat the robot process until failure. Raise *error mode* to manage this situation otherwise (**else**)". In the task execution there are nodes representing simple waiting actions (waitForPlan, installStabilizers...) and other nodes that describe a movement execution according to a specific *Mode* (movePlatform, moveArm…). When moving the platform or the arm autonomously fails, there's always the possibility of operating the robot manually through the connection to a teleoperation node.

4.3.3 Outlook and Future Work to be Carried Out

The content of this deliverable describes the algorithms at different levels or layers developed to manage and optimize the work of the set of multiple robots to minimize the required time to process the construction site surface according to the different constraints, for example the different characteristics of every robot (tool dimensions, reach, etc.) and the minimization of the crossing of tethering lines.

Once planning algorithms has been defined, developed and tested within the simulation environment, next step is to bring them to the real scenario after second robot unit is built and fine-tune and adapt them according to the physical needs and also considering the results of the different tests being conducted with the robot unit V1 at the testing site.

4.3.4 Summary and Conclusion

As main conclusion from this deliverable we have the fact that the necessary planning algorithms and task allocation basis has been set and tested within a simulation environment. That basis is going to be continuously updated and adapted according new inputs from other task in execution and will finally be implemented within the final demonstrator of the project.

Task 6.5 Development of the central process control system

In reporting periods 1 and 2, the planning procedure and the methods implemented therein for the assignment of tasks to individual robots in a robot fleet were developed. An excerpt of the features contained therein is listed here:

- Consideration of different fleet sizes
- Consideration of different robot types
- Consideration of tethered robots
- Solution for different floor plans
- Optimization of the removal trajectory
- Possibility of continuous re-planning during the removal job

Apart from a continuous improvement during the continuation of the planning process, the parameterization for the second prototype could be carried out and successfully integrated into the existing planning process. In the following an exemplary solution for a given apartment is shown, which is solved by the robot V2:

Figure 4.41. Removal plan for single flat by V2 mobile manipulator.

4.4 Mapping and Localization

Task 6.1 Environmental perception and mapping

A comparison of laser-based and RGBD SLAM algorithms has been done. This included a set of experiments developed to evaluate the performance of common SLAM algorithms. We selected GMapping, Hector Mapping, Cartographer and RTAB-Map due to their availability and relatively easy setup in ROS. For all the experiments a Summit XL Steel Robot was used (RB-Kairos base), i.e. the robot mounts 2 Hokuyo UST-20LX laser range finder sensors in the robot edges. The

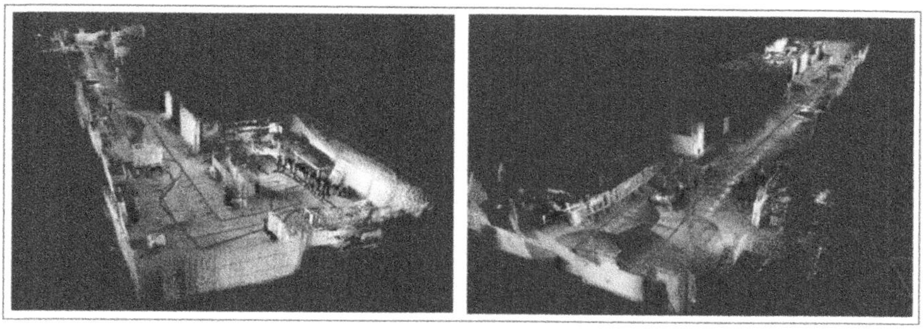

Figure 4.42. Exemplary comparison of RTABMap and the Odometry.

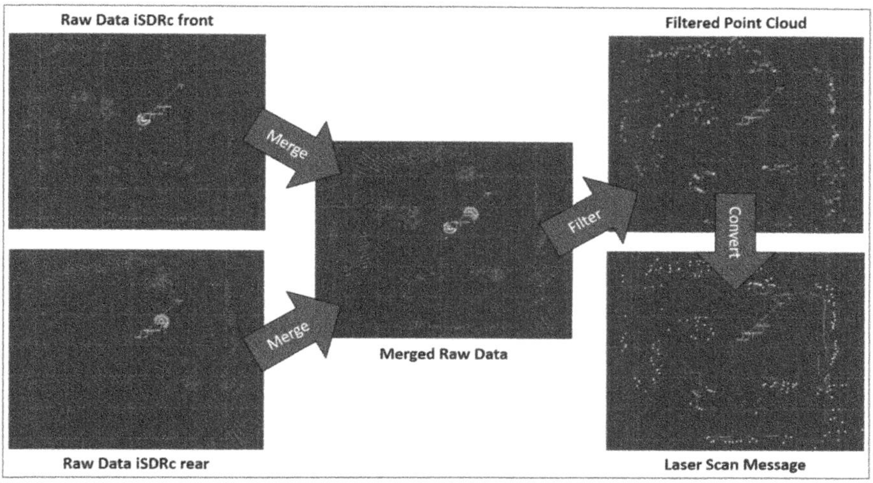

Figure 4.43. Radar Processing steps for localization.

analysis done compares the advantages and limitations of the listed methods and shows the accuracy of the estimated robot pose, as some performance parameters as the CPU or memory usage. Experiments to improve odometry estimation have also been carried out. Different algorithms and devices have been selected to provide alternative odometry sources that can improve the robot wheel odometry. In these experiments the robot standard and non calibrated odometry source was used and the errors of the estimated odometries compared with the ground truth. An example of using RTABMap (red) in comparison to Robot odometry (blue) is given in the following figures.

In addition to the above mentioned findings based on existing algorithms using lidar data, the 2D perception by radar data was further advanced. The consortium had agreed on the redundant solution to localize by radar and lidar in case the problem of the suction quality of V1 could not be eliminated completely. In this way a permanent localization could be achieved even with no 100% extraction despite

Figure 4.44. Mounting positions and orientation of iRTT-AUs and CP-Tag.

dust particles. In this sense, the results for the radar slam have been extended to a pure localization. This means in detail that the common adaptive monte-carlo localization (AMCL) should be carried out with corresponding radar data. This idea fits seamlessly into the semantic mapping step of the overall process, as it automatically provides a lidar-based 2D map of the environment as a result. A later localization by means of radar signals on this map will considerably reduce the setup time of a robotic system and has been targeted accordingly. For this reason, a corresponding fusion and filter algorithm has been developed for the advanced iSDRc from Indurad, which transforms the radar signal into a classical point cloud based on characteristics of the amplitude frequency response. Afterwards the predefined interface in ROS can be used for ACML. In this way, the advantages of the

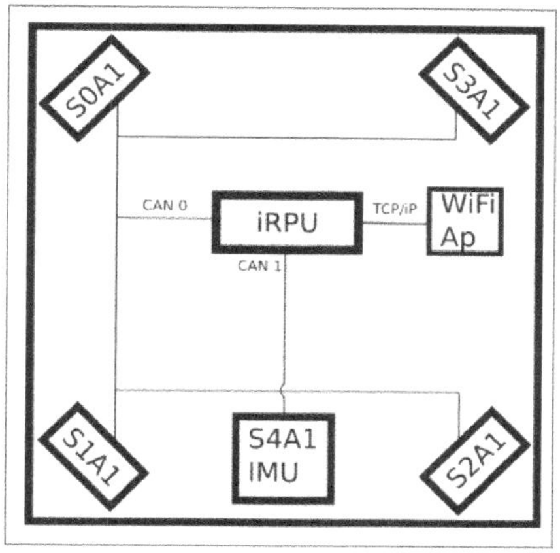

Figure 4.45. Schematic antenna set up.

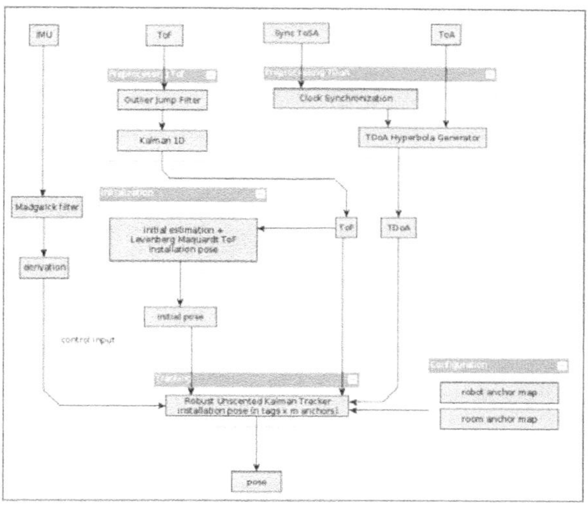

Figure 4.46. Advanced iRTT data flow.

already extensively tested localization method and the penetration of radar signals are efficiently combined.

- Integration of TDoA: measuring reception time differences of radio signals in the 6.4 GHz ultra-wide band with clock synchronization between receptionists over the air.
- New mounting positions on top of the robot as shown in Figures 4.44 and 4.45 with sideward facing antennas for better line of sight to surrounding tag-antennas to reduce multipath effects.

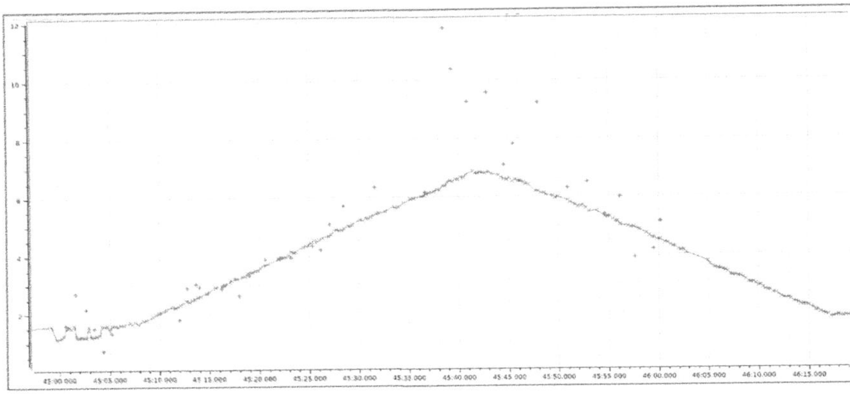

Figure 4.47. Distance points for straight traversal towards CP Tag with front facing antennas. Blue: Distance based on UWB Time-of-Flight measurements. Red: Distance based on ISM Phase Difference.

- Diverse redundant distance measurement system based on detection of radio wave phase differences in the 2.4 GHz ISM (Industrial Scientific Medical) band (at 2,443GHz center frequency).
- Integration of inertia measurement unit (IMU) for supportive pose estimation with Madgwick filter algorithm.
- Robust Kalman Observation Model, with improved convergence of estimated positions and orientations (the pose).
- Integrated Ornstein-Uhlenbeck Process Model (IOU), that deals with stochastic nature of the distance measured with radio ranging systems.
- Improved non-line of sight (NLOS) probability estimation, based on UWB NLOS metrics.

All above mentioned improvements (except for the phase difference measurements in the above mentioned ISM band) and their interconnectivity can also be seen in the following figure.

4.5 Navigation

To verify the correct operation of the robotic unit V1 in terms of navigation and operation, it has been decided to carry out simulations in a virtual environment that recreates the conditions of a real environment. The Gazebo simulator has been chosen, as it offers the ability to accurately and efficiently simulate populations of robots in complex indoor and outdoor environments. It is capable of simulating a population of robots, sensors and objects in a three-dimensional world. It generates both realistic sensor feedback and physically plausible interactions between objects.

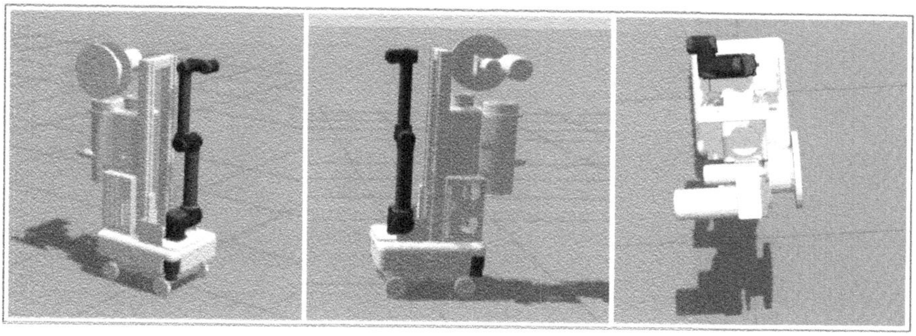

Figure 4.48. Gazebo 3D mock-up model of the robotic unit V1 system in ISO, side and top views.

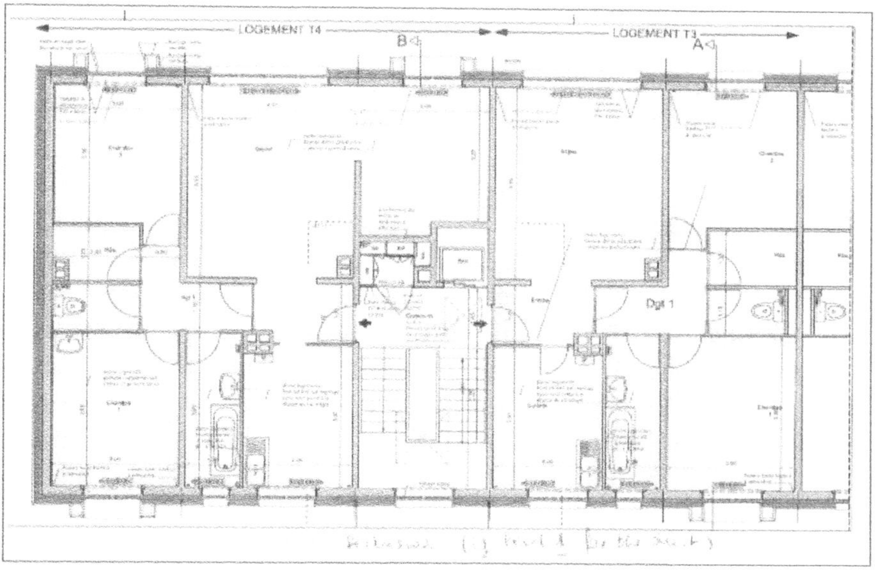

Figure 4.49. Model of a building level with two flats, provided by Bouygues.

A mock-up model of the robotic unit V1 was created with Gazebo to perform simulations in the ROS environment. This model has been generated from the CAD model provided by Telerobot Labs and Robotnik Automation (Figure 4.48).

The environment for this simulation was based on the map of a building level provided by BYCN (Figure 4.49). In order to recreate the simulation environment, it has been used the Gazebo tool building editor. This tool generates a file **.world*, that contains a description of the world to be simulated by Gazebo. It describes the layout of robots, light sources, user interface components, and so on. The world file can also be used to control some aspects of the simulation engine, such as the force of gravity or simulation time step.

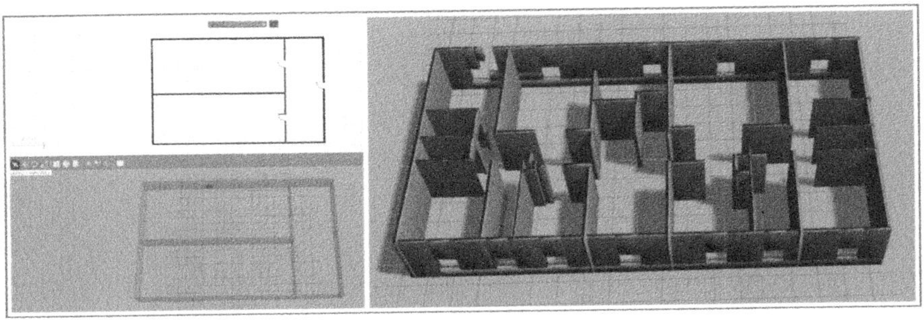

Figure 4.50. Model of a flat created with the Gazebo building editor.

Gazebo world files are written in XML, and can thus be created and modified using a text editor. To simplify those tasks, the Gazebo building editor plug-in has been used (Figure 4.50). This plug-in is very useful to create a scene from scratch. Once the drawing is finished, the world file can be saved for future use into the simulation environment at any time.

To simulate the model in ROS environment, a URDF file (Unified Robot Description File) was generated from the CAD model of the robot. This file contains the physical and mechanical description of the robot, such as the type of joints, the coordinate system of each link, collision elements, and so on. The URDF file has been generated with the SolidWorks to URDF Exporter plug-in of SolidWorks.

Once the robot model for ROS and Gazebo was obtained, a plug-in was added to the Gazebo model to control the movements of the platform within the simulation environment. The chosen plug-in was *libgazebo_ros_planar_move*, for moving the robotic unit in a horizontal plane using geometry_msgs/Twist. The plug-in works by imposing linear velocities in (XY) and an angular velocity (Z) to the controlled bodies at every cycle.

4.5.1 Navigation Tests

In these tests, the robotic unit mock-up model V1 was added into a virtual environment reproducing a flat and, thanks to virtual 2D laser sensors, the proximity of the obstacles around to the mobile platform could be measured (Figure 4.51).

For a better autonomy, the capacity of the robot to re-create a map by itself from its environment was investigated. To generate this map, the *gmapping* ROS package has been used. This package was able to generate a 2D map from the data of the lasers sensors of the platform, as the platform moved through the environment (Figure 4.52). The navigation during the generation of 2D map was tele-operated manually.

Once generated the navigation map for the platform, a simple navigation test has been performed (Figure 4.53). A user-defined goal point was indicated in the

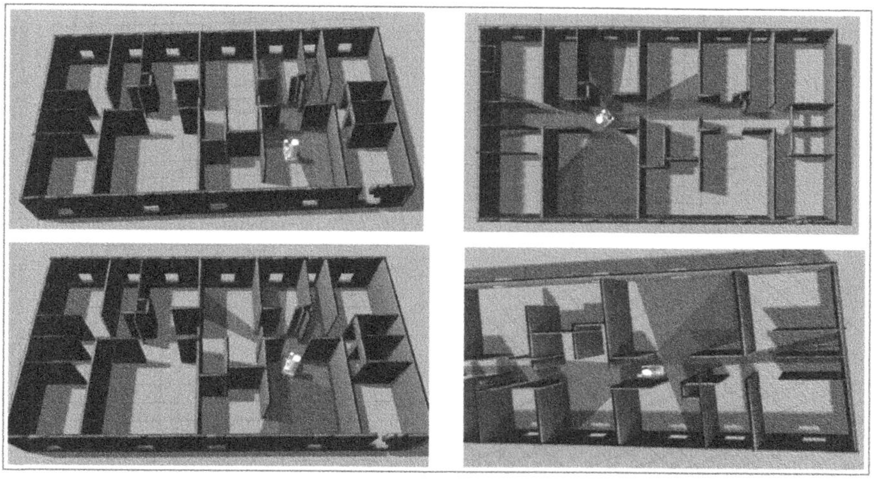

Figure 4.51. Virtual robotic unit (in white) obtaining data from the virtual 2D laser sensors.

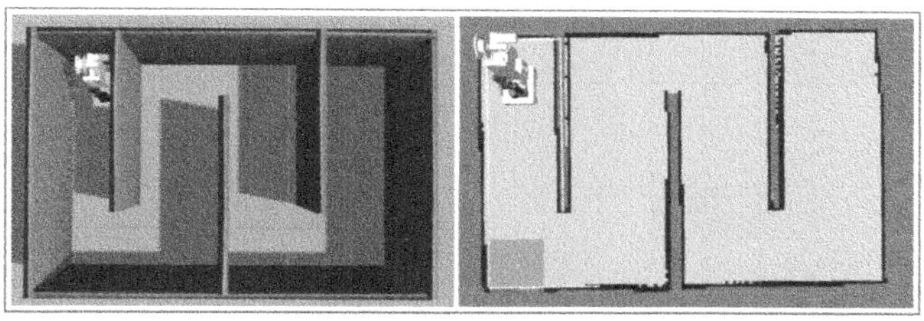

Figure 4.52. Map generated from Gazebo's simulated environment.

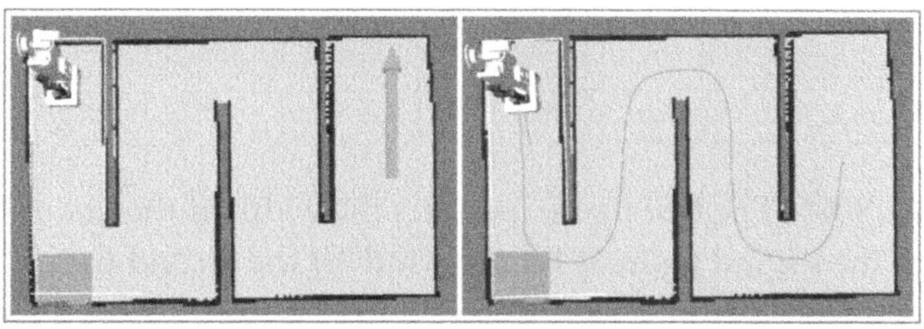

Figure 4.53. (a) Goal for the robotic unit. (b) Path planning by the algorithm of navigation.

map, and then the planner could plan an optimal and free collision path, that was sent to the virtual controller to execute the calculated trajectory.

4.5.2 Bent Corridor Tests

To verify that the robotic unit V1 was able to turn in narrow bent corridors, different tests have been performed with varied sizes of corridors (700x700, 700x800,

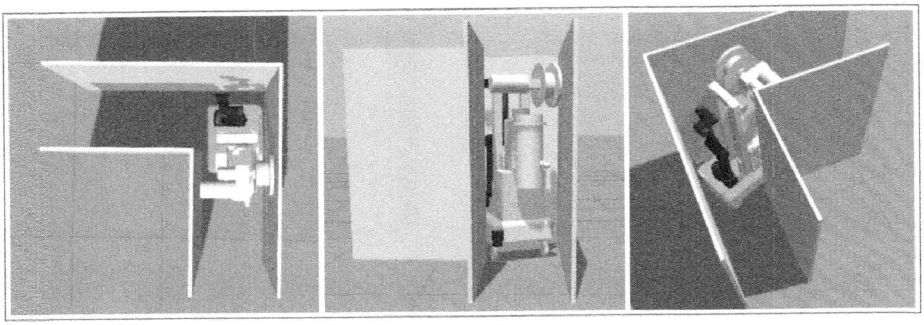

Figure 4.54. Robotic unit turning in a 700x700 corridor.

Figure 4.55. Simplified robotic unit turning in a 700x700 corridor.

700x900, 800x800). The first figure indicates the width of the first part of the corridor, whereas the second figure is for the width of the second part of the corridor that connects to the first one. A Gazebo model has been created for each corridor. The navigation along the narrow bent corridors has been made manually to simplify the complexity of the task.

4.5.3 700x700 Corridor Test

In the case of 700x700 corridor, the robotic unit has not been able to turn in the corridor, getting stuck while turning (Figure 4.54).

It was decided to remove the aspiration unit and the cable reel, as shown in Figure Illustration, since they are outside of the footprint of the mobile platform, and check whether the robotic unit could turn in the corridor.

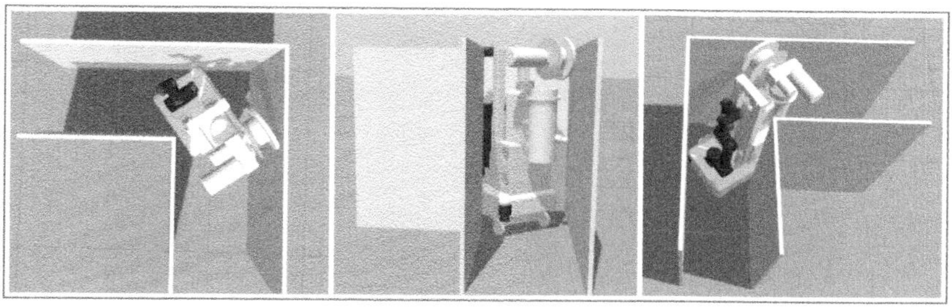

Figure 4.56. Robotic unit turning in a 700x800 corridor.

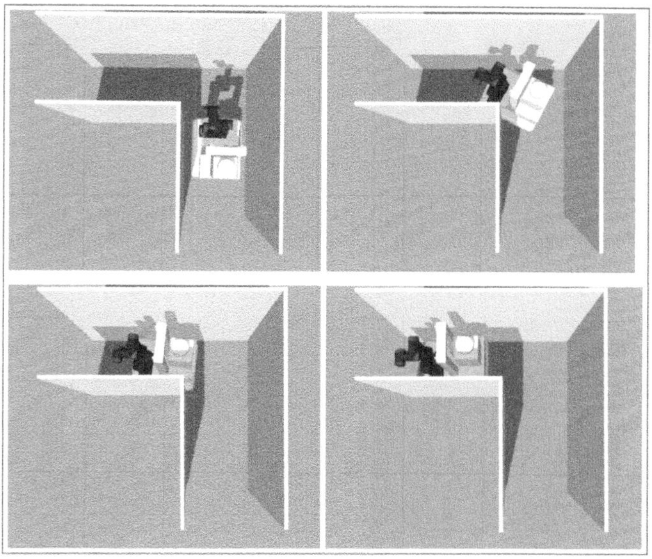

Figure 4.57. Simplified robotic unit turning in a 700x800 corridor.

Without the aspiration unit and the cable reel, the robotic unit has been able to turn the 700x700 corridor, although very tightly along the corridor.

4.5.4 700x800 Corridor Test

With the corridor of 700x800, the robotic unit has not completed turned, because the cable reel stands out of the footprint on the platform, preventing the rotation.

A test without the cable reel worked perfectly (Figure 4.57).

4.5.5 700x900 Corridor Test

In the corridor of 700x900, the robotic unit has been able the turn in the corridor correctly, this time with the cable reel and the aspiration unit as it was in the initial setup (Figure 4.58).

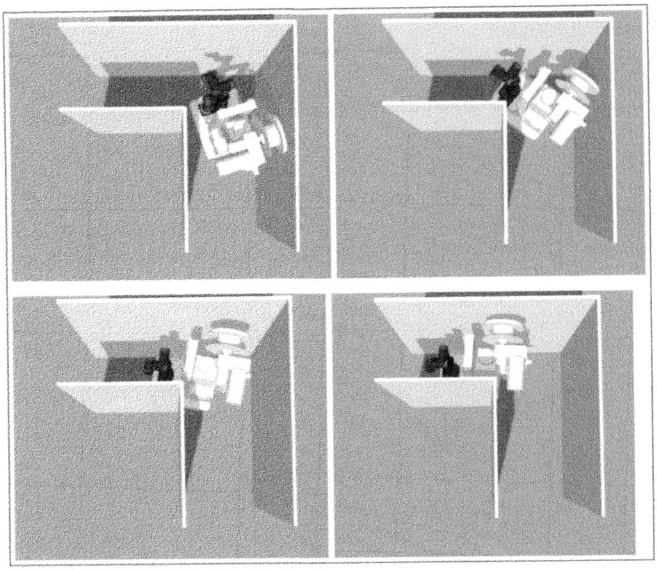

Figure 4.58. Robotic unit turning in a 700x900 corridor.

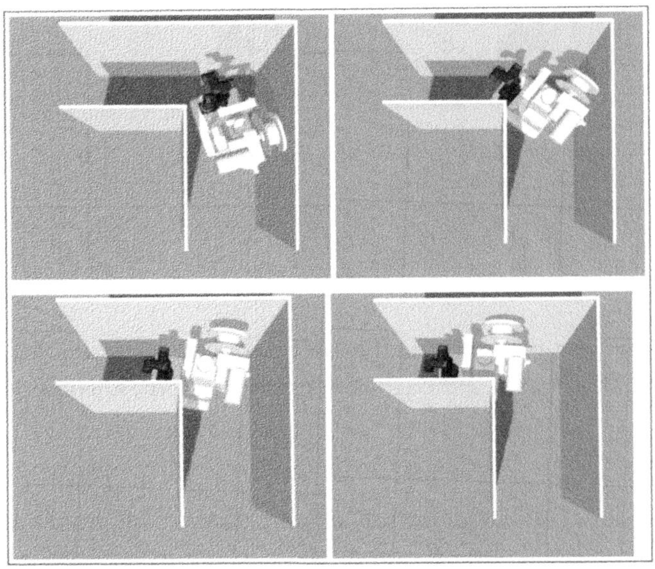

Figure 4.59. Robotic unit turning in a 800x800 corridor.

800x800 corridor test

In the corridor of 800x800, the robotic unit has been able to turn in the corridor correctly, this time with the cable reel and the aspiration unit as it was in the initial setup.

Table 4.2. Summary of bent corridor test results.

Bent corridor Crossing	With robotic unit V1 system	With simplified robotic unit V1
700x700 corridor	No	Yes
700x800 corridor	No	Yes
700x900 corridor	Yes	–
800x800 corridor	Yes	–

4.5.6 Summary of Bent Corridor Test Results

Table 4.2 contains the results of the tests performed in different bent corridors. It is indicated if the platform could turn in the corridors with the complete robotic unit V1 system, or with a simplified robotic unit with less protruding elements.

4.6 Stability Observations

Deliverable 2.1
D2.1 Preliminary simulations of the robotic units

Authors and contributors:
CARRASCO JIMENEZ Ramon, EURECAT
DETERT Tim, RWTH
FAUROUX Jean-Christophe, SIGMA Clermont
MARAJE Siddharth, SIGMA Clermont
MARTIN Daniel, EURECAT
MOHY EL DINE Kamal, SIGMA Clermont

4.6.1 Deliverable in the Context of the Work Program

Work package 2 (WP2) aims to support the other work packages by creating relevant scientific and engineering models in order to bring clear answers to the technical choices to be made during the design process of the robotic unit.

Deliverable D2.1 is the first result issued from WP2, with the ambition to provide simplified models of the robotic unit that is being designed at the beginning of the project. The models constitute the foundations of a design and control simulation environment to be used all along the project.

The deliverable D2.1 takes input from "*Task 1.1 – Specification of asbestos removal tasks*", particularly for the specification of the environment to clean, including narrow corridors, and "*Task 1.2 – Initial trials of operational process*", specifically for the tool requirements and manual productivity.

D2.1 provides a preliminary testing environment with simplified mock-ups of the robotic units implemented in geometric/kinematic/dynamic simulation tools prior to the final design of the robot. Based on the requirements, a tall robotic unit has been considered for preliminary evaluation, made up of a manipulating arm on a mobile platform. A virtual environment was reconstructed from a digital map of the testing area, adding attributes from Task 1.1 to represent the spatial dimensions of the scenario.

The parametrized models developed in D2.1 provide useful design rules for "*Task 2.2 – Stability model and development of mechatronic stabilizers adapted to the task and environment*" and "*Task 2.3 – Virtual scenario of a representative asbestos removal use case*", which will include a virtual mock up and refined models. Further co-simulations will be implemented to address specific calculations like the stability modelling in Task 2.2.

4.6.2 Methods and Work Carried Out

This deliverable required extensive exchanges with the partners for defining and double checking the requirements on the robotic unit and performing the three main following tasks:

- **Extracting a core simplified subset of requirements** critical for feeding the first models. These requirements were gathered during several meetings, including a visit of Bouygues at Clermont-Ferrand (05/04/2016) and a visit of several flats under or after their cleaning process, organized by Bouygues at Clermont-Ferrand (01/06/2016), which showed a realistic environment and its specific difficulties, particularly in terms of confined and restricted space.
- Creating **preliminary models** and extracting results to feed the discussion. The main constraints to consider for the robotic unit were:
  - tool reachability for tall walls of 3 m height, ceiling at 3 m and surface all around the mobile base;
  - non-collision robot-environment in narrow corridors and while passing through doors;
  - singularity avoidance for future control;
  - quasi-static stability in all the workspace
- Exchanging with partners regularly for **banning or confirming architectural choices** during web conferences, phone meetings and brainstorming

sessions. A particular contribution was made for defining the **arm architecture**. Finally, three kinematics were modelled and tested.

The considered approach for analysis was to develop several types of models:

- **Geometrical models**, mostly for assessing tool reachability and collision avoidance. The models can be of two types:
  - Skeleton models, reducing links to a simple average line. The models are represented with the Geogebra software [Geogebra, 2016], an interactive mathematical tool that provides intuitive modelling method both in 2-D and 3-D through algebraic expressions and geometric constructions. Geogebra allows model parametrization and has been proved to be particularly suitable for dimensional synthesis [Prause *et al.*, 2015].
  - 3D CAD models for robot representation, simulating tool paths and avoiding collisions. The links have a volumetric simplified shape. One big advantage of CAD is that dimensional constraints can be used both on the joints (direct geometric model) and on the tool (inverse geometric model). The assembly constraint solver manages them simultaneously, which is particularly useful for redundant models, where the redundant joints can be fixed in the direct geometric model and the remaining joints are positioned by solving the inverse geometric model. The CAD models were made with Catia V5 / V6 (SIGMA) and SolidWorks 2016 (EURECAT)
- **Quasi-static models**, including the masses of the bodies and contacts on the ground, to evaluate the static stability margin of the robotic unit. Algebraic equations were added to a Geogebra model for barycentric calculations and stability margins.
- Preliminary **multi-body models**, including contact forces and inertias, to be re-used in Task 2.3 for evaluating the complete dynamics of the robotic unit during its motion and cleaning tasks. Adams 2013 was used and will take its full interest for Task 2.3 as it allows modelling flexible bodies and as we think the stiffness issues in the robot may compromise its precision.
- **Controllable models**, into a complete realistic environment based on ROS, ready to represent and control the robotic unit V1. This work was performed by EURECAT, as the main integrator of the future prototype V1. A ROS tutorial was also organized by SIGMA in Clermont-Ferrand for improving the ROS competencies of the whole consortium (31/05/2016-01/06/2016).
- **Preliminary simulations of the robotic units**

This section provides a summary about the initial analysis and preliminary models of the robotic unit. It includes requirement extraction, preliminary dimensioning, stability analysis, study of three different arm architectures, force analysis, preliminary dynamic and productivity analysis and finally a controllable model of the robotic unit in its environment.

4.6.2.1 Requirements and constraints on the robotic unit

A core simplified subset of requirements critical for the first models can be found below:

- **Environment** to be cleaned: type of surfaces to clean (plastering, spattling compound, paint, with an option on ceramic tiles and glue), geometry (horizontal/vertical/inclined, flat or curved with a curvature radius of 1.5m) and positioning of the surfaces, minimal door dimensions (70 cm x 2 m), constraints due to seclusion, obstacles and tight environment (particularly 90° angles in narrow 70 cm wide corridors);
- **Robotic unit settings**: maximal acceptable dimensions (Width <0.6 m, Length <0.8 m, Height <1.8 m) and mass (300 kg), approximate masses of the main components, required workspace (reaching a ceiling at 3 m);
- **Cleaning process** and associated tools: typical nature and dimensions of existing tools for manual cleaning (rotating disks of 150-300 mm of diameter, grinding cylinders), involved forces (typically 80 N of normal force, but depends on the tool).

The robotic unit integrates a robotic arm equipped with an abrasive tool mounted on a mobile base will perform asbestos removal operation. The design of this robotic arm will primarily be influenced by the following constraints:

- **Number of Degrees of Freedom (DoF):** For asbestos cleaning scenario, 5 degrees of freedom are needed at least considering the scenario of curved walls: 3 translational DoF and 2 rotational DoF, with axes perpendicular to the tool axis.
- **Reachability:** The dimensions of the room to be cleaned put constraints on the link lengths of the arm. According to the list of specifications, the height wise dimension of the wall to be cleaned is maximum 3 meters. Thus, the robotic arm must be able to reach the top, ground and corner points on the wall.
- **Payload capacity:** The required payload carrying capacity is 50 N maximum since the weight of the tool is around 5 kg. Additionally, the removal process requires at least 80 N of normal force.

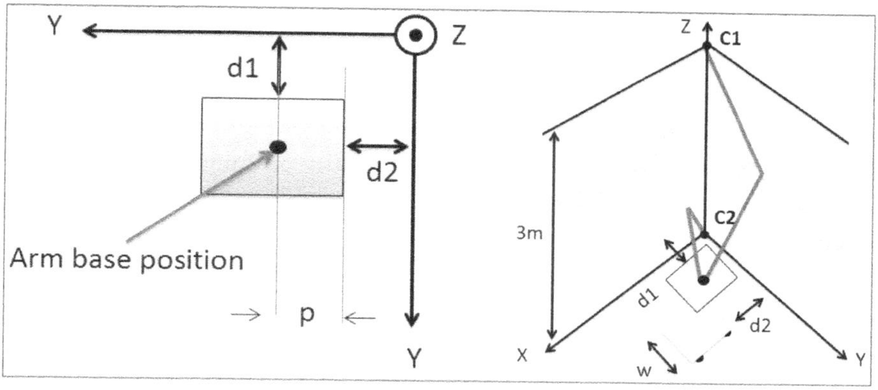

Figure 4.60. Constraints of the arm modelling. (a) Position of the robot with respect to the walls. (b) Constraint points of the arm to reach.

- **Stability of the Robotic Unit**: The robotic unit, while performing removal operation, can lose its stability due to its variable pose and the external forces applied on the robot. Depending on the considered forces, two approaches can be considered:
  - **Static stability:** As the mobile platform has a vertically elongated aspect ratio and is supporting a robotic arm that can overhang, the structural stability of the total system is dependent on the arm structure and pose. This quasi-static model is suitable for slow cleaning motions and mainly considers the wheel-ground contacts.
  - **Dynamic stability**: During cleaning operation, interaction of the robotic unit with the environment generates dynamic forces (normal and tangential reaction forces, frictional forces) on the robotic unit, both on the tool and supporting wheels. Dynamic effects will also be generated due to the high speed cleaning motions with weighting parts. This type of model takes all the forces into account and is suitable for simulating a fast cleaning.

4.6.2.2 Preliminary geometric dimensioning

The first trial of the arm modelling was done based on the reachability approach. The distances of the robot from the left and the front wall, d1 and d2 respectively, are adjusted at their minimum value and considered as constraints (Figure 4.60 Illustration(a)) in order to minimize the lengths of the links of the arm. Assuming the arm base height is h, the constraint for the robot is to reach the farest points in its workspace C1 and C2 is shown in Figure 4.60 Illustration(b). The width of the mobile base is defined as w. Variable p defines the position of the arm base from the front edge of the robot.

According to the robot specifications, the width of the mobile platform is 0.6 m, its length is 0.7 m and the height is 1.5 m. Distances d1 and d2 are fixed at 0.3 m and the arm link lengths are minimal when based on the distance from the arm base to the furthest point, then divided by two since we have two links. A small offset is then added to links to avoid the singularity on fully extended arm. The inverse geometric model validates the lengths by reaching C1 and C2. Hence, the needed link length is 0.95 m for this simplified model. A similar approach will be used for the more detailed models studied in Section 3.3.

4.6.2.3 Stability Analysis of the robotic Unit

The effect of the static and dynamic factors mentioned in section 3.1 on the stability of the robotic unit must be analysed in order to perform removal operation efficiently. However, for preliminary selection of the arm, the main focus is to assess only the static stability of the system.

To perform the static stability analysis, a suitable stability margin should be considered. For the preliminary analysis, a static stability margin proposed by McGhee and Frank – '***projection of the centre of mass in the support polygon***' [McGhee and Frank, 1968] is used as a stability margin. In order to evaluate the stability, robotic unit must be modelled with realistic weight distributions of its components. The Geogebra software is used for building the skeleton model of the robotic unit.

For stability analysis, the design parameters are ***link lengths*** of the robotic arm, ***dimensions*** of the mobile platform, ***positioning*** of the robotic unit with respect to the wall, ***masses*** of the components from robotic arm and mobile platform.

4.6.2.4 3-D skeleton model of the Robotic Unit

A 3-D model of the robotic unit was created in Geogebra and the centre of mass of the total system was evaluated (Figures 4.61–4.62). The objective of the model was to build a **quasi-statically balanced model without external forces for the approximate evaluation of the stability margins and structural design**. The design parameters for this model are arm link lengths (LM, MN, NO), their masses and the mass of the mobile platform. This model allows assessing the effect of design parameters on the variation of the centre of mass of the robotic unit. It revealed that in certain arm configurations, the projection of the centre of mass is close to the edge of the support polygon, as can be seen on the green supporting rectangle on the top left part of Figures 4.61–4.62. This indicates increase in the toppling moments and may cause instability of the system. Thus, the arm configuration and the weight distribution in the arm architecture is a critical factor while achieving stability.

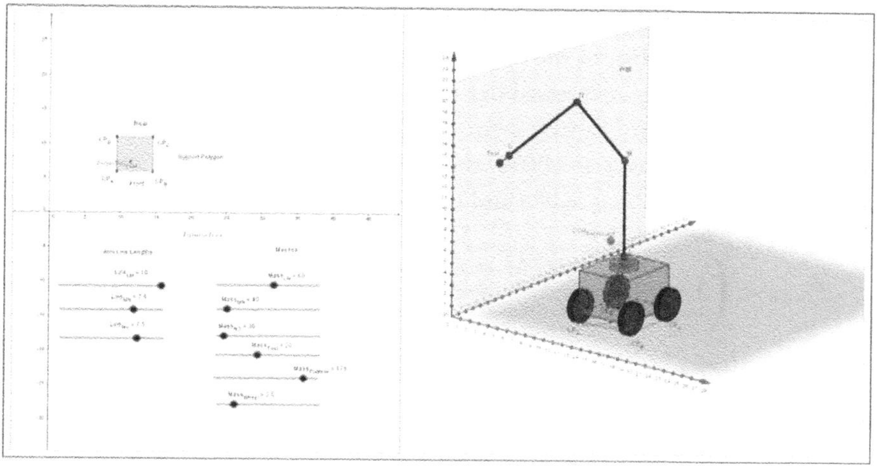

Figure 4.61. 3D Model of the robotic unit (cleaning a wall).

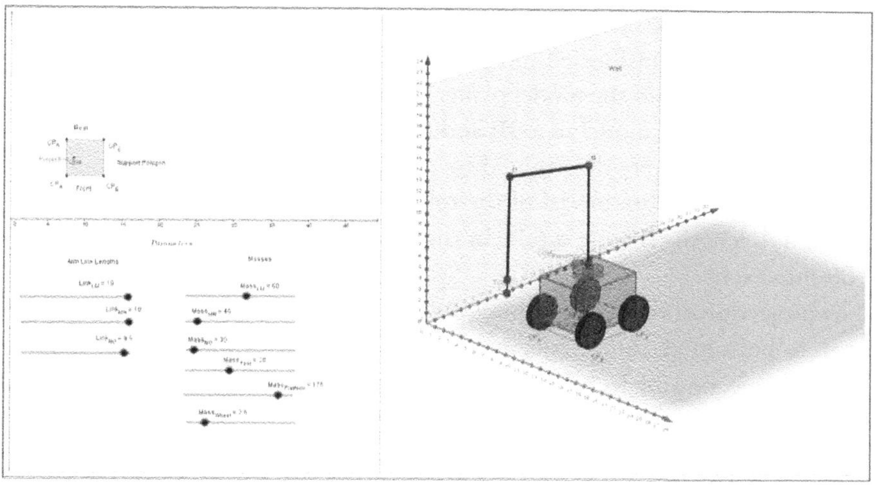

Figure 4.62. 3D Model of the robotic unit (cleaning ground).

The main advantage of this 3D model is that it can cover several cleaning scenarios of the ground/walls/ceiling. If components of the robotic unit change during the project, this model should always be used for a fast evaluation of the stability margin of the modified robotic unit.

4.6.2.5 Stability analysis of robotic unit based on different arm architectures

For the arm dependent stability analysis of the robotic unit, several 2D models of the robotic unit with different arm architectures were built for a simplified and focused analysis of the arm. Stability of the robotic unit was studied during an operation of the robotic unit with the tool moving in the plane of the wall. To quantify stability, the shift in the projection of centre of mass of the overall system from its centre position is expressed in a non-dimensional way, in percentage. Thus, the longitudinal stability can be expressed by the P_{Stab} variable as indicated by equation (1). When P_{Stab} is 100%, the stability is maximal. The robot is at the limit of instability for P_{Stab} = 0% and unstable under 0%.

$$P_{Stab} = \left(1 - \frac{Dist\left(COM - Centre_{Supportpolygon}\right)}{Length_{Supportpolygon}/2}\right) \times 100 \tag{4.1}$$

As the stability depends on the nature of the arm, three architectures A-B-C are proposed for consideration and presented below.

Architecture A) 5 DoF classical arm

The first obvious consideration is a classical 5 *DoF* arm. The 2-joint wrist is made by revolute joints with axes perpendicular to the tool axis. For simplifying the analysis, the 2-joint wrist is transformed into a single revolute joint in the 2D model with an axis perpendicular to the model plane. The base revolute joint of the arm is also ignored, reducing the model to a **3R serial arm** in the 2D model.

The link lengths were determined by considering the requirements on reachability. The 2D Geogebra model of the robotic unit was constructed to obtain the required link lengths (Figure 4.63). In this case, links MN and NO have a length of 900 mm each.

The projection of the centre of mass of the overall system was also determined for the top to bottom cleaning trajectory. The robotic unit shows a stable behaviour since the projection of its centre of mass always lies well within the supporting polygon. For example in Figure 4.64(a), the length of the support polygon is 600 mm and the distance between the projection of the centre of mass and the centre of the robotic unit is 77 mm. Thus, for the configuration shown in Figure 4.64(a), P_{Stab} can be calculated by equation (2).

$$P_{Stab} = \left(1 - \frac{77}{350}\right) x100 = 78\% \tag{4.2}$$

Similarly, for the configuration shown in Figure 4.64(b), P_{Stab} is found to be 71%.

Even if architecture A is capable to generate the cleaning motions along the walls with sufficient quasi-static stability, it cannot be used in our case because of serious collision issues. As can be seen in Figure 4.63, the mobile platform has to be put at a distance x_A = 350 mm from the wall to let the end effector pass between the mobile platform and the wall. This safe operational distance from the wall also

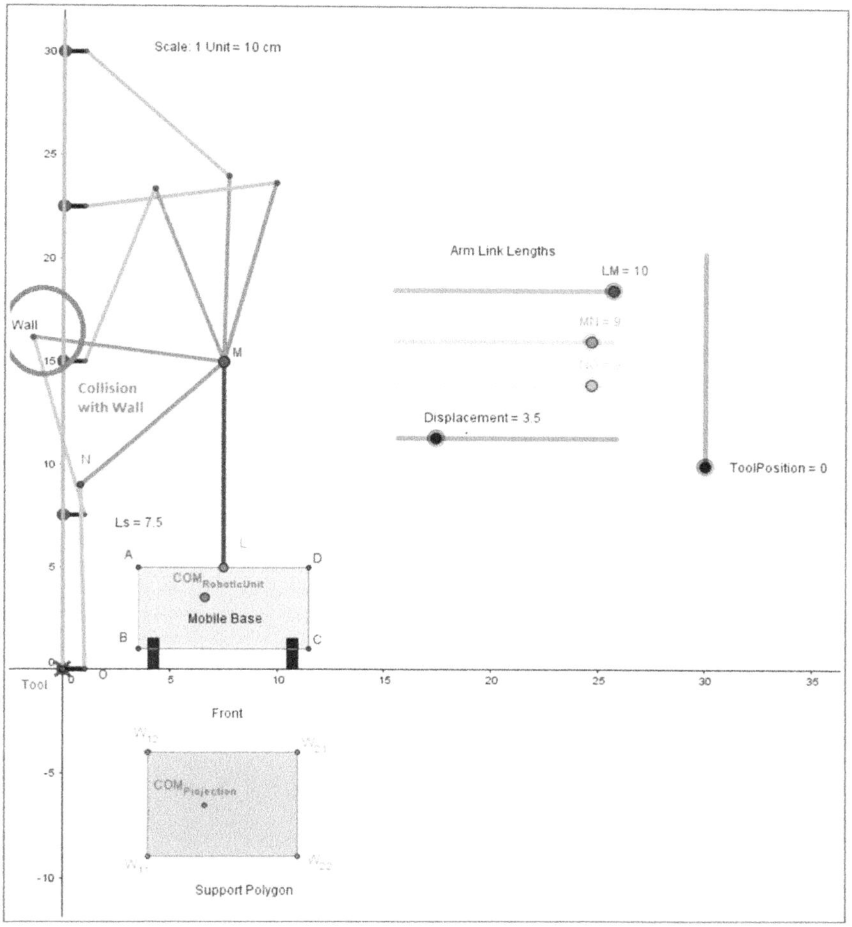

Figure 4.63. Collision with the wall while cleaning from a close distance (x_A = 350 mm).

serves to avoid collision with the wall. Figure 4.63 shows the elbow of the robotic arm colliding with the wall. This means either the safety distance x_A is not long enough or the link lengths MN = NO = 900 mm are too long, even if they must be long enough so the ceiling at 3 m high can be cleaned.

Architecture A has even more collision issues in the case of a narrow corridor, a common feature in the flats to be cleaned. Figure 4.65 shows a 100 cm wide corridor, which is a large value compared to the narrowest corridors of 70 cm. Even in this optimistic case, the model shows not only the elbow collision with the left wall when the tool altitude is $y_O = 120$ cm but also a second elbow-collision with the right wall when $y_O = 240$ cm. This second case shows clearly that architecture A cannot be used for our project.

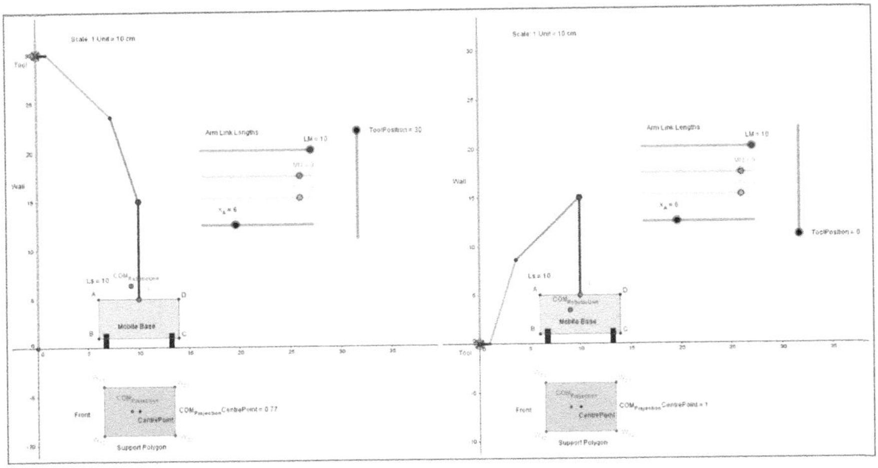

Figure 4.64. 2-R architecture: Wall cleaning scenario. (a) Topmost point. (b) Ground point.

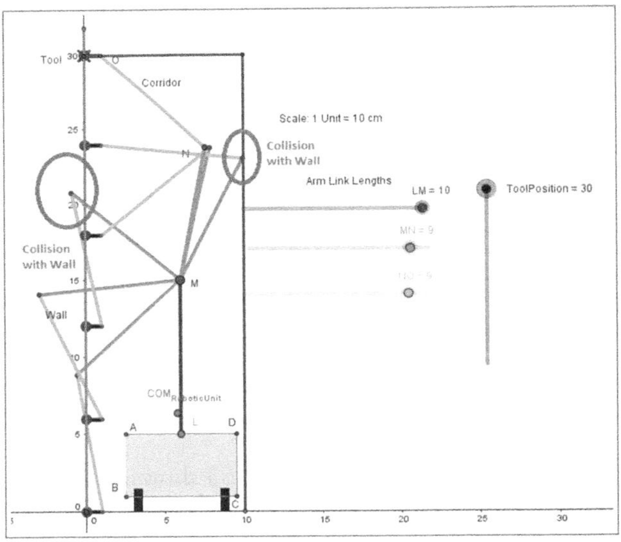

Figure 4.65. Cleaning in a narrow corridor with architecture A leads to arm collision with the wall.

Architecture B) 6 DoF redundant solution with a 5 DoF arm mounted on a vertical slider

This redundant architecture is considered in order to avoid the use of two long link lengths required in case of a classical anthropomorphic arm, as discussed for architecture A. To study the planar case, the architecture B is reduced to a 3R serial arm with a vertical slider at the base. A Geogebra model of this architecture was built to determine link lengths and study static stability. The link lengths obtained in this

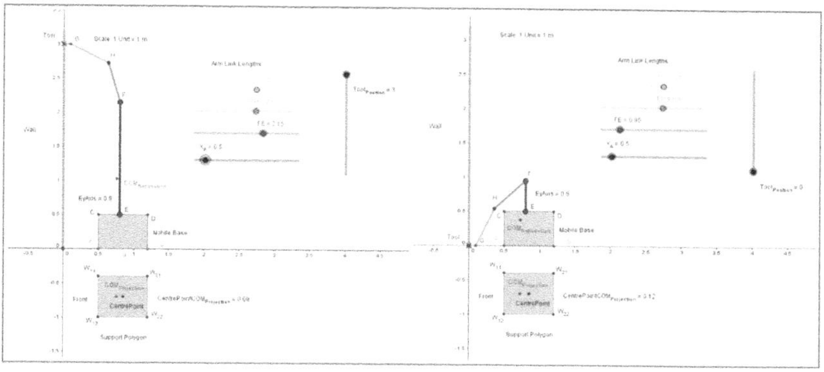

Figure 4.66. 2-R architecture with slider: Wall cleaning scenario. (a) Topmost point. (b) Ground point.

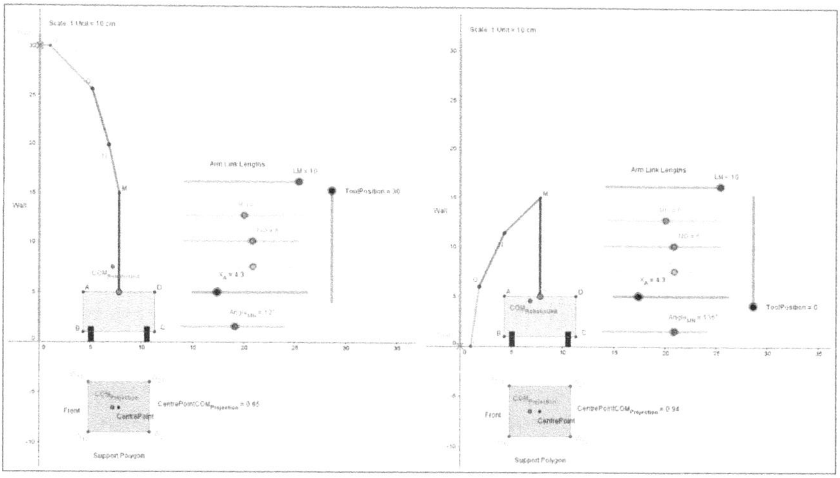

Figure 4.67. Wall cleaning scenario with architecture C. (a) Topmost point. (b) Ground point.

case are around 60 cm each. Also, the structure shows a stable behaviour during cleaning trajectory and the longitudinal stability margin P_{Stab} for the configurations shown in Figure 4.66(a) and 4.66(b) are 74% and 66% respectively.

Architecture C) 6 DoF redundant solution with a 6 DoF arm having three shorter moving segments

Ignoring the spherical wrist and the base revolute joint, this architecture can be reduced to a 4R plane serial arm. A Geogebra model of architecture C was built to determine link lengths and study its static stability. One important advantage of architecture C is that it can clean in narrow corridors, whereas architecture A fails because its arm collides with walls of the corridor (Figure 4.65).

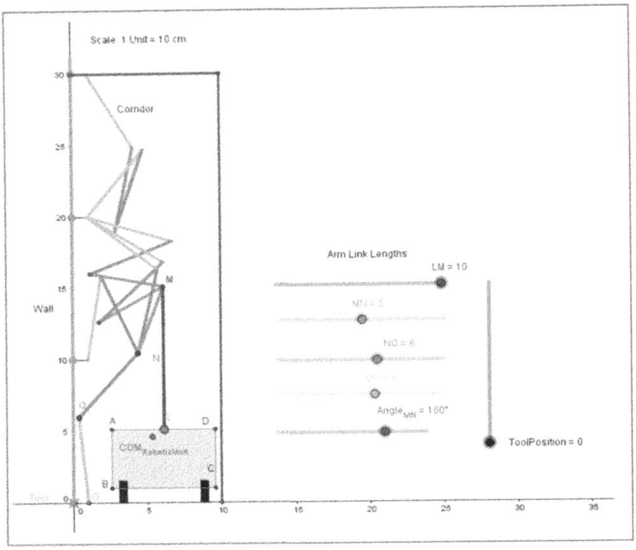

Figure 4.68. Cleaning in a narrow corridor with architecture C.

The static stability of this architecture was studied (Figure 4.67) and the values of P_{Stab} in (a) and (b) configurations are 81% and 73% respectively. This means **architecture C seems to be more stable than the two others** in these two reference positions.

Figure 4.67 also illustrates the notion of redundancy. The idea is to adjust the angle of the revolute joint M so that the NQO double-segment arm can move vertically without crossing a singularity. The optimal angle of joint M could be chosen by maximizing the distance to singularity during the sweeping motion. Three positions of joint M can be seen in Figure 4.106 during the path planning of the arm for a vertical cleaning motion in a narrow corridor. It can also be seen that no collision occurs with the left or right walls.

In order to demonstrate the potential interest of architecture C, a CAD model was also built, showing a possible implementation (Figure 4.107).

An interesting advantage of architecture C over architecture B can be seen in Figure 4.107(a) and Figure 4.108: the first segment can be made horizontal so that the tool can reach all the periphery of the mobile base, even in tight corridors.

Two simulations of path planning for a straight line can be seen in Figure 4.109. A vertical motion on a wall can be seen in Figure 4.109(a). When the end-effector is ascending, the first segment is progressively lifted. Then, a singularity is avoided by deliberately bringing the first segment downwards so that the two remaining segments can find their natural position. In a second time, a horizontal motion on a ceiling at 3 m can be seen in Figure 4.109(b). Thanks to the lateral shifting of the arm, no collision can occur between the arm and the central stabilizer.

Figure 4.109(b) also shows a possible issue during the arm control: two possible configurations of the wrist joints J5 and J6 allow pointing the tool axis towards the ceiling. The first one, represented in Figure 4.107(b), brings the tool actuator towards the right side of the picture. This wrist configuration was not possible in Figure 4.109(b) because of a collision risk between the tool actuator and the terminal wall of the corridor for the right part of the trajectory. Collision was avoided by wrist reversal to the second possible configuration, which points the tool actuator towards the left of the figure. Passing from the first to the second configuration is achieved by adding a rotation angle of 180° to joints J5 and J6. The robot should be always capable to switch from one configuration to the other, according to collision risks of the tool actuator with the environment. This collision avoidance should be considered for control.

As a conclusion on architecture C, it has four advantages over the two others:

- Three shorter segments instead of two long segments avoid any collision with the walls in narrow corridors
- Its first segment can be oriented horizontally and allows the tool to reach the complete periphery of the mobile base
- Its stability at reference poses seems to be the best of the three architectures
- Its redundant additional mobility can be used to keep the arm far from singularities

4.6.2.6 Force analysis of the actuators

Model A1

The first arm model is shown in Figure 4.110, where **J1** and **J2** are two passive rotational joints. The arm is similar to the architecture A presented in section 3.3. The joints could be actuated either by a rotational or translational actuator. In this model A1 was considered the second case, using for instance two screw jacks **P1** and **P2**. This architecture is inspired from the digging machines or earth movers which provide high stiffness and big load. The mass of the arm is 80 kg, based on the project specifications divided on two links of 0.95m each. The arm is simulated in the planar case while cleaning the wall and then cleaning the ceiling. The interaction forces with the wall along x and y directions are defined as **Rx** = -100 N and **Ry**= -100 N respectively while in the ceiling case only **Ry** = -200 N is considered (100 N of normal force + 100 N of tool mass).

Case 1: Wall cleaning

The arm is simulated during tangential wall cleaning from top to bottom with a speed of 3 m/min which is realistic for such kind of work and enough to keep the

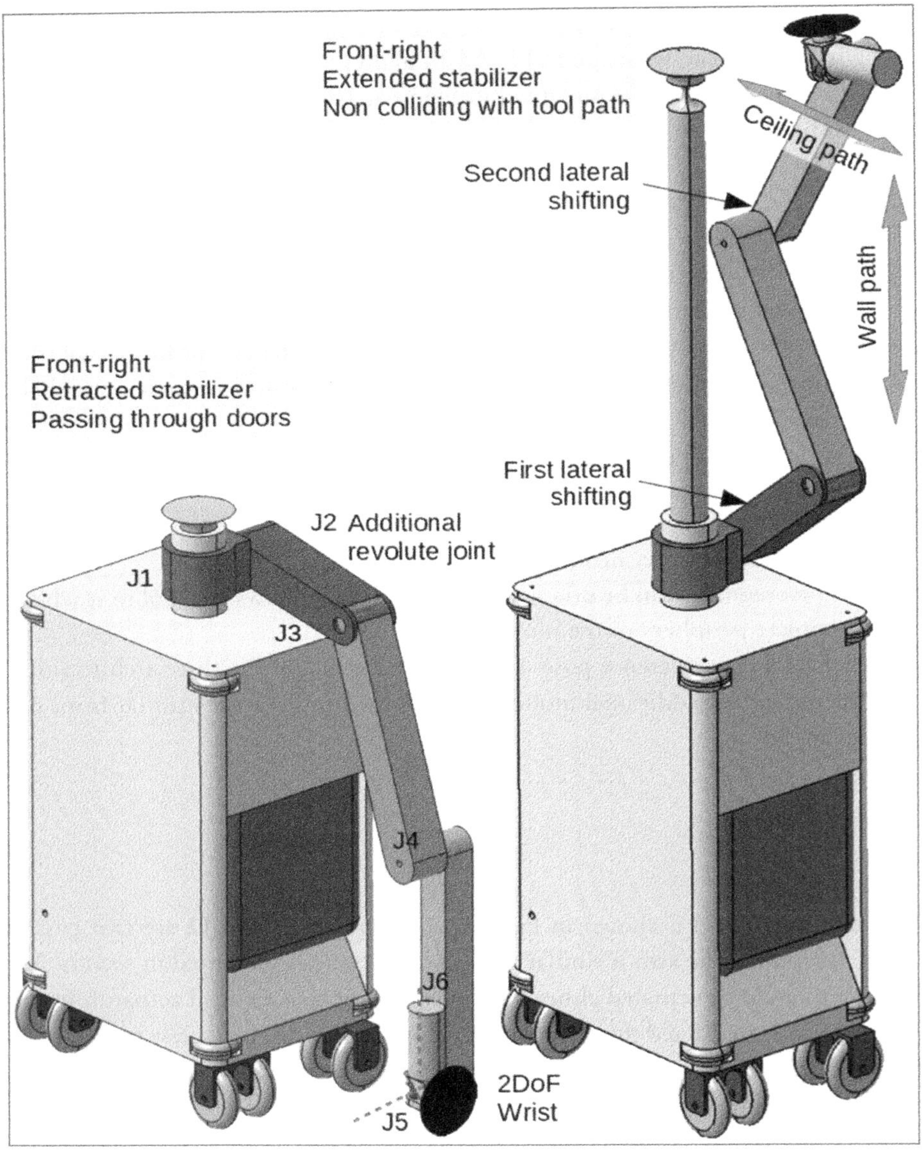

Figure 4.69. CAD model and possible implementation for an arm with architecture C. (a) Tool cleaning the lower part of a wall. (b) Tool cleaning the ceiling above the mobile platform.

stability of the platform to avoid falling down because of dynamics (quasi-static model). The results of the simulation are analysed and shown in the figures below. Figure 4.111 shows that the maximal lengths needed for **P1** and **P2** prismatic joints are around 0.9m. Figures 4.112 and 4.113 show that the maximal torques needed on the rotational joints **J1** and **J2** are 543 and 259 Nm respectively.

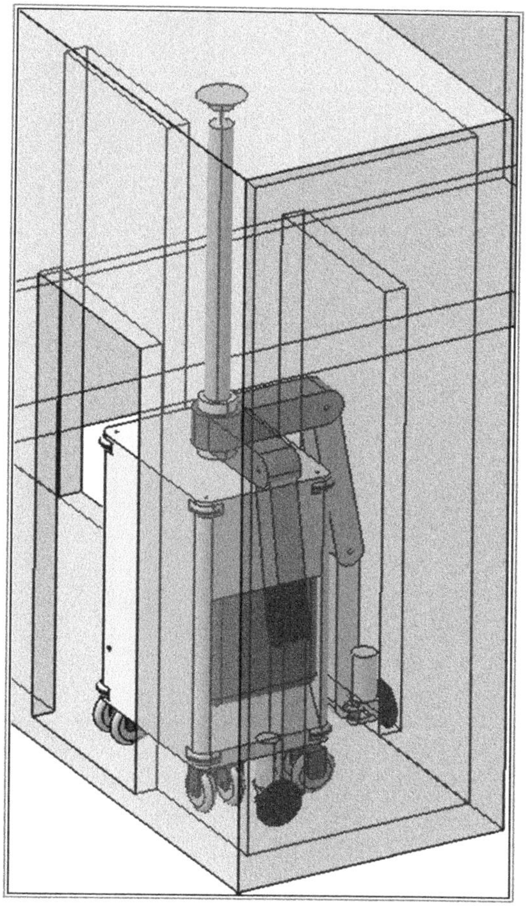

Figure 4.70. The first segment can be made horizontal so that the tool can reach all the periphery of the mobile base, even in tight corridors.

The maximal forces needed on the translational actuators **P1** and **P2** are 11606 N and 819 N respectively as shown in Figures 4.76 and 4.77. Hence, the smaller the angle b between the force direction and the radial direction from the joint, the maximum the torque, based on what Figure 4.78 shows. Thus, working near a fully extended arm leads to the high actuator forces shown in the graphs. These singular positions should be absolutely avoided as the actuation forces tend towards infinite in this case.

Case 2: Ceiling cleaning

The robot arm moves tangentially to the ceiling with a speed of 3m/min. The results of the simulation are analysed and shown in the figures below. Figures 4.79 and 4.80 show that the maximal torques needed on the rotational joints **J1** and **J2**

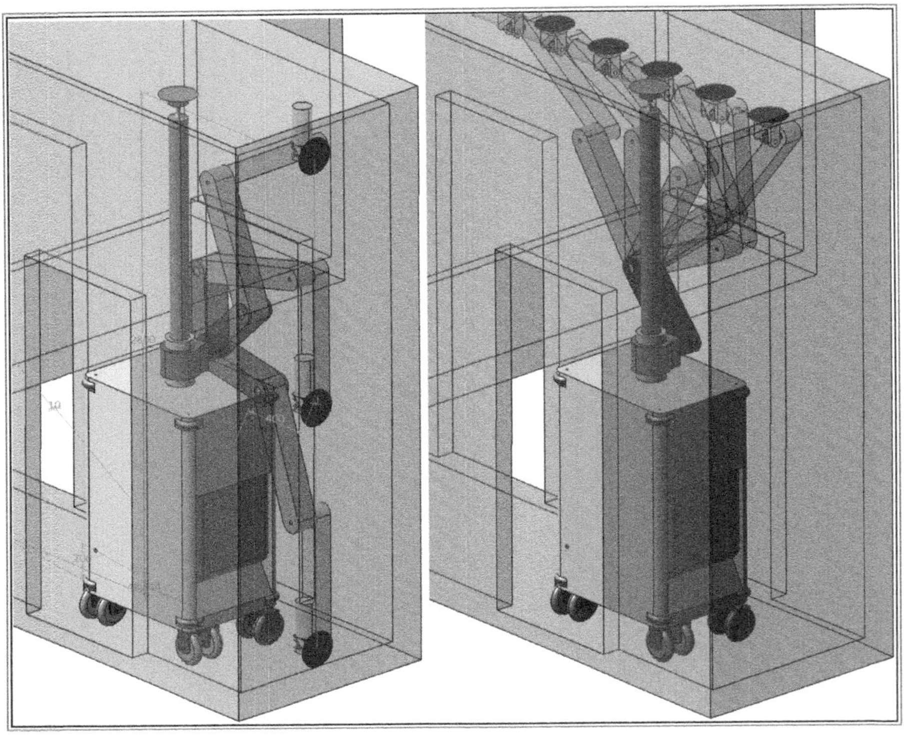

Figure 4.71. Path planning of a straight line. (a) Vertical line on a wall with singularity avoidance thanks to J3 reversal. (b) Horizontal line on the ceiling with no collision arm-stabilizer thanks to the lateral shifting of the arm. Collision avoidance between the tool actuator and the terminal wall required switching to a second wrist configuration with respect to Figure 4.107(b).

are 550 and 380 Nm respectively. The maximal forces needed on the translational actuators **P1** and **P2** are 1980 N and 1545 N respectively as shown in Figure 4.81 and 4.82.

Summary about model A1:

The model A1 is a special implementation of architecture A with translational actuators. The interest of using, for instance, electric screw jacks as actuators would be their non-reversible behaviour and compatibility with the electric power supply chosen for the robotic unit. The drawback of this implementation is that it has a limited workspace and that the actuation forces tend to become very high at full arm extension.

Model A2

A second actuation model as shown in Figure 4.83 is studied to reduce the limitations of the first model. It is still based on architecture A with two revolute joints

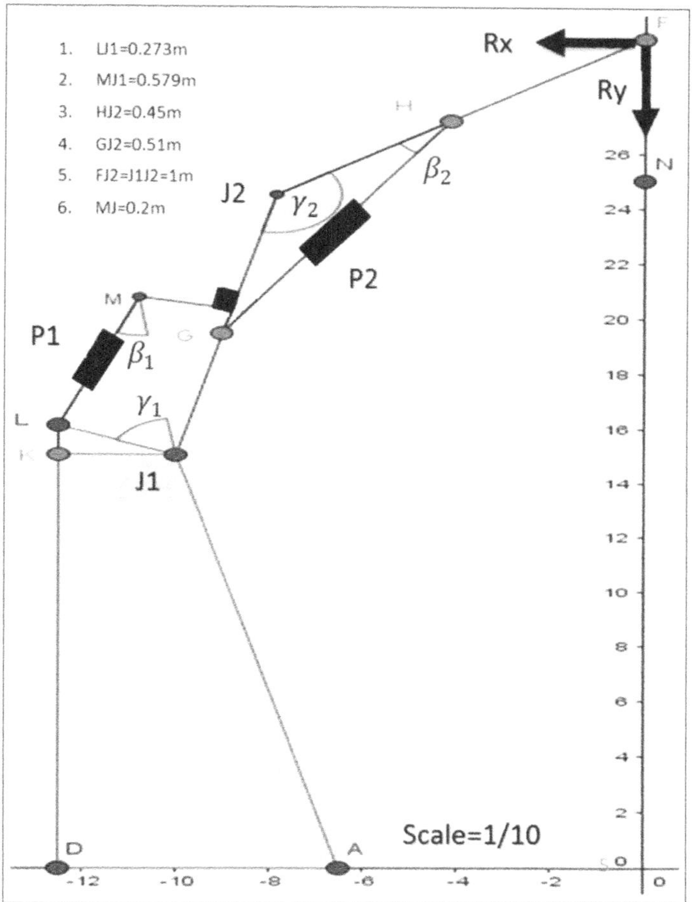

Figure 4.72. model A1 for force analysis, based on architecture A motorized by two translational actuators P1 and P2.

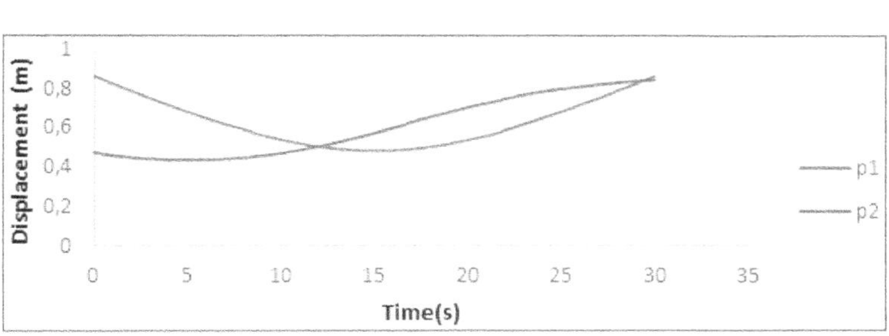

Figure 4.73. Evolution of translational actuators P1 & P2 positions for model A1.

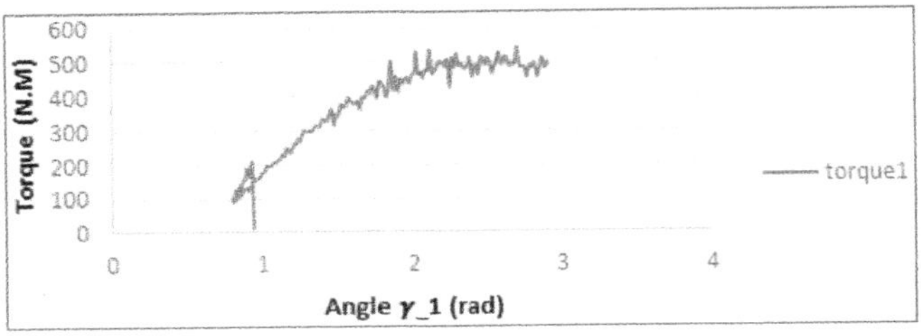

Figure 4.74. Torque on joint J1 vs. angle g_1 for model A1.

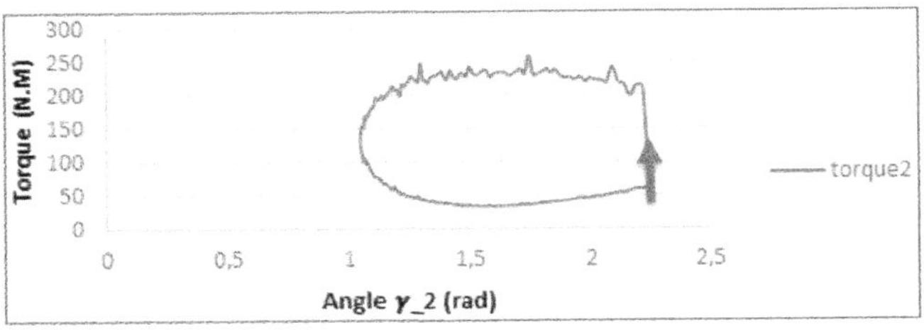

Figure 4.75. Torque on joint J2 vs. angle g_2 for model A1.

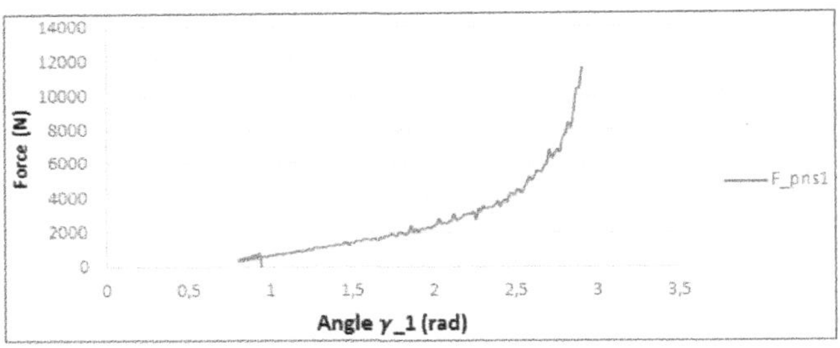

Figure 4.76. Force applied by translational actuator P1 vs. angle g_1 for model A1.

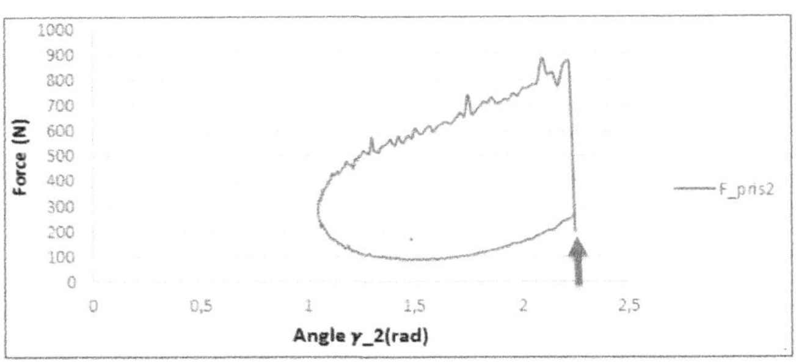

Figure 4.77. Force applied by translational actuator P2 vs. angle g_2 for model A1.

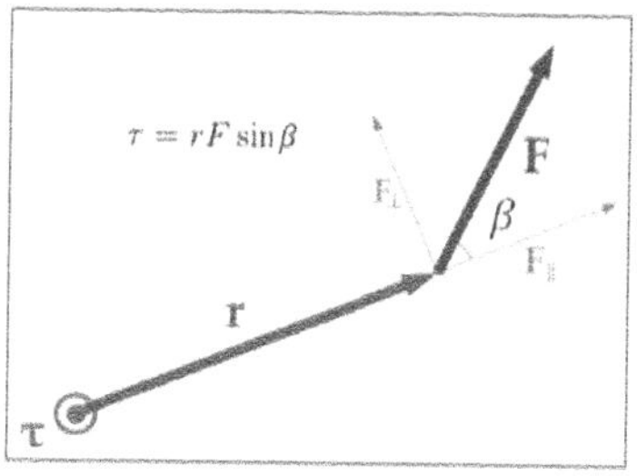

Figure 4.78. Relation between torque t and force F.

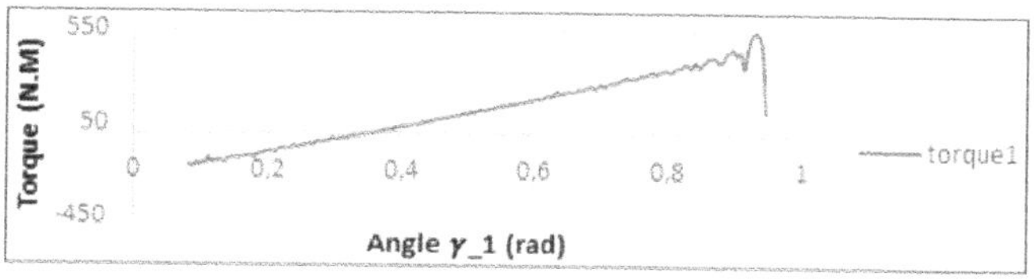

Figure 4.79. Torque on J1 vs. g₁ for model A1.

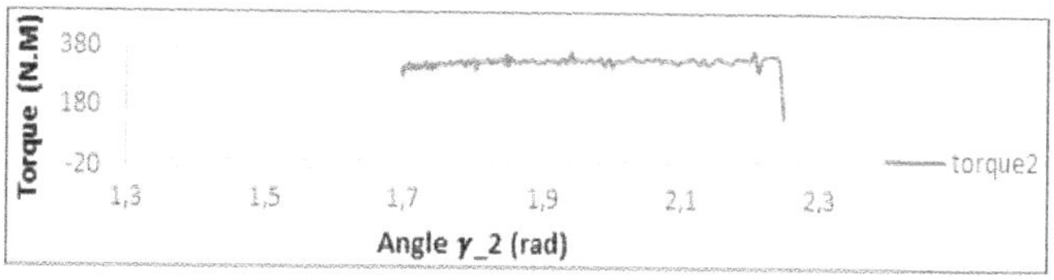

Figure 4.80. Torque on J2 vs. g_2 for model A1.

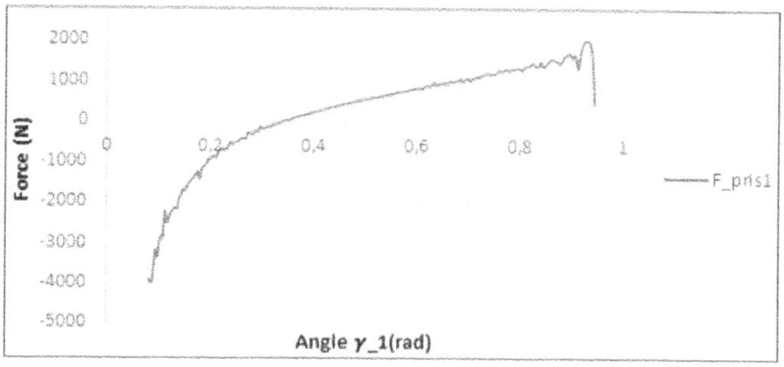

Figure 4.81. Actuator force of P1 vs. angle g_1 for model A1.

J1 and J2. This time, J1 is actuated by a rotational actuator whereas J2 is controlled by an indirect linkage connecting the fore-arm to the base. Joint **J2**, **J3** and **J4** are passive rotational joints, whereas **P1** is a translational actuator. The interest of this structure is that all the actuators are supported by the base, which could bring to lighter links and a higher dynamics. This model is similar to the well-known 5-bar mechanism which is famous for its stiffness and high acceleration performance,

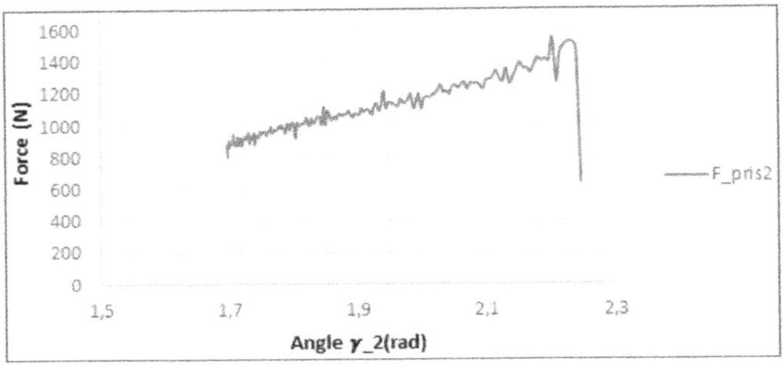

Figure 4.82. Actuator force of P2 vs. angle g$_1$ for model A1.

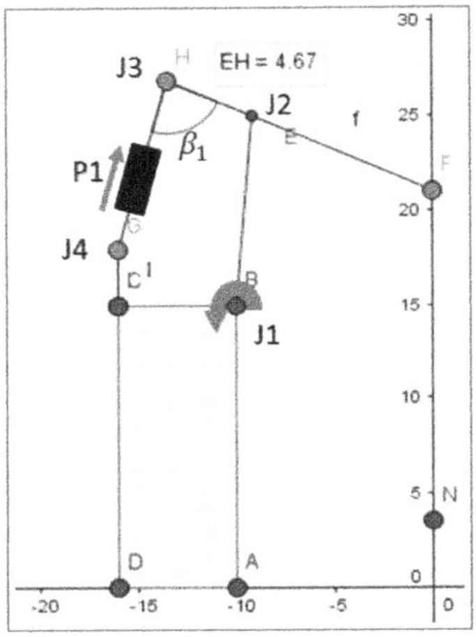

Figure 4.83. Model A2 for force analysis, based on architecture A motorized by a rotational actuator J1 and a translational actuator P1. The overall setting has a parallel architecture, suitable for high dynamics.

typical of parallel architectures. The arm weighs 80 kg and is simulated for wall and ceiling cleaning as previously for model A1, with the same interaction forces.

The conclusion about the simulation results showed that, in order to reach point **A** (bottom right of the robot in Figure 4.83) the required length of the prismatic joint **P1** is 1.7 m. Taking into account the torques estimated before on **J1** and **J2** (Figures 4.76 and 4.77), if **J1** and **P1** are actuated only, then the maximal force to be applied by **P1** will be roughly 2858 N.

Summary about model A2:

The forces and torques needed to actuate the arm for the asbestos removal task are feasible although they are high. But the main drawback of this arm is its big size for our specifications and the singularity when J4, J3 and J2 are collinear which limits operation on the ceiling.

Summary about force analysis of actuators:

Overall, the recommended solution for actuation is to keep rotational actuators on all the joints and not use translational actuators, as it is less limiting in terms of workspace and as forces in the jacks may become very high at full arm extension. Electric screw jacks could have the interest of non-reversibility and high stiffness, but cannot be used because of the limitations shown above. The designers should be careful to find a rotational actuator that can be locked or braked for better stiffness.

4.6.2.7 Workspace of the manipulator

This section investigates the workspace for architecture A. The intersection between the arm workspace and the wall or ceiling is shown in Figure 4.84 and represents the attainable area in the wall/ceiling for a fixed position of the arm base. The workspace was constructed by discretizing the joint space and using the direct geometric model (DGM) of a 3R arm with architecture A and two links of 0.95 m as discussed before. The inputs of the DGM are the joint interval values. Iterating the joint value between the maximum and minimum joint limits gives a 3D mesh, that can be intersected with the wall and ceiling to obtain two 2D meshes shown in Figure 4.84.

The workspace, and consequently the arm, must be dimensioned to reach the top and bottom of the wall, the ceiling and the furthest points discussed before. For the walls, the usable workspace is the biggest rectangle included into the 2D projected workspace on the wall (Figure 4.84(a)) with a Z height covering the complete wall height (3 m). Its width LY is the maximal length that can fit in the 2D projected workspace. This allows minimizing the displacements of the mobile platform in Y, as the idea is to cover the wall surface by jointly tiling several rectangles separated by mobile platform displacements of length LY.

The approach is a bit different for the ceiling, where there is no limitation to 3m for one side of the rectangle. The two dimensions LX and LY of the rectangle included into the 2D projected workspace on the ceiling (Figure 4.84(b)) result from a compromise and should be optimized, depending on the motion capacities of the mobile platform.

4.6.2.8 Preliminary dynamic model of the robotic unit

A parametrized 3-D model of the robotic unit based on arm with architecture A was built with ADAMS to study the effect of dynamics during a vertical trajectory in the plane of wall. This model will be adjusted to the chosen arm architecture in the future. Thus, it should be considered as preliminary and useful for demonstrating the model possibilities. The masses of the components of the robotic unit were adjusted according to the latest assumptions.

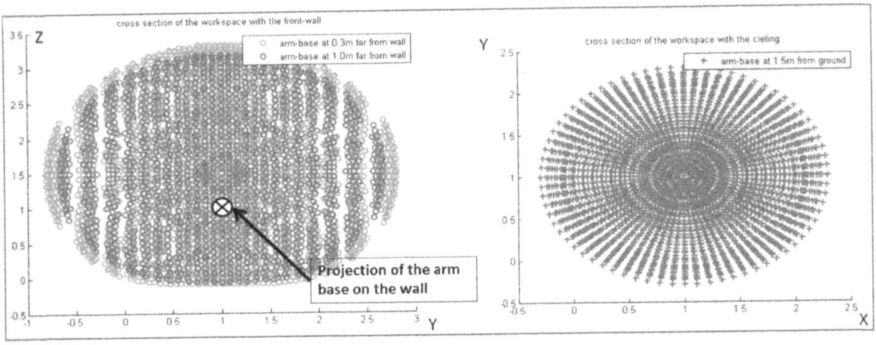

Figure 4.84. Arm workspace of architecture A.

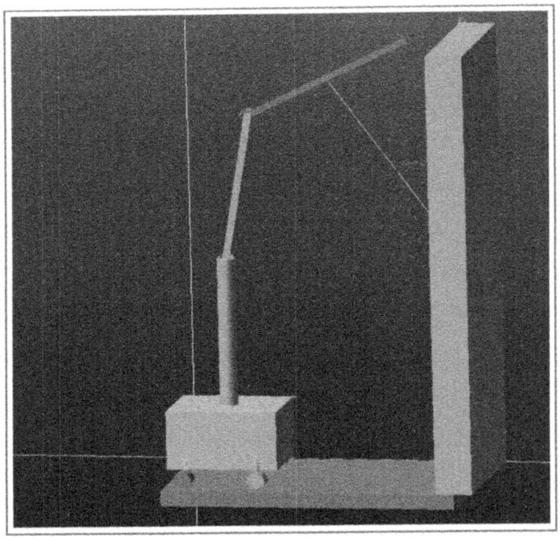

Figure 4.85. ADAMS model of the robotic unit with arm architecture A.

The inverse geometric model of the 2R robot was programmed in Matlab to get the joint angles corresponding to a vertical trajectory. These values were then imported to the ADAMS model as a spline to generate the trajectory.

A contact between link 2 (red) and the wall was created. Each of the four wheels was also constrained to move in the plane of the floor by creating contacts. This model showed a stable behaviour throughout a vertical trajectory.

This preliminary dynamic analysis shows a stable behaviour of the robotic unit for the given speed of the end-effector. However, this model must be further developed by using the chosen arm architecture and introducing realistic external tool forces in order to have a better understanding of the dynamics.

4.6.2.9 Productivity of the robotic unit

It is important to evaluate the productivity of the robotic unit and to bring it as high as possible with respect to the productivity of a human operator.

The current productivity of a human worker has been measured around 6.4 square meters/hour (source: Bouygues). The approximate productivity of a robot with a tool of 150 mm diameter moving at 3 m/min is $0.15mx3m/minx60min/h = 27m^2/h$. However, the productivity is related to the tool size and its translational speed. Even at a speed of 3 m/min, which is relatively low for machining, the expected productivity of a single robot could be four times the productivity of a human operator, provided the cleaning can be made in a single pass. But this has to be confirmed by removal tests. The path planning strategy has also to be defined. This important figure of productivity should be assessed as precisely as possible all along the project.

References

[Geogebra, 2016] Geogebra, the graphing calculator for functions, geometry, algebra, calculus, statistics and 3D maths. https://www.geogebra.org. Retrieved online in 2016.

[McGhee and Frank, 1968] R. B. McGhee and A. Frank, "On the stability properties of quadruped creeping gaits," Mathematical Biosciences, vol. 3, pp. 331–351, 1968.

[Prause *et al.*, 2015] Isabel Prause, Jean-Christophe Fauroux, Mathias Hüsing and Burkhard Corves. "Using Geometry Sketchers and CAD Tools for Mechanism Synthesis", paper OS3-032, in Proc. of IFToMM 2015, The 14th World Congress in Mechanism and Machine Science, October 25–30, 2015, Taipei International Convention Center, Taiwan, 11p.

Dynamic Modelling of the Robotic Unit for Stability Evaluation

Process of asbestos removal involves dynamic interaction between cleaning environment and robotic unit. This interaction occurs at two levels. Firstly, grinding tool mounted on the robotic arm interacts with the cleaning surface to generate reaction forces. Secondly, wheels of the mobile base are subjected to ground reaction forces due to:

- weight of the robotic unit
- inertia forces generated due to the motion of robotic arm.

In Figure 1 a typical interaction model existing within the process of asbestos removal is detailed. Reaction forces generated by tool-surface interaction (ground,

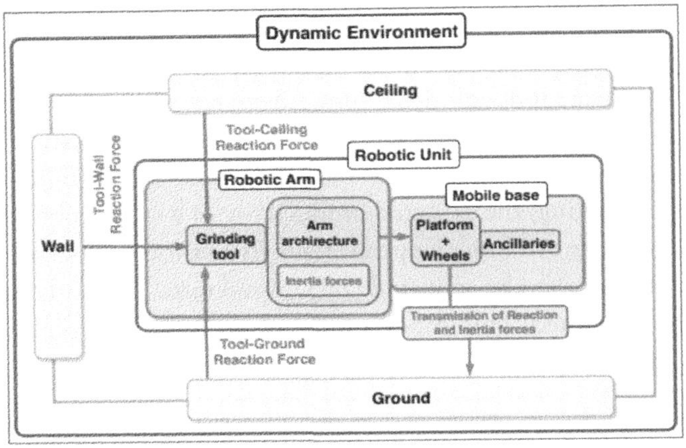

Figure 4.86. Interaction model of asbestos removal use case.

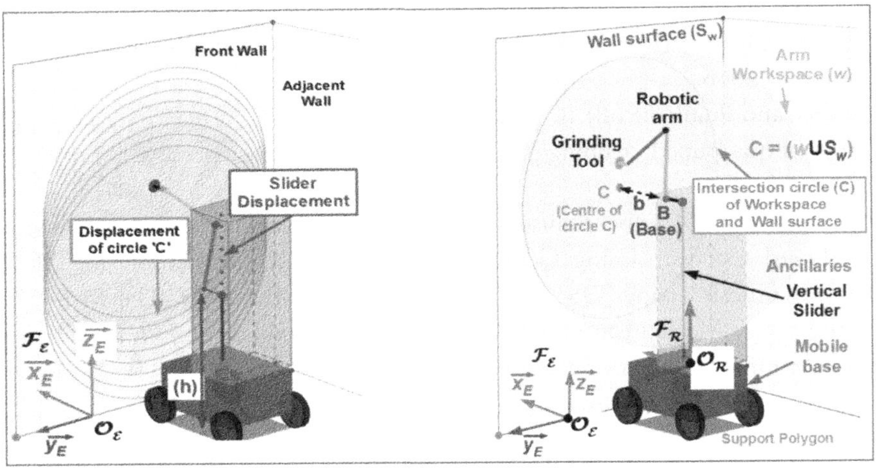

Figure 4.87. Workspace (left) and GeoGebra representation (right).

wall and ceiling) are transmitted to the arm architecture through tool-arm connection. Since, arm performs accelerated motions, varying inertial forces are generated in addition to grinding reaction forces. Connection of arm to the mobile base transmits these forces to the mobile base which are further passed on to the ground surface through wheels. Due to these transmissions, stability of the robotic unit is significantly affected. Therefore, in order to study effects of arm motion and grinding reaction forces on the stability of robotic unit, a dynamic simulation model capable of simulating realistic asbestos removal scenario should be constructed.

Representation of asbestos removal use case

Figure 2 shows robotic unit placed at a distance 'b' from wall surface (Sw). At this distance, the 3D workspace (w) of the robotic arm intersects with surface (Sw) in circle C. The area of C is the area available for cleaning at a given base placement such that collision free (arm-cleaning environment) continuous trajectories are feasible. However, since the robotic arm is mounted on a vertical slider (figure 3), circle C can be translated along a vertical axis to a resulting workspace with a geometric shape called 'stadium' (rectangle with semicircles on either of the opposite sides). Thus, even at fixed base position, workspace of the robotic arm is extended due to presence of architectural redundancy (P-joint).

Stability Evaluation

Two methods of stability evaluation are adapted during this study.

a) Numerical Method

Cartesian trajectory input: Targeted operational surfaces in construction applications can be 3-dimensional in nature (curved walls, ceiling or ground). However, in this work, focus is limited to non-curved surfaces. Thus, 2-D Cartesian trajectories are defined while performing simulations. Cartesian pose of the end-effector frame corresponding to trajectory coordinates serves as an input to the algorithm.

IKS: Inverse kinematics solution is needed to generate arm motion. Since arm is kinematically redundant, IKS is determined using redundancy resolution method that uses pseudo-inverse of the kinematic Jacobian matrix.

Recursive computation of joint velocities: Angular velocities and accelerations of moving links contribute in generating dynamic effects on the robotic unit. Newton-Euler recursive algorithm is used to compute these entities.

b) Stability Evaluation: Zero moment point for the instantaneous dynamic state of the robotic unit is computed using equations 3.2a and 3.2b. Stability is then evaluated as per equation 3.3

Method of co-simulation

In this section, a procedure for constructing ADAMS-MATLAB co-simulation is described in detail (Fig. above).

1. Parameterized multibody ADAMS model of the robotic unit
   (a) Direct geometric model (DGM) of the robotic arm is found using SYMORO. DGM provides pose of the each link frame in Cartesian space coordinates

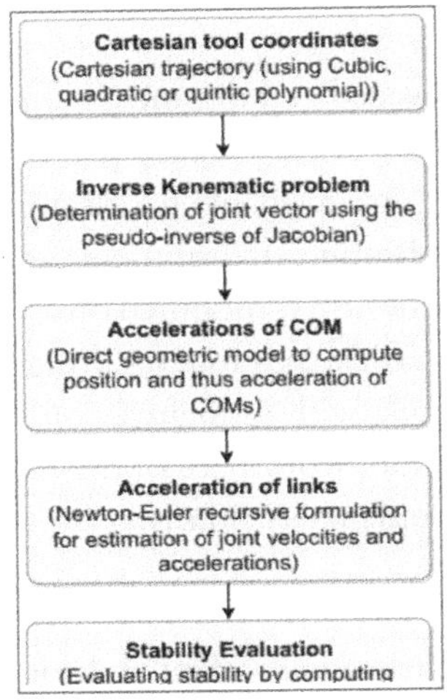

Figure 4.88. Numerical evaluation of stability.

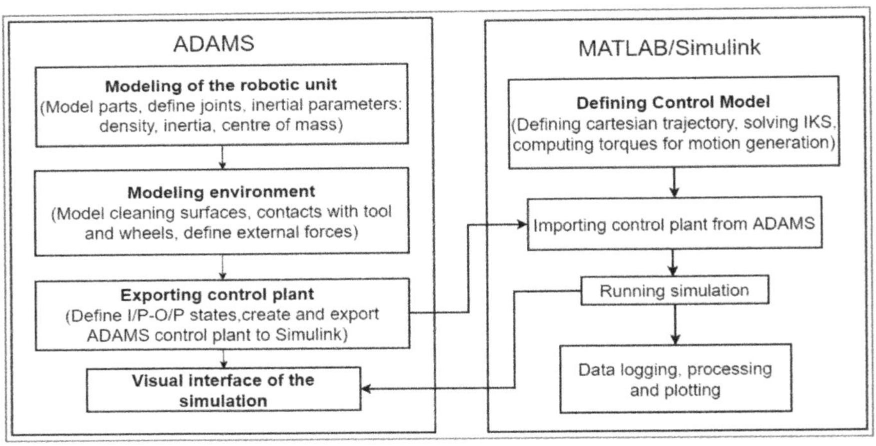

Figure 4.89. Flow-chart of ADAMS-Matlab co-simulation.

   (b) Using these co-ordinates, construction points are created in ADAMS environment
   (c) Individual links are then built with the help of these construction points

2. Input-Output control plant
   (a) State variables are assigned to each joint angle as well as C.o.M of each link to track position, velocity and acceleration. These are output variables of the control scheme.
   (b) Input variables consist of torques given to the joints of the robotic arm
   (c) ADAMS control plant is created, then exported to Matlab-Simulink environment co-simulation purpose.

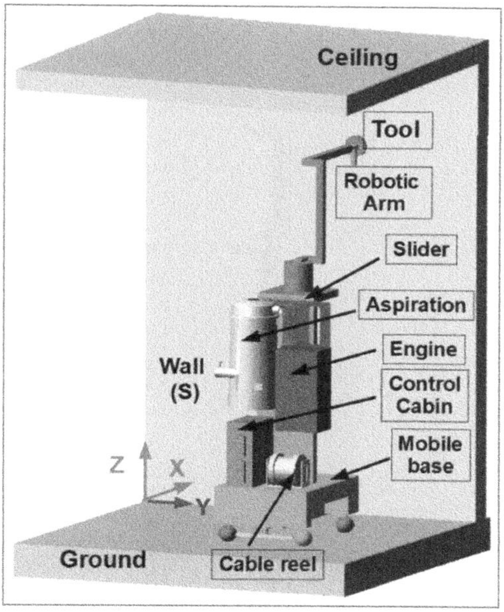

Figure 4.90. Multibody dynamic model of the Robotic unit - V1.

3. Control Scheme in Simulink environment
   (a) PID controllers are implemented at each joint angle to compute required input torques for desired motion
   (b) Gravity compensation torques are added to PID generated torques to account for the effect of gravity
   (c) Output coming from ADAMS i.e. joint velocities and accelerations are used for PID feedback
4. Estimation of ZMP
   (a) All the necessary components to calculate ZMP are obtained in run-time through ADAMS output whereas angular moments can be calculated from available information
   (b) ZMP criteria is then implemented to compute stability at each pose while performing desired motion

Results of stability evaluation

The results of the stability analysis by this approach is shown in the following section.

Acceleration in +ZE, h = 1.675 m: Here, tangential reaction forces act along -ZE. Moment responsible for longitudinal stability decreases from top to bottom

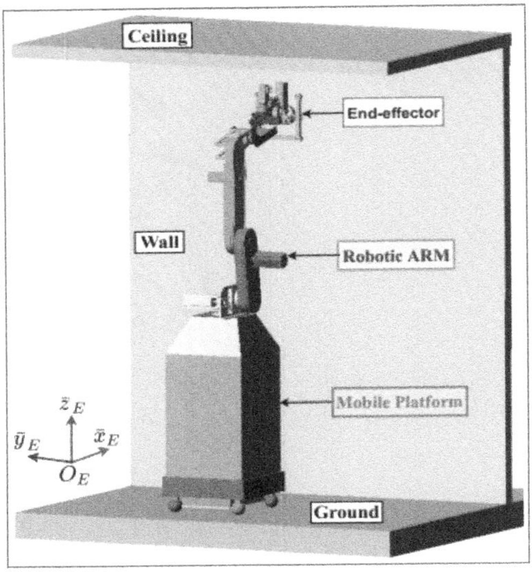

Figure 4.91. Multibody dynamic model of the Robotic unit - V2.

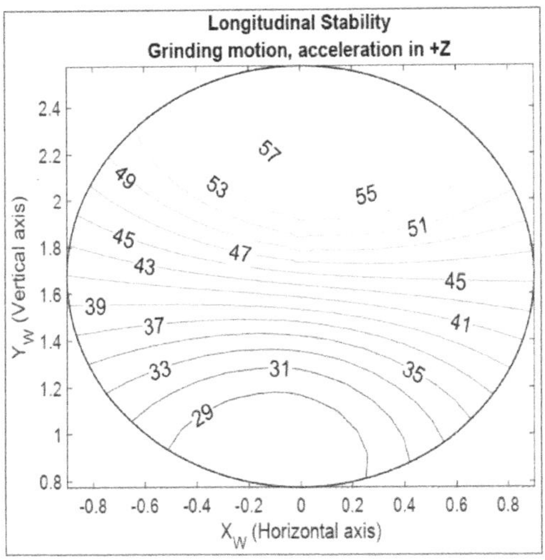

Figure 4.92. V1-Longitudinal stability contour for acceleration in +ZE.

due to decreasing distance along ZE. This is reflected in the longitudinal stability maps shown in figure 9.

Acceleration in −ZE, h = 1.675m: Here, tangential reaction forces act along −ZE and moment now pushes ZMP along −XE. Thus, longitudinal stability at respective points for −ZE motion is greater than that for +ZE motion (Fig. 10).

Acceleration in +YE, h = 1.675m: Tangential forces act along -YE pushing ZMP towards lateral edge of the polygon. Area on the right of red line shown in figure

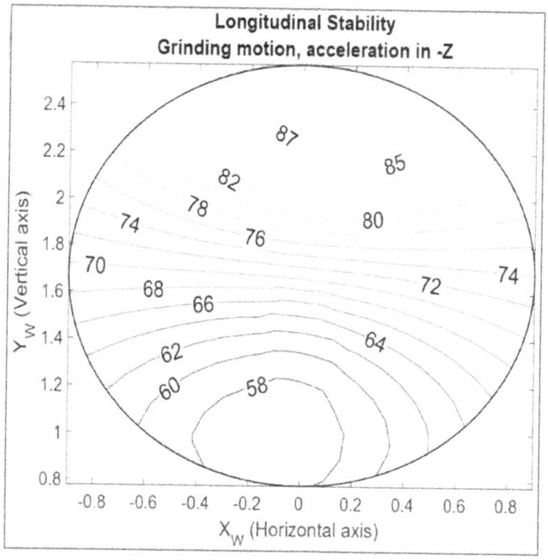

Figure 4.93. V1-Longitudinal stability contour or acceleration in −ZE.

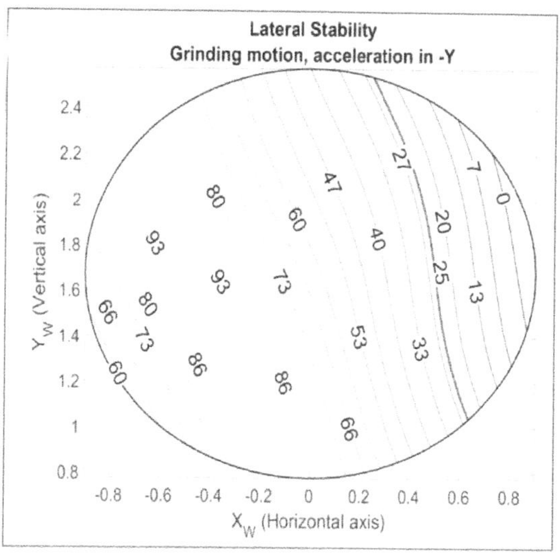

Figure 4.94. V1-Lateral stability contour for acceleration in −YE.

10 is stable. However, area on the left side of the line has stability is below 25% as shown in figure 11.

Acceleration in -YE, h = 1.675m: Tangential forces act along +YE direction creating moment that tends to move ZMP towards other lateral edge of the support polygon. In general, the following important conclusions can be made from stability maps:

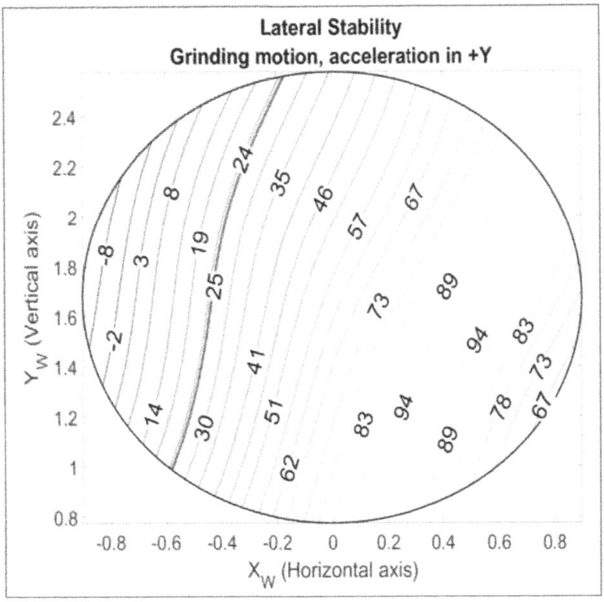

Figure 4.95. V1-Lateral stability contour contour for acceleration in +YE.

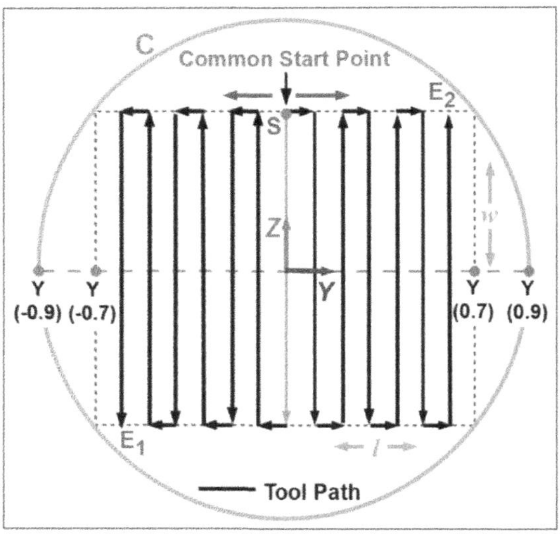

Figure 4.96. Path traced by tool on the wall.

- Longitudinal stability is always above critical limit of 25% for motion along Z direction.
- Lateral stability is affected based on the positive or negative direction of the cleaning along Y.

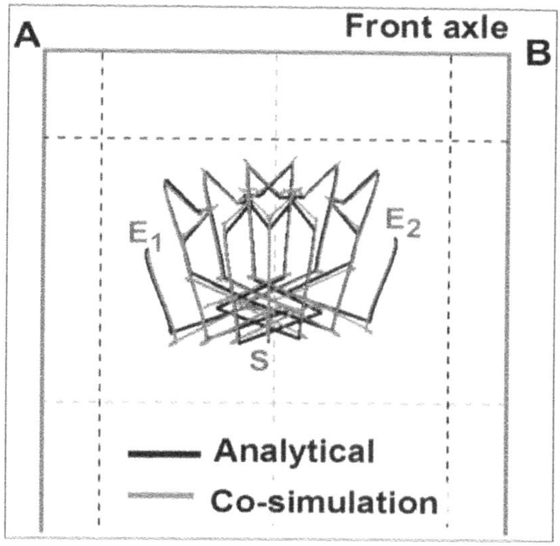

Figure 4.97. ZMP estimation through numerical and co-simulation.

Stability based trajectories

Asbestos removal grinding trajectories are illustrated based on the conclusions of stability analysis (Fig. 13). Since, most cleaning walls are rectangular, we assume a rectangle of dimensions (1.38 × 1.13) m inside C. A boustrophedon path is tracked separately from point S to E1 and S to E2 instead of one single path from E1 to E2. The tool motion along -Y while cleaning inside Y< 0 zone and along +Y while cleaning inside Y> 0 zone results in a stable manipulator operation. Fig. 14 shows variation of ZMP obtained by MATLAB simulation and co-simulation during the two grinding trajectories. It can be safely concluded that analytical model can be trusted for simulating grinding trajectories in future.

Conclusions of stability-based computations

Important conclusions of the work are listed below:

- Dynamic interaction model of the asbestos removal scenario was studied
- Dynamic simulation of the use case was created using two methods namely-numerical and analytical.
- Dynamic stability of the robotic unit was studied for identifying reaction wrench during cleaning scenario
- Stable directions for cleaning, starting from the centre of the workspace and moving laterally towards the outside were identified, and a global boustrophedon cleaning trajectory in two parts was proposed to maximize stability.

References

Cooperation

[1] Zine Elabidine CHEBAB, "Conception et Commande Collaborative de Manipulateurs Mobiles Modulaires (C3M3)", PhD dissertation. December 08th, 2019

[2] Zine Elabidine CHEBAB, Jean-Christophe FAUROUX, Grigore GOGU, Laurent SABOURIN, Nicolas BOUTON and Youcef MEZOUAR. "Structural Analysis of Mobile Manipulators", Session 3-Robotics, in Proc of 15th IFToMM World Congress, June 30-July 4, 2019. Krakow, Poland. 10 p.

[3] Siddharth MARAJE, Jean-Christophe Fauroux FAUROUX, Chedli-Belhassen BOUZGARROU and Lounis ADOUANE. "Stability analysis of an asbestos removal mobile manipulator for safe grinding trajectories", Session 3-Robotics, in Proc of 15th IFToMM World Congress, June 30-July 4, 2019. Krakow, Poland. 10 p.

Grinding

[1] QUENTIN, Trebot, PHILIPPE, Vaslin, JEAN-CHRISTOPHE, Fauroux, et al. Robotized grinding experiments of construction materials for asbestos removal operation. In : IFToMM World Congress on Mechanism and Machine Science. Springer, Cham, 2019. p. 2621-2630.

[2] https://www.numatic.fr/project/hzd900/

References

[Geogebra, 2016] Geogebra, the graphing calculator for functions, geometry, algebra, calculus, statistics and 3D maths. https://www.geogebra.org. Retrieved online in 2016.

[McGhee and Frank, 1968] R. B. McGhee and A. Frank, "On the stability properties of quadruped creeping gaits," Mathematical Biosciences, vol. 3, pp. 331–351, 1968.

[Prause *et al.*, 2015] Isabel Prause, Jean-Christophe Fauroux, Mathias Hüsing and Burkhard Corves. "Using Geometry Sketchers and CAD Tools for Mechanism Synthesis", paper OS3-032, in Proc. of IFToMM 2015, The 14th World Congress in Mechanism and Machine Science, October 25–30, 2015, Taipei International Convention Center, Taiwan, 11p.

Deliverable 2.2
D2.2 Simulation environment and model of the representative asbestos removal use-case with achievable precision

Authors and contributors:
MARAJE Siddharth, SIGMA Clermont
FAUROUX Jean-Christophe, SIGMA Clermont

4.6.3 Deliverable in the Context of the Work Program

Deliverable **D2.2**, simulation environment and model of the representative asbestos removal use-case with achievable precision is a part of task **T 2.3** – Virtual scenario of a representative asbestos removal use case. In deliverable **D2.1**, three arm architectures were analyzed for evaluating static stability. The second of the three solutions – 6 DoF redundant solution with a 5 DoF arm mounted on a vertical slider was selected for version 1 prototype while the third one – a 6 DoF redundant solution with a 6 DoF arm having three shorter moving segments was kept for consideration for version 2 prototype. After assuming a 3 DoF wrist for each of the architectures, the second one becomes a redundant 7 DoF (***P-6R***) architecture while the third one becomes redundant ***7-R*** architecture. In this deliverable a dynamic simulation environment was built in order to create a platform for overall system layout and analysis of the two robotic units resulting from the integration of the two redundant arm architectures, Version 1 – robotic unit integrated with P-6R arm and version 2 – robotic unit integrated with 7R arm. Preliminary investigation of the two robotic units is carried out to test the multibody dynamic models as well as preliminarily investigate static and dynamic stability. The objectives of this deliverable are:

[1] **Preliminary investigation of robotic units**: This investigation involves simulating simple cleaning trajectories e.g vertical or horizontal cleaning motions to test the models as well as to preliminarily investigate the dynamic stability by recording ground reaction forces.

[2] **Characterization of Workspace**: Evaluation of the intersection of the 3D workspace of the robotic arm on the cleaning surfaces is necessary to anticipate the movement of the mobile platform. Thus, the 3D workspaces are visualized to ensure the accessibility to the cleaning surfaces while performing simulation.

This task takes input from tasks 1.1 (use-case specifications), 3.2 (technical requirements), 1.4 (operation process parameters), 2.1 (results from preliminary simulation of the robotic unit) and 2.2 (preliminary proposition on mechanical stabilization).

4.6.4 Methodology

This section explains a methodology adapted for building and testing a Simulation environment consisting of realistic dynamic model of the robotic unit and the cleaning environment.

[1] **Multibody Dynamic Simulation (ADAMS Software)**:

a. **Modelling of the Robotic unit:** A parameterized multibody dynamic model of the two robotic units will be built. The models will serve for simulation of trajectories and estimation of stability. The typical characteristics of this model are:

[2] Dimensions of the mobile manipulator:

a. Realistic dimensions of the mobile platform and stabilizers
b. Parameterization of the robotic arm based on the DH-model providing ease of accommodating changes in link lengths, joint offsets and wrist design.
c. Pose of the arm base mounting on the mobile platform

[3] Pose of the mobile manipulator within environment:

a. Distance of the mobile platform from the cleaning wall
b. Different types of cleaning environments like large rooms, corridors, rooms with obstacles like bath tub or kitchen sink
c. Different ceilings with variable heights and structural resistance

[4] Contacts of the mobile manipulator with the environment:

a. Tool-surface
b. Wheels-ground
c. Stabilizer-environment(ground)

[5] Trajectory generation: Inverse geometric model of the robotic arm is solved in matlab to get the joint variables. These variables are then imported to ADAMS to feed to each of the joint.

a. **Locomotion into environment** to cover all surfaces inside the flat. A retracted configuration of the arm has to be found for optimal stability during locomotion. The optimal trajectory in constrained environments such as narrow corridors has to be planned.

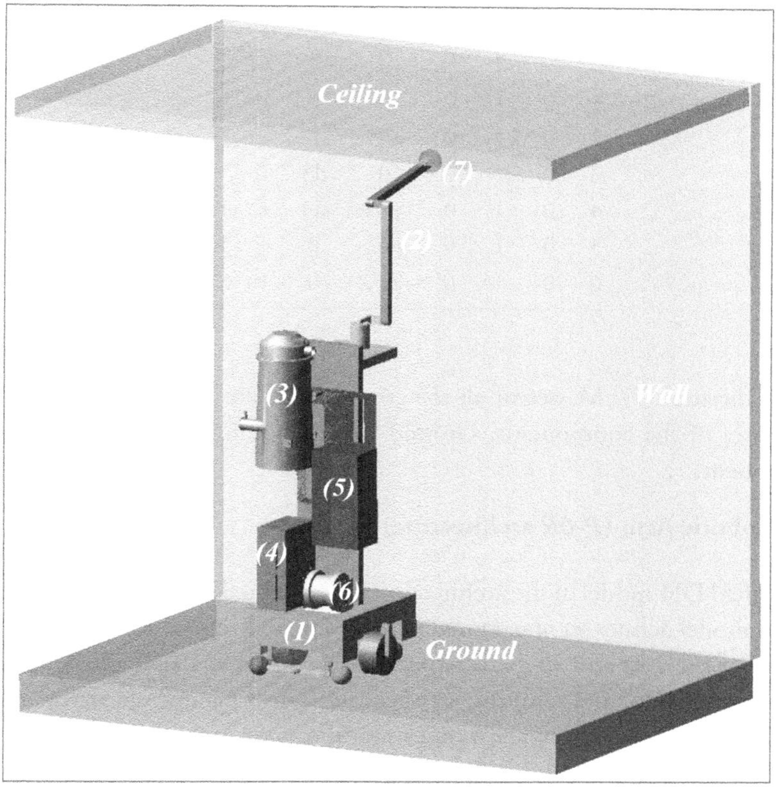

Figure 4.98. ADAMS model of the Robotic unit Version 1.

b. **Simulation of precise tool trajectories for fixed pose of the mobile base:** The attempt is to identify trajectories for cleaning wall, ground and ceiling. Cleaning productivity is directly related to joint accelerations; however, it is limited by process quality. Thus, the simulation environment will be used to identify maximum joint acceleration in future.

4.6.5 Multibody Dynamic simulation

To evaluate the dynamic behavior of the robotic unit inside cleaning environment, ADAMS rigid body model of the two robotic units is built. Some sample cleaning trajectories are generated to assess the dynamic performance of the robotic unit. The description of the model goes below.

[1] **Description of the version 1 ADAMS Model**:

The main components of the robotic unit indicated in figure 1 are: mobile platform (1), robotic arm (2), aspiration shell (3), cabinet (4), engine (5), cable reel

Table 4.3. Modified DH parameters of the *P-6R* architecture.

j	σ	μ	γ	A	d	θ	R
1	0	1	0	0	0	th_1	r1
2	0	1	0	*pi*/2	d2	th_2	r2
3	0	1	0	0	d3	th_3	r3
4	0	1	0	-*pi*/2	d4	th_4	0
5	0	1	0	*pi*/2	0	th_5	0
6	0	1	0	-*pi*/2	0	th_6	0

(6) and the tool (7). Masses of all the components are adjusted according to the real masses of the components. Ground wall and ceiling constitute the cleaning environment.

a. **Robotic Arm (*P-6R* architecture)**

Modified DH model of the architecture is formulated using SYMORO-software [1]. The model defines set of parameters, joint variables and constants, which decide the pose of the arm and its geometry respectively. Table 1 lists all the parameters while figure 2 shows the resulting architecture. These parameters are incorporated into ADAMS model so as to allow quick modifications in the pose as well as dimensions of the lengths to evaluate changes in dynamic performance. Here, parameters d_3 and d_4 determine two link lengths of the arm and values of these parameters are 0.6m each. Parameters r_2 and r_3 take equal values so that the offset between joints th_1 and th_3 along axis Z_3stays zero.

b. **Mobile Platform:**

Model of the platform with dimensions equivalent to the original dimensions is built in ADAMS. Platform supports the mounting of Cabinet, Cable reel, Vertical slider and aspiration unit. Platform has six wheels as shown in the figure below. Three axles are named 1, 2 and 3 respectively from front to rear. The wheel naming respects the W_{as} Convention with a an axle number (1 for front, 2, for middle, 3 for rear) and s a side number (1 for right, 2 for left).

c. **Constraints:**

To model the interaction between the robotic unit and the cleaning environment, contacts are added into the model. These 3D contacts evaluate contact forces based on the inter-penetration of meshed parts. Four wheels of the platform plus the two additional center wheels have contacts with the ground.

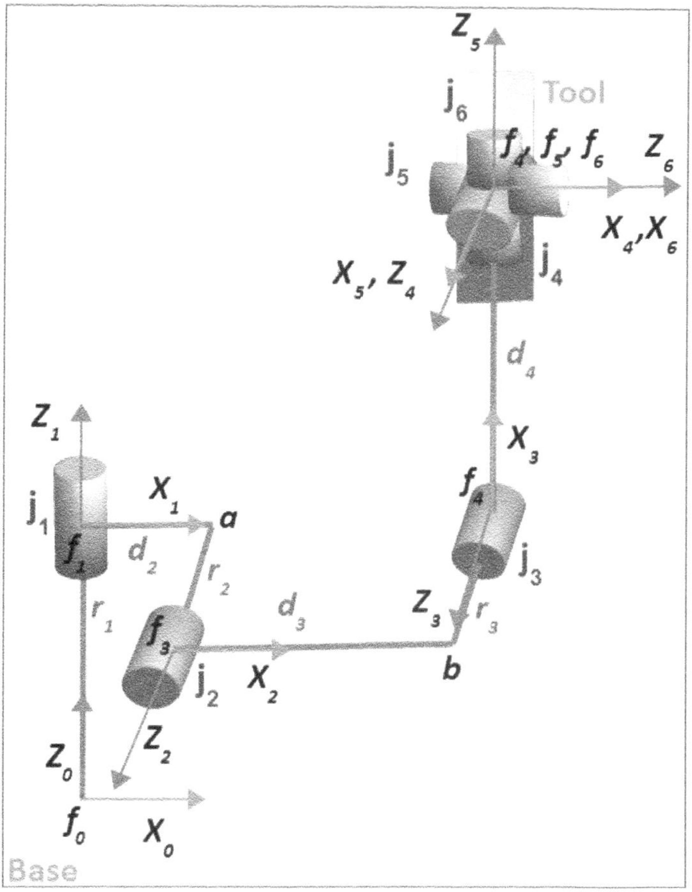

Figure 4.99. Visualization of the P-6R architecture.

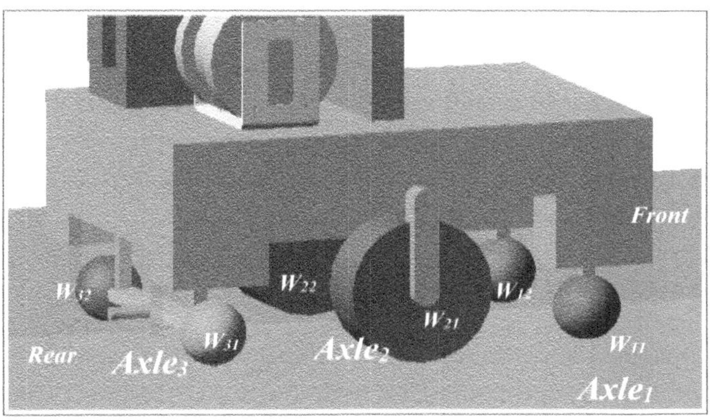

Figure 4.100. Nomenclature of the wheels.

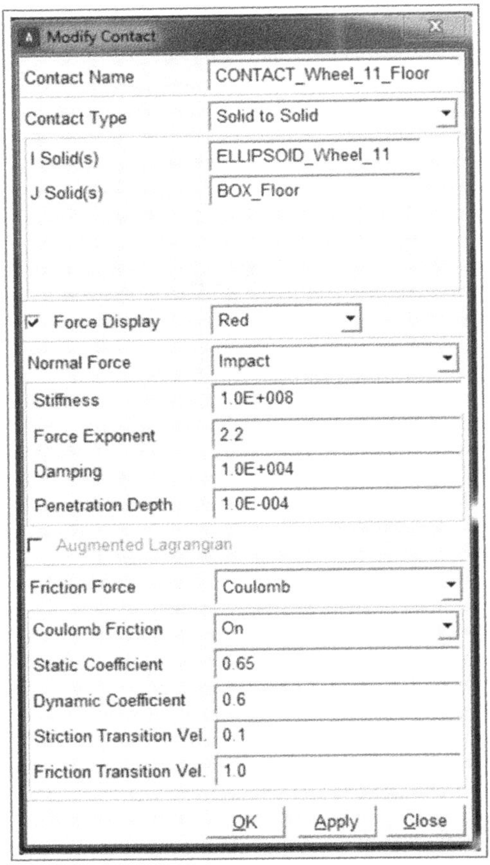

Figure 4.101. Wheel-ground contact in ADAMS.

Figure 4.101 shows contact parameters for wheel W_{11} with the ground. Values of parameters – stiffness, force exponent, damping, and penetration depth are kept unchanged (ADAMS default values). Coefficient of friction values are approximated as 0.65 and 0.6 for static and dynamic friction respectively. Also, a contact is provided between the tool and the wall to be cleaned (Figure 4.102) so as to apply desired cleaning force and to experience reaction forces.

d. **Workspace of the Robotic Arm**

In order to imagine the workspace of the robotic arm at a given configuration of the robotic unit, 3D geogebra model is built. Since ADAMS can't construct exact geometrical volume of the workspace, geogebra is preferred over ADAMS. Pink box represents the space occupied by ancillaries (cabinet, engine, aspiration unit, cable reel etc.)

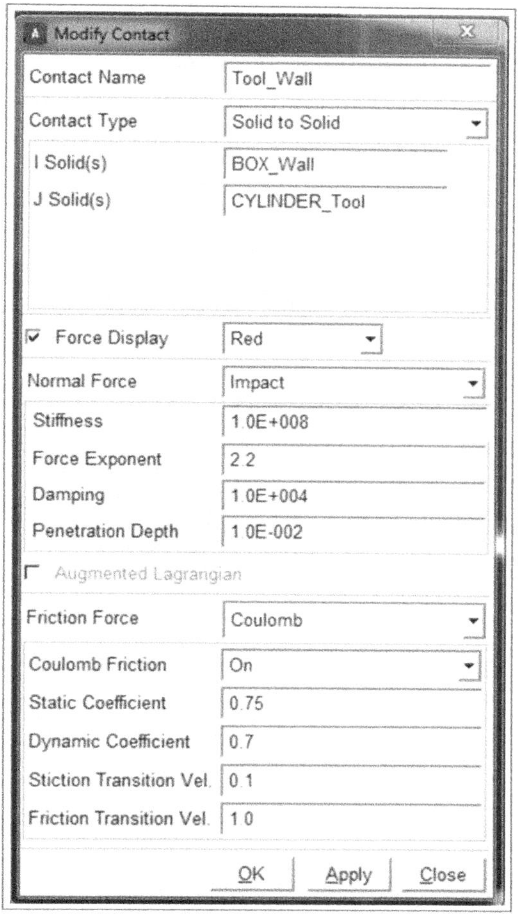

Figure 4.102. Tool-wall contact in ADAMS.

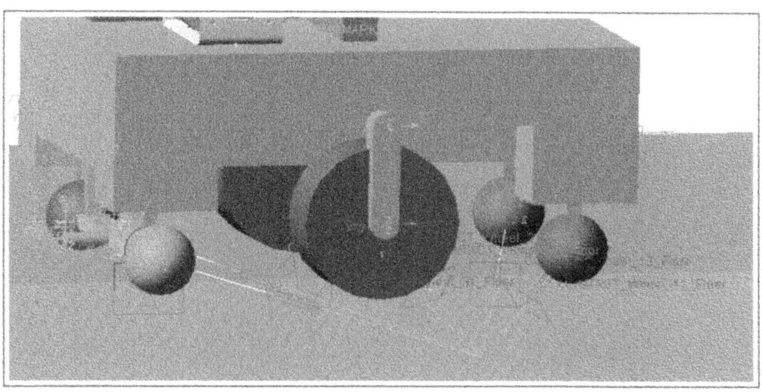

Figure 4.103. Contacts between wheels and ground.

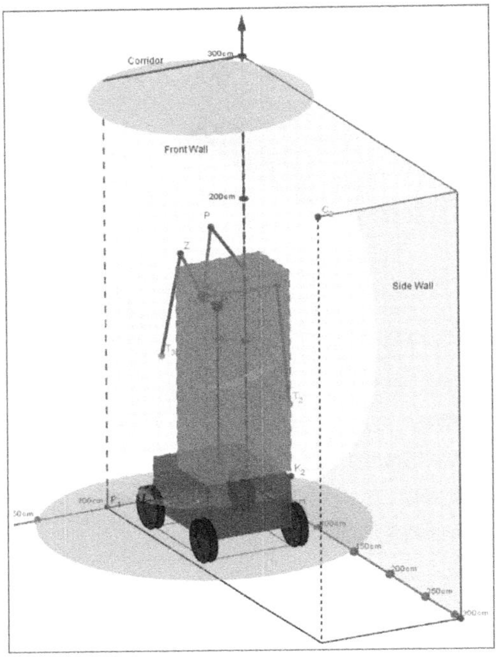

Figure 4.104. Intersection of the workspace with ground and ceiling.

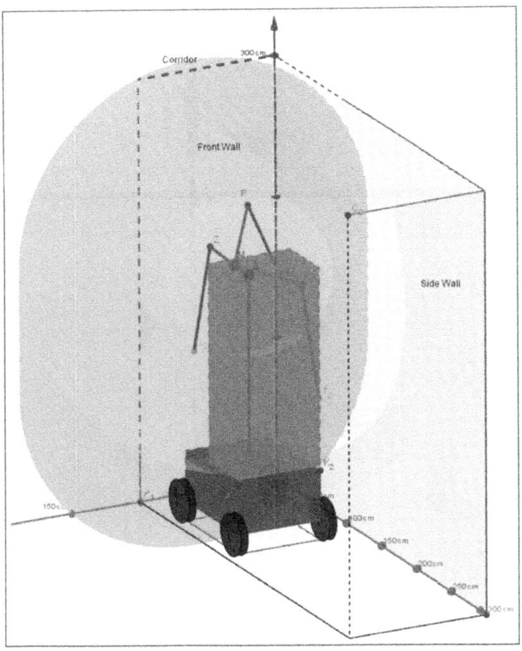

Figure 4.105. Intersection of the workspace with side and front all.

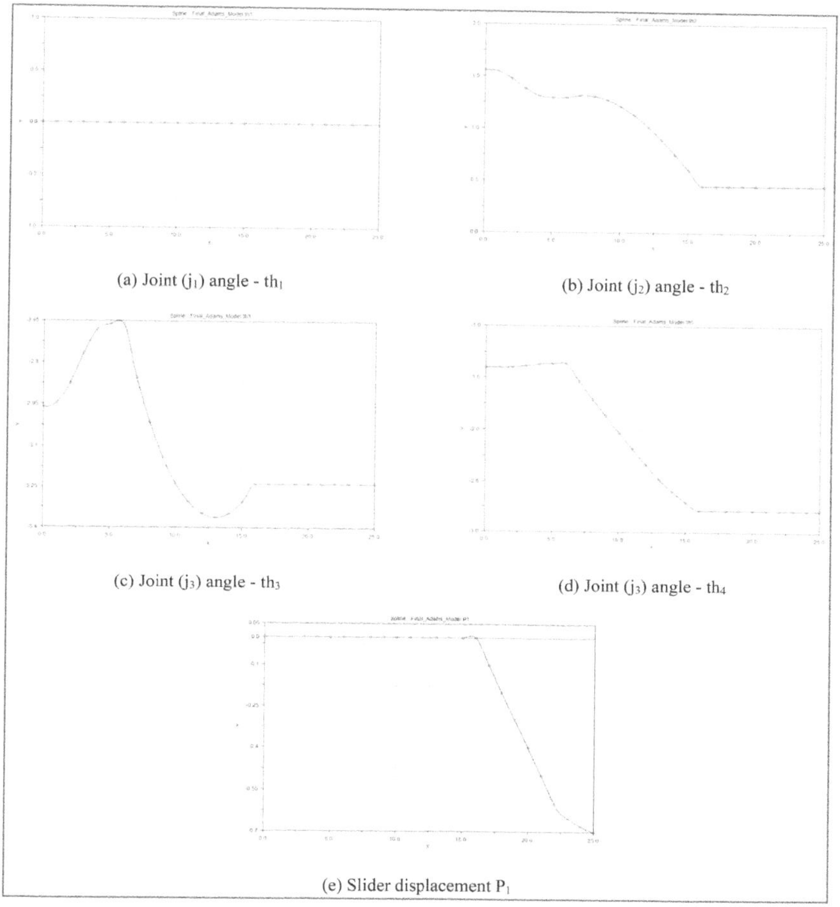

(a) Joint (j_1) angle - th_1

(b) Joint (j_2) angle - th_2

(c) Joint (j_3) angle - th_3

(d) Joint (j_3) angle - th_4

(e) Slider displacement P_1

Figure 4.106. Joint angle curves of the joints actuated during vertical trajectory.

e. **Dynamic Simulation**

The aim of this simulation is to test the functioning of the model by simulating a simple vertical cleaning trajectory. Ground reaction forces also have to be evaluated in order to assess the stability of the robotic unit.

i. Vertical wall cleaning trajectory

As already described in the section 2.c, an inverse geometric model of the robotic arm is built in matlab. A vertical cleaning trajectory is generated to get the joint angles for the corresponding end-effector positions. These joint angles are imported to ADAMS to feed the respective joints of the dynamic model.

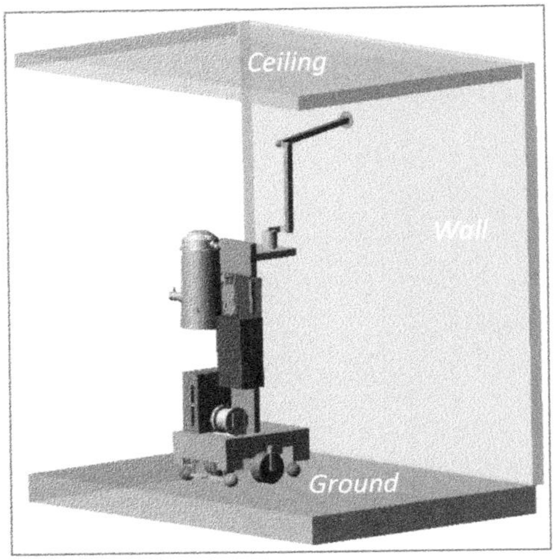

Figure 4.107. Initial position of the robotic unit.

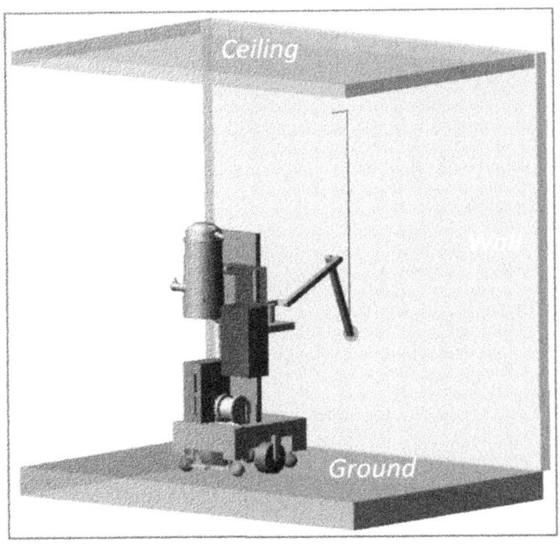

Figure 4.108. Final position of the robotic unit.

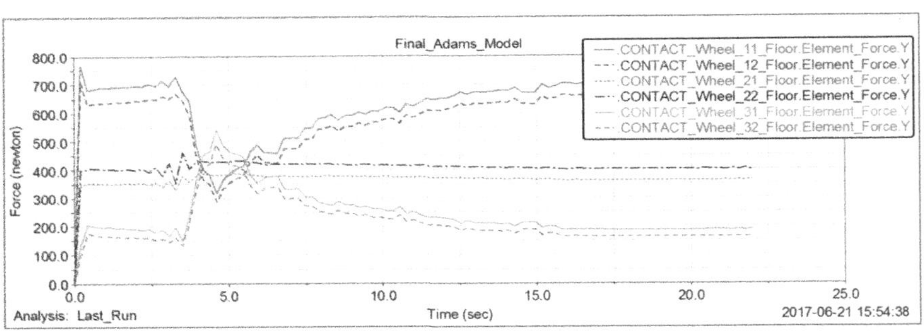

Figure 4.109. Ground reaction forces while performing vertical trajectory.

Ground reactions obtained during this simulation are shown in the figure below. Form Figure 4.109, it can be seen that none of the ground reaction forces approach to zero during the complete cleaning motion. Thus, the robotic unit has a stable behavior. Ground reactions on front wheels (W_{11} and W_{12}) initially have higher values since the robotic arm operates in the frontwards. Once the robotic arm starts pushing on the wall, ground reactions on the front walls decline whereas, those on the rear wheels increase.

ii. Conclusions on Version 1

a. Parameterized multibody model of the robotic unit was built in ADAMS. Parameterzation of the robotic arm was achieved through DH parameterization
b. 3D geogebra model helped in evaluating intersection of the workspace of the robotic arm with cleaning surfaces which is dependent on the configuration of the mobile base (distance of the robotic unit from the wall).
c. The model was successfully tested by implementing a vertical cleaning motion. Model showed a realistic behavior while performing the trajectory. Static and dynamic stability was observed that gives us the confidence for actual testing of the prototype.
d. Limited workspace: Figure shows the collision of the Link M_1I_1 with the pink box. Thus, the workspace of the robotic arm is limited only frontwards. Limited workspace implies restriction on the workspace to be cleaned.
e. Issues due to vertical slider: To enable the movement of the vertical slider, plenty of space is consumed (shown with the green box in figure 11). This space can't be used for mounting components and hence the big pink box sized space is needed at the back side to allow all the mountings.

However, a robotic unit integrated with a 6 DoF arm having three shorter moving segments takes eliminates these limitations and allows all around cleaning. In deliverable **D2.1** the idea of such a robotic unit has been already elaborated. In the present deliverable the attempt will be to build the multibody model of the robotic unit and evaluate stability. This robotic unit will be referred as version 2 in this draft.

2. **Description of the Version 2 ADAMS Model:**

The components of the version 2 robotic unit are: mobile platform, 7R redundant robotic arm, aspiration shell, cabinet, engine and the tool and stabilizers. Masses of all the components are adjusted according to the real masses of the components.

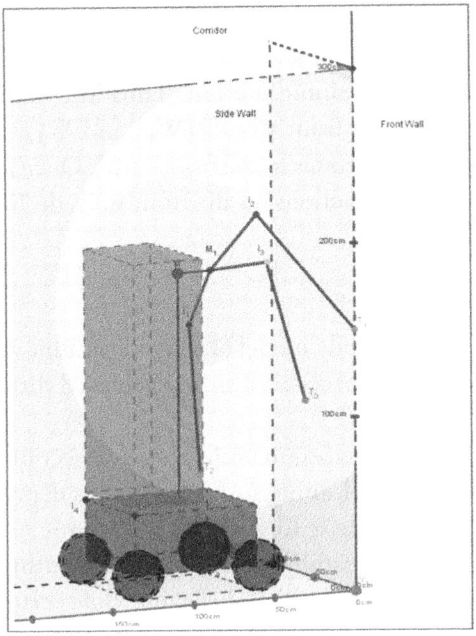

Figure 4.110. Collision of the link with the slider.

a. **Robotic Arm**:

Similar to version 1, robotic arm is parameterized according to the DH parameters. Table 2 lists the modified DH of the *7R* redundant architecture. Parameters d_3, d_4 and r_5 determine the three link lengths of the architecture. In ADAMS model, d_3 is 0.4m while d_4 and r_5 are 0.7m each. These values are obtained from the CAD design provided by TLabs based on the idea proposed by SIGMA. Offsets r_3 and r_4 collectively compensate for the offset r_2 such that the offset between joints j_1 and j_4 along z_4 is zero.

A tree diagram explaining the architecture of the robotic arm is presented in Figure 2. The figure presents joints, frames attached to each joint as well as the parameters that determine the link lengths and joint offsets. DH-parameters define the architecture of the arm. These parameters are incorporated in modelling to modify the pose and dimensions of the robotic arm and study the effect of their variation.

b. **Wheels Nomenclature**

Wheel nomenclature follows the same methodology as that of version 1. In addition to wheels, stabilizers are also added to the mobile platforms that are named using same nomenclature.

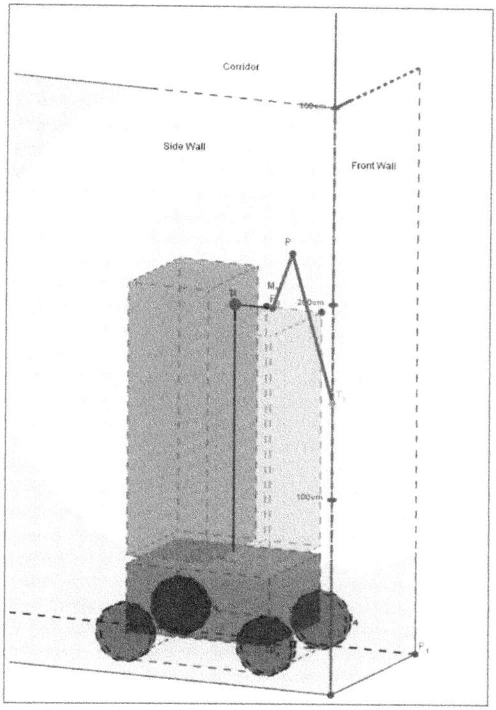

Figure 4.111. Space consumption in Version 1.

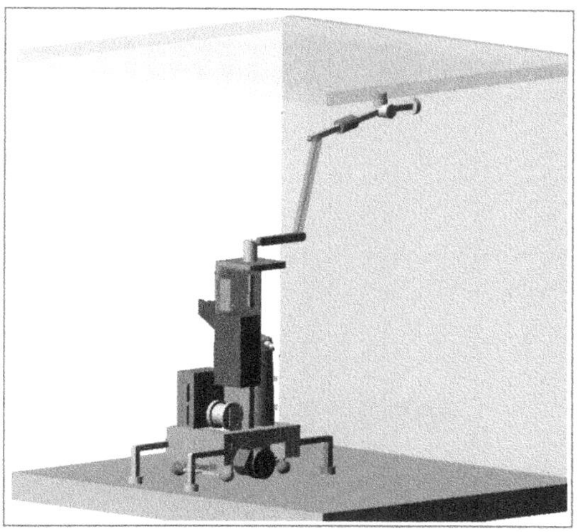

Figure 4.112. ADAMS model of the Robotic unit.

Table 4.4. DH parameters of the robotic arm.

j	σ	μ	γ	α	d	θ	r
1	0	1	0	0	0	th_1	r1
2	0	1	0	*pi/2*	d2	th_2	r2
3	0	1	0	0	d3	th_3	r3
4	0	1	0	0	d4	th_4	r4
5	0	1	0	*-pi/2*	0	th_5	r5
6	0	1	0	*pi/2*	0	th_6	0
7	0	1	0	*-pi/2*	0	th_7	0
8	2	0	0	0	d8	- *pi/2*	0

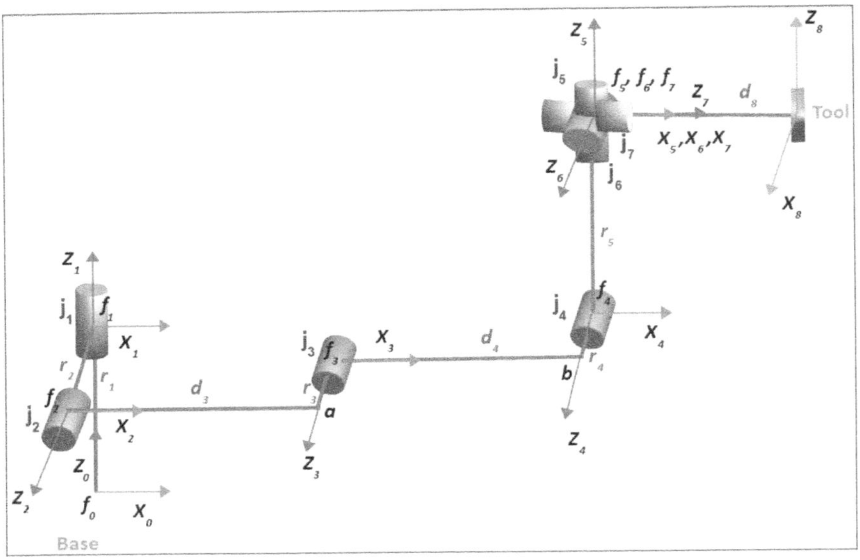

Figure 4.113. Visualization of the 7R architecture.

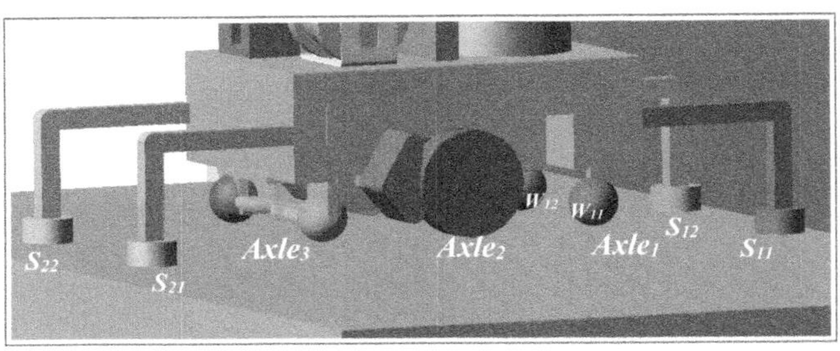

Figure 4.114. Wheels Nomenclature.

c. **Workspace of the robotic arm**

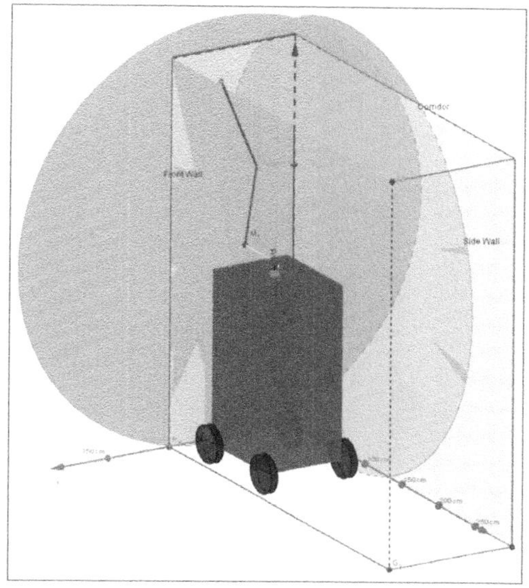

Figure 4.115. Intersection of the workspace with side and front wall.

d. **Dynamic Simulation:**

In case of presence of obstacles, robotic arm has an overhanging configuration and hence the stability can be a critical issue. Currently, to investigate a worst case scenario dynamic stability of the robotic unit is tested for two cases – 1) Without stabilizers 2) With stabilizers. Top to bottom vertical trajectory is simulated to study the dynamic response of the system. The objective of this simulation is also to study the stability in worst case and evaluate the need of stabilizers.

1. Vertical Trajectory without stabilizers:

Robotic unit is places as close to the obstacle as possible. Width of this obstacle is 1m. A vertical trajectory is traversed to study the stability of the robot. Figure 4.114 shows initial and the final position of the trajectory performed by the robotic unit. Figure 4.118 shows ground reaction forces in Y direction on all the wheel contact are presented.

Front wheels (W_{11} and W_{12}) experience higher reaction forces as the robotic arm is operational on the front side. Middle wheels (W_{21} and W_{22}) have moderate reaction forces while the rear wheels (W_{31} and W_{32}) are subjected to low reaction forces.

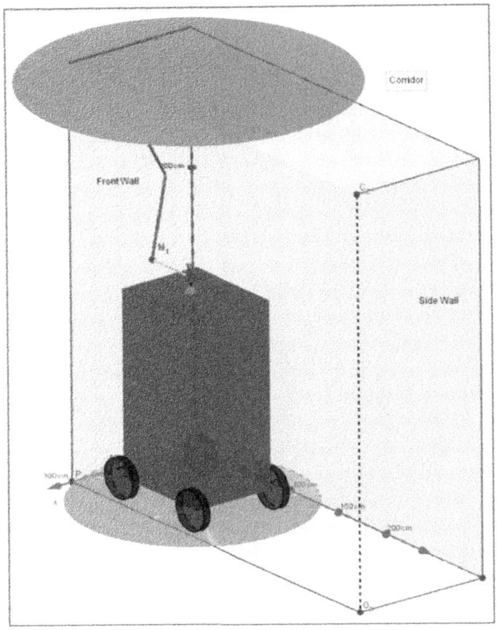

Figure 4.116. Intersection of the workspace with ground and ceiling.

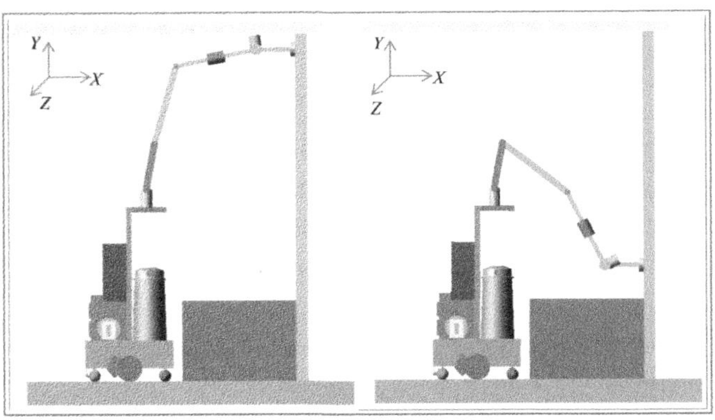

Figure 4.117. Initial and final positions of Robotic unit for vertical trajectory.

In Figure 4.116, a graph of lateral stability margins against time for the vertical trajectory is shown. Stability margin is calculated by a formula:

$$\frac{(W_{11} + W_{21} + W_{31}) - (W_{12} + W_{22} + W_{32})}{(W_{11} + W_{21} + W_{31}) + (W_{12} + W_{22} + W_{32})}$$

A good stability can be seen during the entire operation. Longitudinal stability margin is difficult to calculate here as we have six wheels. However, the rear wheels

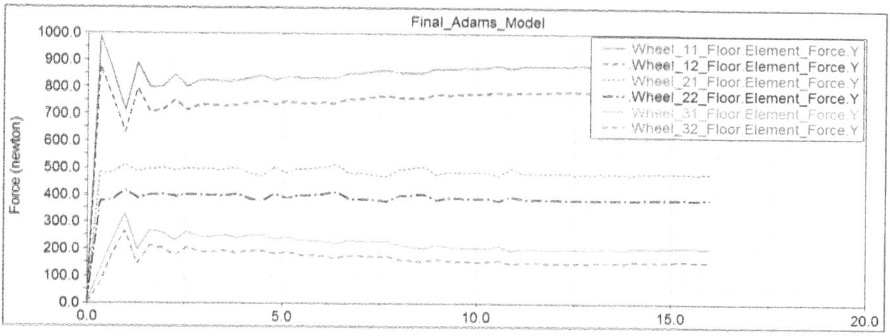

Figure 4.118. Ground reactions on wheel contact points.

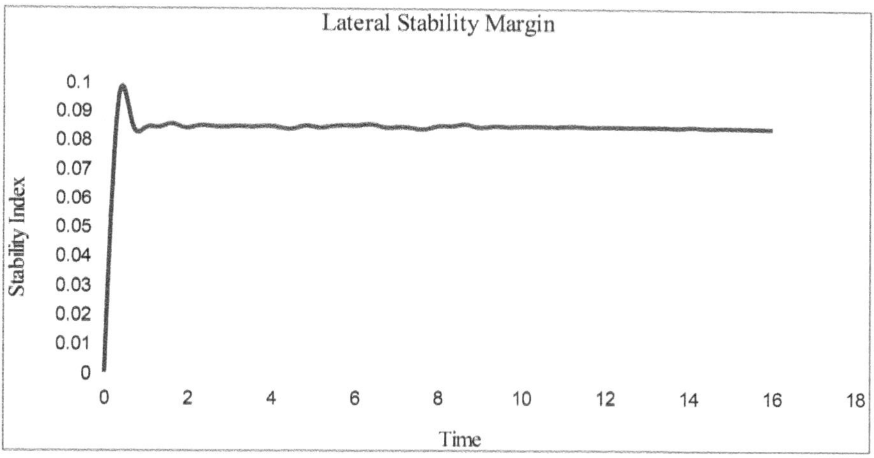

Figure 4.119. Lateral stability margin.

show a reaction force in the range of 200–250 N and hence we can say that the robotic unit is not close to instability.

2. Vertical Trajectory with Stabilizers:
   a. Stabilizers aligned in front

During this simulation, robotic unit was lifted up by 4 – stabilizing legs, hence no wheels are in contact. Stabilizers deployed on the ground carry the entire load of the robotic unit and a larger support polygon is capable of imposing higher stability to the robotic unit. Figure 4.121 shows ground reaction forces on 4-stabilizing legs.

The longitudinal stability margin is calculated as:

$$\frac{(S_{11} + S_{21}) - (S_{12} + S_{22})}{(S_{11} + S_{21}) + (S_{12} + S_{22})}$$

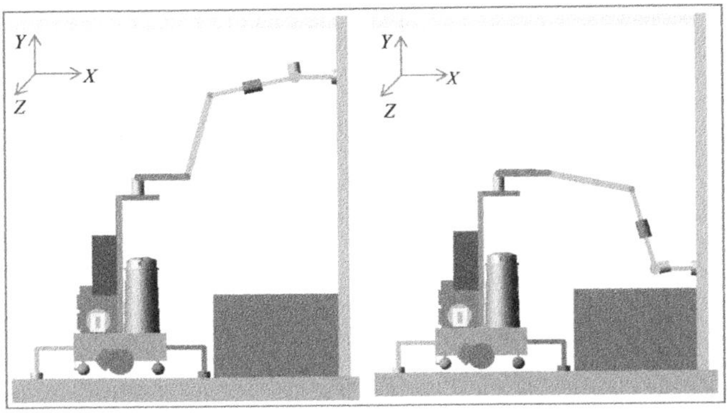

Figure 4.120. Initial and final positions of Robotic unit for vertical trajectory.

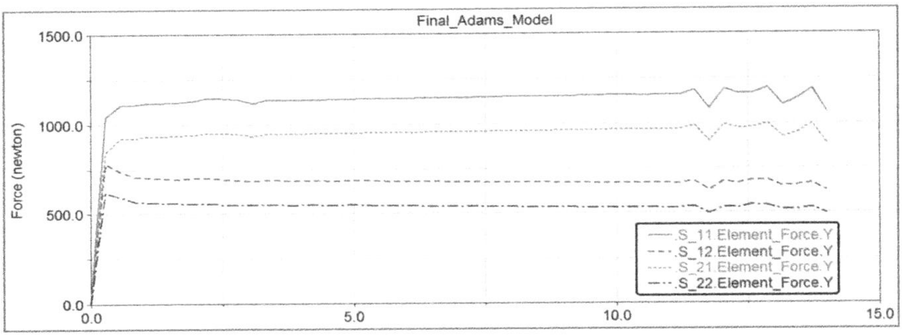

Figure 4.121. Ground reactions on stabilizer contact points.

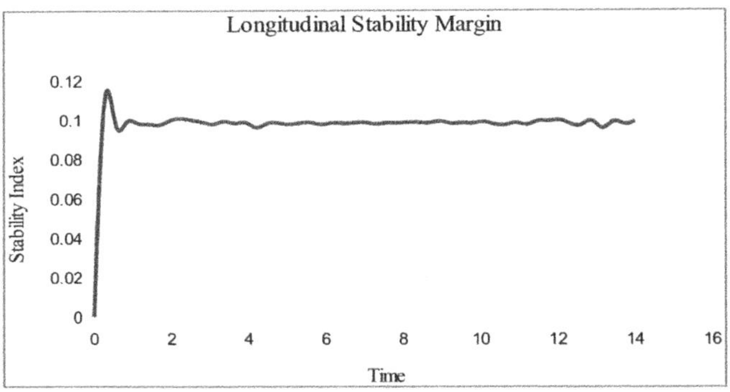

Figure 4.122. Longitudinal stability margin for vertical trajectory.

Figures 4.119 and 4.120 show lateral and longitudinal stability margins for the vertical trajectory with stabilizers extended in the front of the robot. Using stabilizers improvement in the lateral stability can be seen by comparing graphs presented in Figures 4.116 and 4.120.

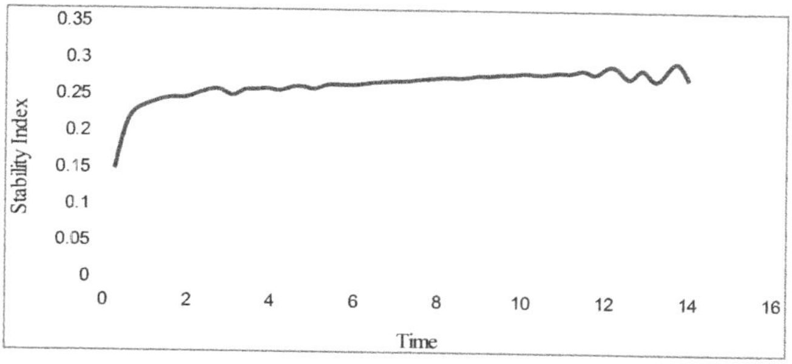

Figure 4.123. Lateral stability margin for vertical trajectory.

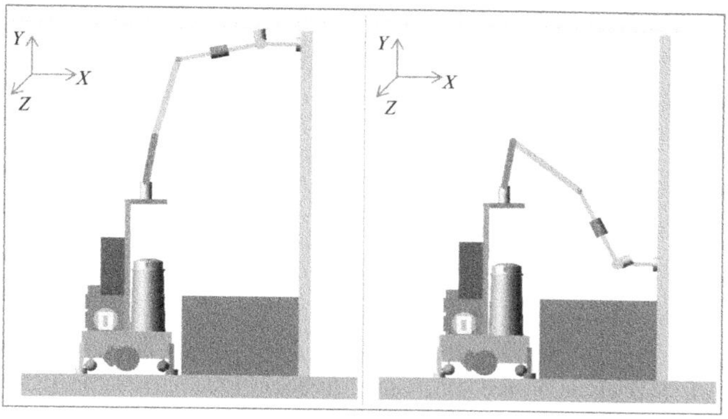

Figure 4.124. Initial and final positions of Robotic unit for vertical trajectory.

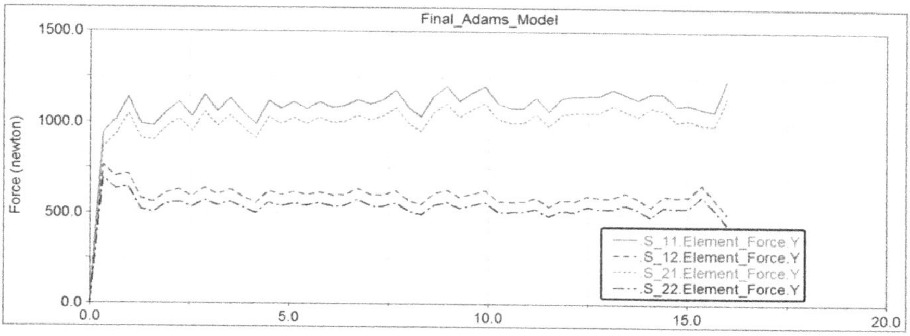

Figure 4.125. Ground reactions on stabilizer contact points.

b. Stabilizers aligned laterally:

In this scenario, stabilizers are aligned laterally, hence increasing the support polygon in lateral direction.

The ground reactions are similar to the one obtained in case of stabilizers aligned frontwards (Figure 4.122). However, the ground reaction curves in this case are not as smooth as they are in Figure 4.118 indicating more vibrations. Thus, a strategy of putting stabilizers in front is recommended.

4.6.6 Outlook and Future Work to be Carried Out

The simulation environment developed in during task 2.3 will now serve for rigorous investigation of efficient cleaning trajectories. As version 2 is identified to be a better solution, future investigation will mostly concentrate this version.

The detailed Workflow of the simulation goes as belows:

[1] Wall case:

- 1.1 Vertical Trajectory and front-rear stability
  - 1.1.1 Bottom Stabilizers
    - 1.1.1.1 Ascending
    - 1.1.1.2 Descending
    - 1.1.1.3 Comparison
    - 1.1.1.4 Vertical dynamic efforts vs gravity – influence on front-rear stability margin
    - 1.1.1.5 Removal of front stabilizers for improving stability margin
    - 1.1.1.6 Bottom-top stabilizers
      - 1.1.1.6.1 Ascending
      - 1.1.1.6.2 Descending
      - 1.1.1.6.3 Comparison
      - 1.1.1.6.4 Vertical dynamic efforts vs gravity – influence on front-rear stability margin
      - 1.1.1.6.5 Interest of bottom-top co pushing
  - 1.1.2 Horizontal Trajectory and lateral stability
    - 1.1.2.1 Low altitude of end effector, low slider
    - 1.1.2.2 Mid-altitude of end effector, high slider
    - 1.1.2.3 High altitude of end effector, extended arm
    - 1.1.2.4 Collision avoidance with stabilizer feet
  - 1.1.3 Influence of the distance from the wall
    - 1.1.3.1 Collisions
    - 1.1.3.2 Singularities
- 1.2 Ground
  - 1.2.1 Hollow workspace
  - 1.2.2 Preferable trajectory

- 1.2.3 Collision avoidance with the base and stabilizer feet
- 1.2.4 Path planning of the mobile platform

1.3 Ceiling

- 1.3.1 Limited plane workspace
- 1.3.2 Worst case scenarios for stability
- 1.3.3 Collision avoidance with stabilizer feet

1.4 Productivity evaluation of the mono-robot system
1.7 Interest on Productivity of using multi-robot system

Summary and Conclusion

Expected results: It is expected that through these realistic simulations, precise and efficient trajectories could be obtained. This task will further provide input to task **T2.4 for further improving stability through collaboration**. The results from this section would also be helpful for motion planning and task planning.

References

[1] Wisama Khalil, et al. OpenSYMORO: An open-source software package for Symbolic Modelling of Robots. IEEE/ASME International Conference on Advanced Intelligent Mechatronics, Jul 2014, Besan¸con, France. pp.1206-1211, 2014.

4.7 Hybrid Control

Deliverable 6.3
Semi-autonomous execution of a of a single asbestos-removal-task implemented and process control system documented

Authors: Tobias Haschke, Oriol Orra, Diego Fogliacco, Tim Detert, Claudia Cornely

4.7.1 Deliverable in the Context of the Work Program

The deliverable 6.4 is marked as the forth milestone of the Bots2ReC project. As written in the deliverable name the project plan scheduled a semi-autonomous execution of a single asbestos removal task. This means in form of robot actions that the asbestos removal robot can move towards the desired execution point, can bring the grinding disk towards the wall and start to clean a plain wall element.

In this case, the definition of "semi-autonomous" execution is chosen as the user can always interact with the robot through tele-operation via the user interface. As an example the operator can drive the robot via joystick to the desired operational point.

According to these functional parts, this deliverable strictly depends on different pre steps that are mentioned below. The first important input for this deliverable are the 3D environment data, which are delivered by the **Task 6.1**. The **Task 6.2** is in a quite close connection to the environment data as it uses the environment model for localization and mapping during the task execution. As this deliverable is meant to be "semi-autonomous", the connection to **Task 6.2** is not as close as to **Task 6.1**. Beside these tasks, this deliverable depends on the implemented control approaches for the mobile platform and especially for the robotic arm. These control approaches are implemented within the **Task 4.3** and **Task 5.3** and require robust components due to the aspired use case. As the process of asbestos can vary due to changing environment properties (e.g. wall thickness or subsoil) some data sets of process parameters need to be extracted from the removal tests of **Task 1.4**.

After this deliverable the next steps are to widen up the size and the number of asbestos removal tasks. According to this the current solution needs to be adapted to the second version of the prototype which will also bring the side-effect adding multiple inhomogeneous robots to the robot team that should execute the removal tasks in parallel. These steps are going to be solved in **Task 6.5** and **Task 2.4**.

4.7.2 Methods and Work Carried Out

The Central Process Control System (CPCS) is fully implemented in C++ within the Robot Operating System (ROS) framework. Due to this, the main elements as e.g. the task planner or task_manager of the CPCS are written as node-based units, which can be distributed over the whole ROS network. Beside the ROS-based implementation, the control algorithms are extended to use Linux real-time kernels to take care of the necessary control frequencies. According to Bots2ReC, the RT Preempt Kernel is used for the self-developed control algorithms. The complete development process is related to the overall asbestos removal process.

4.7.3 Overall Asbestos Process

The overall asbestos removal process can be separated into main processes and their sub-processes that automatically create an easy-understandable workflow for the user and for the programmer as well. All in all the five main processes *Construction site preparation*, *Data Mining*, *Task Planning*, *Task Execution* and *Construction site completion* are defined and individual sub-processes are listed within the table below.

#	Main Processes	#	Sub-Process	Part of CPCS
1	Construction site preparation	a	Measurement of asbestos contamination	No
		b	Seclusion of removal area	No
		c	Robotic setup on construction site	Yes/No
2	Data Mining	a	Create environment model	Yes
		b	Specify removal jobs	Yes
3	Task Planning	a	Create complete removal data set	Yes
		b	Generate robot tasks	Yes
4	Task Execution	a	Reactive task management	Yes
		b	Navigate to execution position	Yes
		c.i	Asbestos removal (Open Loop Control)	Yes
		c.ii	Asbestos removal (Closed Loop Control)	Yes
5	Construction site completion	a	Removal of robotic setup	Yes/No
		b	Removing seclusion	No
		c	Official acceptance	No

The first and fifth main processes and their sub-processes are mostly done by external experts and therefore do not need to be described in detail for this deliverable. The only robotic relevant element of these main processes is the setup and the removal of the robotic infrastructure. This contains the CPCS with its computers and displays for the User Input and the local network setup with Ethernet switches and Wi-Fi routers.

The second main process is *Data Mining* that is an all-important step within the robotic process chain. Due to the non-existing 3D environment sensors during the removal process, a correct and exact model of the refurbishment site has a significant impact on the removal efficiency. If the environment model is too inaccurate, the CPCS might not find the necessary overlaps between model and current sensor data. This can have the effect that the robot will also need tele-operated input by the user for simple geometries as plain walls. A part of the environment model is the task allocation towards geometrical elements as walls, floors and ceilings.

Task Planning is the next main process after mining the environment data. It contains the conversion from environment and process data into a complete data set that is called the *removal problem*. Within the planning step, this problem is solved in an optimal way to generate robot-specific commands that can be executed directly.

During the *Task Execution* process, the commands are executed to perform the desired job in the real world. Therefore, it is necessary to implement a task management system that can actuate the mobile base for the navigation towards the operational points. In addition, the task management system also needs the ability to control the robotic arm for the asbestos removal. Especially the multi-modal control of the robotic arm is a key step within the removal process for handling all the different sensor inputs for the optimal task execution.

4.7.4 Realization of sub-processes

This chapter describes the sub-processes in detail and can point out the technical details for each step. Some elements are already documented in other deliverables and therefore these explanations are skimped.

4.7.4.1 Data Mining

Data mining is a typical pre-process that need to run before the asbestos removal robot is running on site. As it is not possible to run these steps in the necessary accuracy and resolution as there are no sensors planned for this task.

a) Create environment model

The root of the environment data set is a point cloud, which can be created via two different ways. The first one is to use the so-called BIM files, which are often used for modern refurbishment sites to document all the details about the removal job. However, as BIM is not used for every removal site another solution for the point cloud generation is necessary. For these cases, it is planned to use a low-cost 3D mapping robot that can scan the already secluded rooms beforehand. The point cloud is processed in two consecutive steps from bounding boxes into polygons. The complete workflow and the geometric mechanisms are explained in detail in the **Deliverable 6.2**. In the following table the three processing steps are summarized with three figures taken from the translating algorithm.

Point Cloud	Bounding Boxes	Polygons
	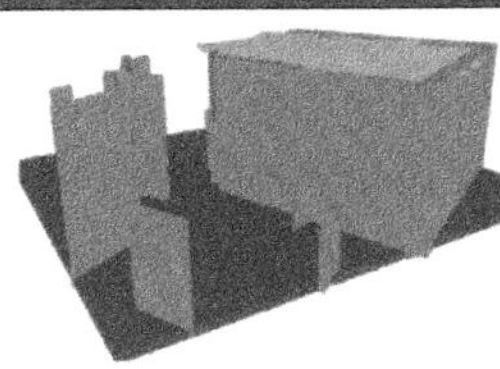	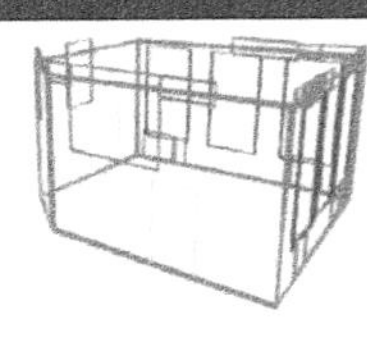

The outcome of polygons is the pre-defined interface for the task planning routine as all of the polygons contain more information than just the points for their geometry. Out of the shape it is possible to find elements like doors or windows and after a short computation the normal vector can distinguish polygons into walls, ceilings and floors. This property comes in handy for the planning phase as the polygons can be connected to pre-defined task parameters as e.g. maximal lateral speed or nominal tool pressure efficiently.

b) Specify removal jobs

In addition to the properties extracted from the polygons the information about the removal process are not granted as they are related to the working surface and its quality. There are obvious process differences as plaster and tiles which automatically need different tools. But there are also non-obvious differences that can occur between the grinding process on painted and plastered walls.

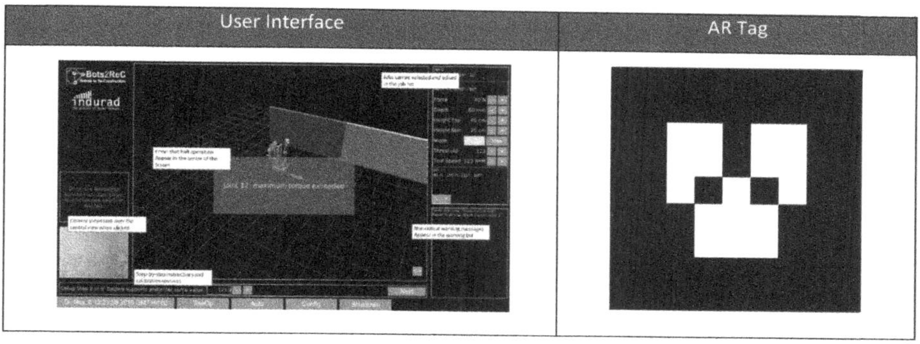

For handling this input about the surface quality, two options are taken into account for the Bots2ReC prototypes. On the one hand, the user can select wall elements within the user interface and setup pre-defined values for grinding and tiles removal processes. The other option is the automatic detection of the surface with AR tags that can be used during the robotic 3D mapping routine mentioned in 4.1.a. This means that the experts, which are doing the asbestos analysis in the run-up to the removal process, will mark the different wall with corresponding AR tags. The information that is given by an AR Tag can contain information such as "Plaster – contaminated" or "Not contaminated" to ease the task definition for the user. The definition of tasks via AR tags is tested at the IGMR at Aachen University together with the 3D mapping approach.

c) Create complete removal data set

Finally, both information, the polygons and the jobs, are combined together in a so-called semantic map object in C++. This one is the final step of the data mining process and the basis for the *Task Planning*.

4.7.4.2 Task Planning

With this complete definition of the asbestos removal problem, the plan for the robotic-based removal process can be generated. Due to the complexity of the problem an optimal way for the *Task Execution* is aspired. The solution should answer questions such as:

- Where does each robot start its removal job?
- How does the user manage problems or unforeseen delays in the best way?
- How many robots are best for this removal problem?
- Which tools should be mounted for this removal problem to finish it as fast as possible?

The details how these questions are answered are worked out in **Task 2.4**. Within this deliverable report, only a brief description shall be granted.

b) Generate robot tasks

The main idea of the *Task Planning* step is driven by the benchmarking process for being as fast as possible for reaching a profitable task fulfillment. Because the robots are wired, the flexibility of each robot is limited by the position of the other robots. Having this in mind, saturation effects exist for the number of robots per room and flat which of cause are related to condition of the removal flat. As an example, two flats with the same base area can be processed with different number of robots if one contains many small rooms and the other one only a few big ones. Beside this question of robots quantity the processing way for one wall by one robot can be optimized as well. As the robot‘s reach is limited due to the size of the arm and the stability margin, walls can be spitted up into small sub-segments that can be processed by the robot from one fix operational point.

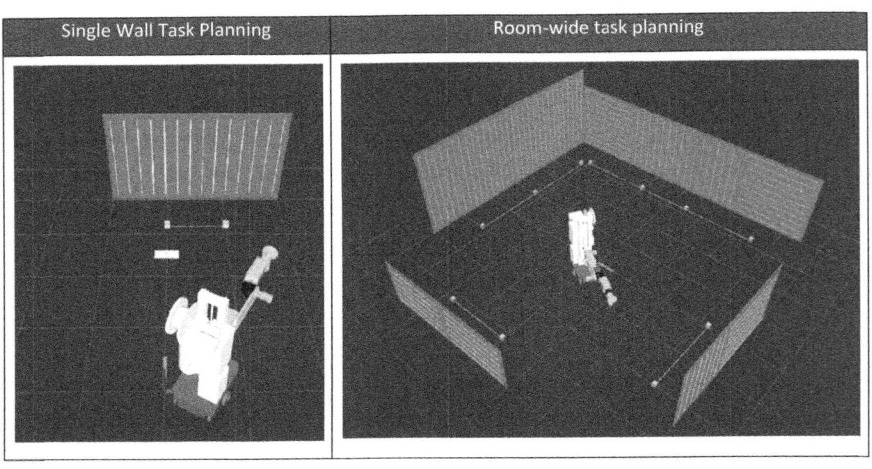

As it is shown in the table above for a *Single Wall Task Planning* situation in this case two operational points (small yellow boxes in front of wall) are necessary to process the complete wall. Combining multiple walls for a *room-wide task planning* will create solutions, as they are shown in the table in the right figure. Within the CPCS software, the solution is separated into some understandable elements for monitoring the systems state and to be executed by the robots:

- *removal_problem*
  One semantic map plus the available robot_team
- *robot_team*
  Collection of available robots and their capabilities
- *removal_plan*
  Solution of the removal problem that contain all removal_jobs for each robot
- *removal_job*
  Collection of all removal_tasks that one robot need to fulfill
- *removal_task*

Removal area of a wall, ceiling or floor that can be removed by one robot from one operational point

Only the removal_job with the embedded removal_tasks are forwarded to the robot for the execution. The data within these objects can be executed directly by the robots hardware without costly computational effort.

b) Navigation to execution task

The navigation towards the operational point given by the *robot_job* is done with typical SLAM approaches based on Lidar sensors and new developed algorithms for the radar-based SLAM if the air is or the sensor surface is polluted with dust. The planning of local paths and the subsequent execution are interwoven into the reactive control architecture that was explained above. Further details about the operating modes of the radar-based SLAM are explained within the **Deliverable 6.3**.

c.i) Asbestos removal (Open Loop Control)

According to the progress of the project, it was necessary to run the first grinding tests in an open control loop. This means, that the robotic arm was fully controlled via the joint encoders without additional information from the process. The planned input by the process are the distance measurements by the iLDRs from Indurad and the Force/Torque feedback of the tool. The Force/Torque feedback is available over two different ways where the one way is to measure it directly with a force/torque sensor at the end effector. The other way is to use the compliant tool and to compute the forces out of the spring stiffness and the measured travelling distance of the compliant tool. With this setup the user finds the initial position of a grinding movement via tele operation. The situation of aligning the tool to the wall is shown in the figure below.

Moving towards the processing wall with a PS3 Controller (Open Loop)

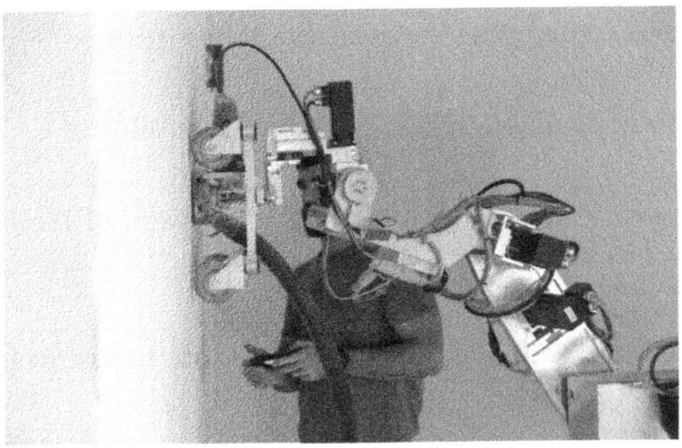

After aligning the tool, some pre-defined trajectories can be executed. In the figure below some rudimentary vertical grinding movements are shown. According to the open loop control the lack of process information, cause some problems for the grinding execution. For example, it is quite difficult for the robot to keep touch to the wall during turns or other non-linear movements with the compliant tool. Often the turning motion of the castor wheels initiate shear forces, which oblige the grinding disk to dive deeper into the walls surface. During this non-desired motion, the force limits of some motors are exceeded and the robotic arm throws an emergency stop.

Single Grinding Operation	Multi-Grinding Operations	Grinding near walls and objects
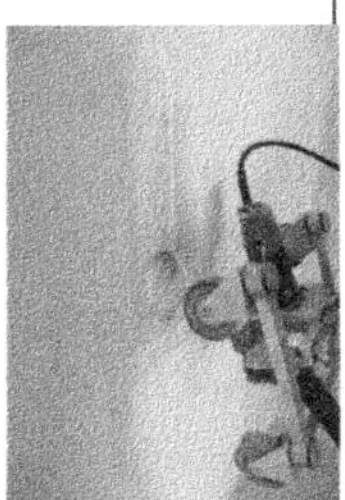	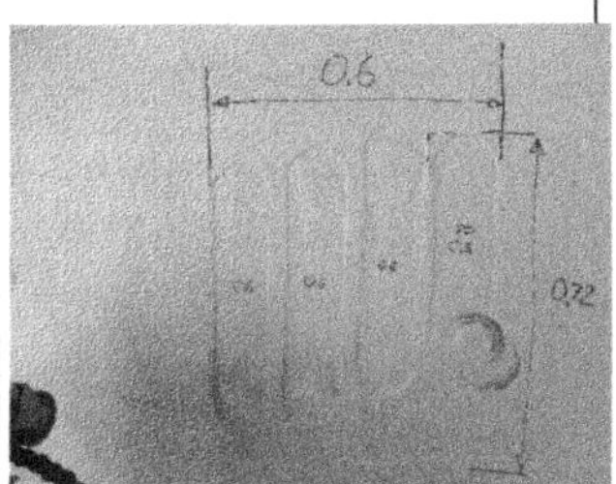	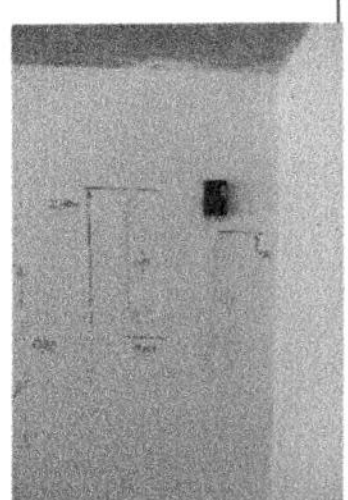

Overall, the open loop control approach is just an intermediate solution for gaining first test data of the grinding process. The first tests have shown that adding the 1D radar sensors and the force/torque feedback is inevitable for a robust removal process.

c.ii) Asbestos Removal (Closed Loop Control)

In parallel to the open loop control tests the closed loop control approach was developed and implemented. As the control approach represents a multi-modal control approach which is not build-in into the Beckhoff Controller of the robotic arm, the self-written controller is running on a real-time preempted Linux NUC. This configuration allows high control frequencies because of the guaranteed computation of the implemented control algorithm by the operating system. Basically there are five building blocks of this control architecture that are shown in the figure below. While the *CPCS* is playing a tangential role the main elements are the *robot*, the *beckhoff_controller*, the *indurad_nuc* and the *arm_nuc*. All the interfaces marked in orange are typical ROS drivers which connects the different elements through pre- or self-defined ROS messages. The only exception is the connection towards the *beckhoff_controller* which is implemented with an ADS Client and Server structure.

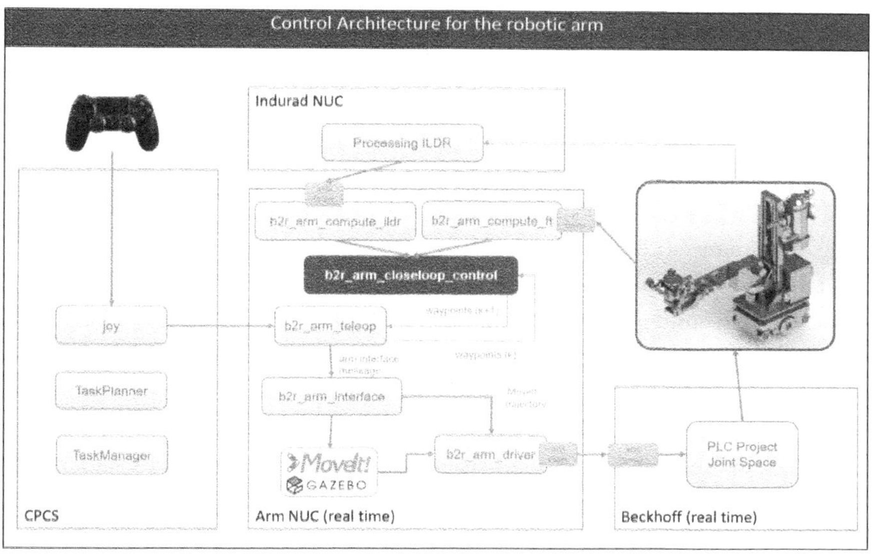

The compensation of process influences through motions of the robotic arm are computed in an interwoven setup of MoveIt, an analytical inverse kinematic solver and the self-written multi-modal controller *b2r_arm_closeloop_control*. A special feature of this setup is the independency of position and force/torque control. This means that there are two different controllers written by the Bots2ReC consortium where the first one is controlling the motors of the arm by position and velocity and the other controller is only using forces and torques. For changing between the different controllers the developer just need to change the PLC Project on the *beckhoff_controller* to address the motor drivers in the correct way. These two approaches are implemented in parallel for running comparable tests for defining the better option for the asbestos removal process.

4.8 Removal Control Approaches

Task 5.3 Control and sensors system of the robotic arm

For the first implementation of the control of the arm while performing the grinding task, a smooth position-force hybrid controller was proposed. This controller changed smoothly between free space (while approaching the wall to be grinded) and contact (while beginning the grinding) modes, thus reducing the impact force of previous similar approaches. The controller was tested not only in a simulation based on Matlab+Adams but also with the Kuka LWR robot at Sigma Clermont by using a camera (for wall distance measurement) and a wrist force-torque sensor. This controller presented good position and force tracking performances and the impact factor was small. Nevertheless, the lateral forces had an important effect on the controller because of the large offset between the disc tool center (in contact with the wall) and the wrist center (where the force-torque sensor is installed), which generated large tilting wrenches that involved tool contact loss when the surface of the wall was irregular. In addition, the controller was dependent on the wheels installed around the tool and these wheels also lost contact. Finally, the cutting parameters (i.e. feeding velocity and grinding force) were not estimated accurately, hence leading to inaccurate results in the grinding quality.

Thereby, a second version of the arm controller is proposed for overcoming all these shortcomings of the first version. The grinding model is identified and included in the controller and the wheels around the tool are removed. In addition, the controller is unified by commanding the arm and the wrist as one unit (and not separately, as in version 1). This controller employs a hybrid hybrid velocity-force control strategy expressed in a desired frame in the end-effector. Addition to that, the proposed controller makes the disc of the grinding tool adapt to the curvity change of the wall surface and corrects the desired directions of force and velocity in real-time. The controller is tested again on the KUKA LWR 7-dof robotic arm programmed to grind different planar and curved walls with unknown geometry. The KUKA arm is again equipped with a six-axes ATI Gamma 3 force-toque sensor and a grinding tool to treat a piece of wall covered with resurfacing concrete. As shown in Figure 4.126, there are no wheels supporting the tool.

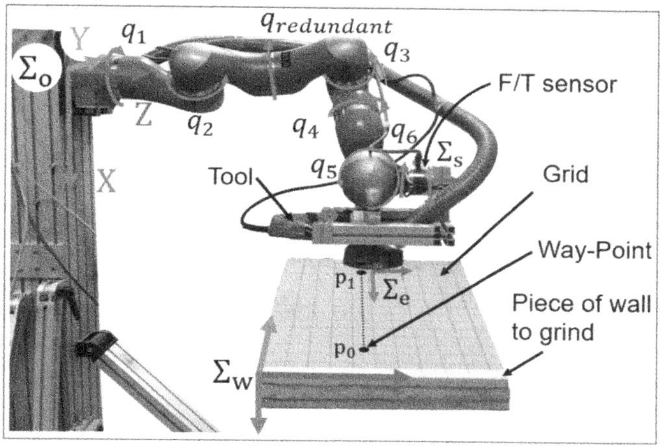

Figure 4.126. The robotic system equipped with grinding tool and force/torque sensor to perform grinding on a piece of wall. The grid is to evaluate the wall before and after grinding.

The three main new features of this new controller strategy are going to be detailed:

- The definition of new compliant frames with regard to the tool, for more precise velocity and force control while contact
- The modeling of the lateral grinding forces, that provoked the tool contact loss with the previous controller
- The non-contact forces elimination from the force-torque sensor readings, so that a more precise contact force control can improve the grinding quality

Choice of the compliant frames

The choice of the velocity and force control frames in the grinding controller implemented here is crucial. The chosen frame for force control is $\{cf\}$ on the center of the disc, it coincides with the end-effector frame origin $\{e\}$ and it is rotated by $\pi/2 - \varphi_v$ around e X, while the control frame for velocity and motion control is $\{cv\}$ fixed at an offset $-R$ along the z direction of $\{cf\}$. The frames are depicted on the disc in Figure 4.127.

There are several reasons why these frames are chosen like this. The force control frame must be fixed to the center of the grinding disc to have good measurements of the lateral forces acting on the disc, and good force control of the normal force at the wall along the direction ${}^{cf}x$, hence there is no velocity control along this direction while the disc is in contact with the wall. This released direction allows the robot to move along the surface of the wall while maintaining the axis of force control normal to the surface. Hence, a desired force can be controlled between the

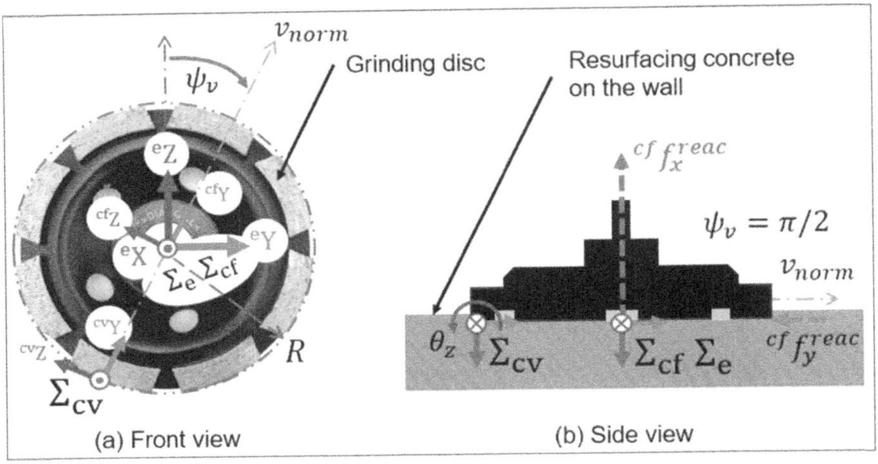

Figure 4.127. The choice of the frames for simultaneous force and motion control.

disc and the wall along this direction. Velocity control is applied in the frame $\{cv\}$ along the remaining linear directions, namely ^{cv}y and ^{cv}z. Hence, position control is applied in the same frame to the remaining axes which are not controlled in force or velocity: the three orientations around ^{cv}x, ^{cv}y and ^{cv}z. The reason why $\{cv\}$ is chosen as in Figure ??, is that the proposed controller needs to adapt automatically the surface based on special orientation controller that will be explained in the next sections. Hence, if the disc is moving along the wall with velocity v_{norm} along ^{cv}y and it needs to orient, applying rotation θ_z around ^{cv}z will generate no reaction force along ^{cv}x which is natural and important for the grinding quality. However, if $\{cv\}$ was chosen to coincide with $\{cf\}$, applying the rotation θ_z around ^{cv}z will generate reaction force along ^{cv}x and the wall can be damaged.

Modeling of the lateral grinding forces

The key idea of this controller is to adapt the orientation of the disc to the wall surface based on keeping the lateral reaction force $^{cf}F^{lateral}$ around a desired value that should be appropriate for grinding. However, for the case of wall grinding in the construction industry, there has been no published model that identify the nominal grinding lateral force $f^{lateral\_nom}$. Moreover, the desired normal grinding force $^{cf}f^{reac}$ and the translational velocity of the disc on the wall $^{cv}v_{norm}$ are not known as well.

Hence, in order to explore the relation between $^{cf}f^{reac\_x}$, $^{cv}v_{norm}$ and the nominal grinding lateral force $f^{lateral\_nom}$, it is necessary to firstly have an idea about the values of $^{cf}f^{reac}$ and $^{cv}v_{norm}$ that should be used as input desired values for the controller. Thus, assisted manual grinding tests are carried out with the experimental setup shown in Figure 4.128. These tests were done by setting the

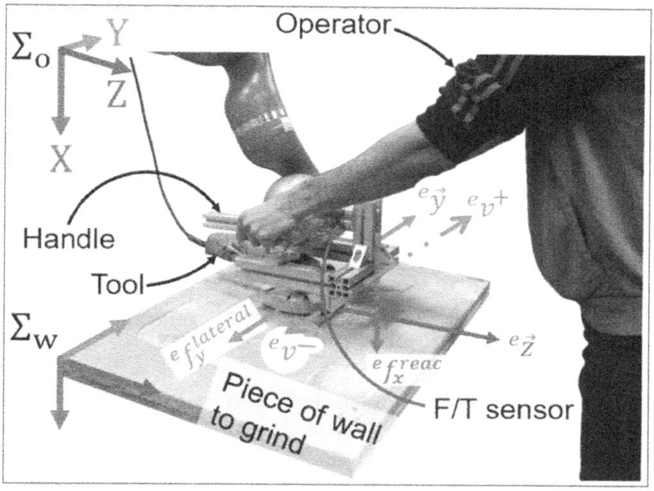

Figure 4.128. The manual grinding setup used to determine the normal grinding force f_{reac} and the translational velocity of the disc on the wall v norm. It is equipped with grinding tool, force/torque sensor and a handle.

robot controller to Cartesian impedance mode, this mode allows to constraint the motion in some directions and release it for others. The wall used for the data collection is assumed planar and its frame $\{w\}$ is set parallel to the frame $\{o\}$. To this end, the robot is set to move freely by hand along the ${}^{o}x$ and ${}^{o}y$ directions while the direction along ${}^{o}z$ is locked. The three orientations around ${}^{e}x$, ${}^{e}y$ and ${}^{e}z$ are fixed such that the frame $\{e\}$ is parallel to $\{w\}$ in order to keep the disc parallel to the wall. Then, the operator performed three grinding tests by moving the wrist manually along ${}^{e}y$ and pressing along ${}^{e}z$ to grind the piece of wall.

The average value of the normal forces in the three tests is 26.21 N. Concerning the lateral velocities, they are obtained from the robot kinematics and their mean square average is 0.0243 m/s. This speed is considerably high for the KUKA LWR to execute the grinding task on the wall. For this issue, the lateral velocity is scaled down to 0.015 m/s and consequently the force to 15 N by respecting the approximate proportional value ($f^{reac}/v_{norm} \approx 1000$). It is assumed here that the normal force is proportional to the translational velocity because of the collision between the disc and the wall. The grinding tool used in the experiment is made of an electric spindle rotating in a high speed with the abrasive disc attached to its end. The description of the grinding parameters that are identified with these experiments are detailed in Table 4.5.

The disc used in the tests has the outer radius $R = 0.0625$ m and the inner radius $r = 0.0575$ m. The spindle rotates 11000 rounds per min (rpm), this is equivalent to $\omega = 1.1519 \times 10^3$ rad/s. F^{reac} and v_{norm} can be obtained from the

Table 4.5. Parameters of the grinding model.

Variable	Units	Explanation
μ_e	(W/s)/m^3	Energy gain
ω	rad/s	Rotational speed of the disc
f^{reac}	N	Applied force
t	m	Thickness of the cut ahead
R, r	m	Contact radii on the disc
d	m	Depth of the cut
b	m	Width of the cut
v_{norm}	m/s	Translational velocity of the disc on the wall
v_{tan}	m/s	Tangential velocity of the disc
μ_f		Friction constant

force-torque sensor and the robot kinematics respectively. In order to determine μ_e, μ_f and d, two automatic grinding tests were performed using the setup in Figure 4.129. Similar to the manual setup, the wall used is planar and its frame $\{w\}$ is set parallel to the frame $\{o\}$. Also, the orientation of the end-effector is fixed such that the disc is parallel to the wall. Then, the robot is commanded using hybrid control in the operational frame $\{o\}$ to execute normal grinding force f^{reac} along $^o x$ and translational velocity v_{norm} on the wall along $^o y$. The values of f^{reac} and v_{norm} are the ones deduced previously from the manual grinding tests as 15 N and 0.015 m/s respectively. The lateral and normal forces (f^{reac} and $f^{lateral}$) are recorded from the force-torque sensor measurements, the value of v_{norm} is obtained from the robot kinematics and the depth of the cut d is measured using a coordinate measurement machine and found to be 0.002 m. The surface's profile of the automatic grinding tests is evaluated using the profilometer (as shown in Figure 4.129), their average roughness is 16.7 μm and they were smooth enough by tactile sensing as well. Accordingly, the values of μ_e and μ_f are obtained by linear least square method.

Non-contact forces elimination from the force-torque sensor readings

Non-contact forces compensation in force-torque sensor readings is important in the grinding task. It is necessary to decouple the forces caused by the weight, inertia, Coriolis and centrifugal effects of the grinding tool from the pure contact forces with the environment. The force-torque sensor provides the readings of wrench W acting on the sensor frame $\{s\}$ with respect to the frame itself. This wrench not only contains the pure contact wrench $^s W_c$ but also the previously described

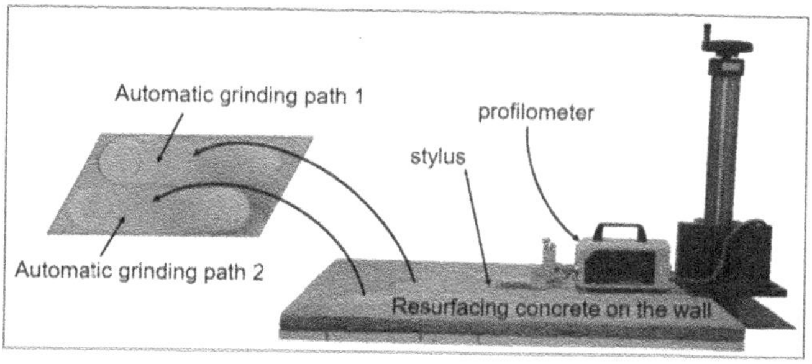

Figure 4.129. Surface roughness measurement setup.

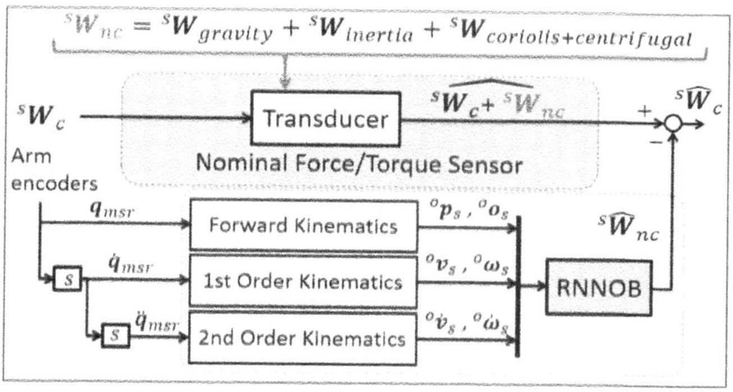

Figure 4.130. Recurrent neural network observer to estimate non-contact forces.

non-contact wrenches ${}^sW_{nc}$ (gravity, inertia and Coriolis-centrifugal). The sensor is activated when the robot is not moving and the disc is facing down, that is, the sensor frame $\{s\}$ is parallel to the operational frame $\{o\}$. Thus, it is necessary to first estimate the non-contact forces and then subtract them from the force-torque sensor output in order to obtain the pure contact forces.

In order to estimate these non-contact forces (${}^sW_{nc}$), a Recurrent Neural Network observer (RNNOB) is used since it outperforms analytical based methods. The recurrent neural network model is trained using the force-torque sensor pose (${}^sp_O$, ${}^so_O$), twist (${}^sv_O$, ${}^s\omega_O$),and its accelerations (${}^sv?_O$, ${}^s\omega?_O$), obtained from direct kinematics, to estimate ${}^sW_{nc}$. Fig. 5 depicts the block diagram of the RNNOB used here to estimate the contact forces.

The data was collected by *manual*, *automatic* and *rotational* trajectories with the grinding tool attached to the force-torque sensor as shown in Fig.3. The *manual* data was collected by setting the robot controller to gravity compensation mode and then moving the wrist manually to various poses in the workspace with random velocities and accelerations. The *automatic* data was generated by turning the

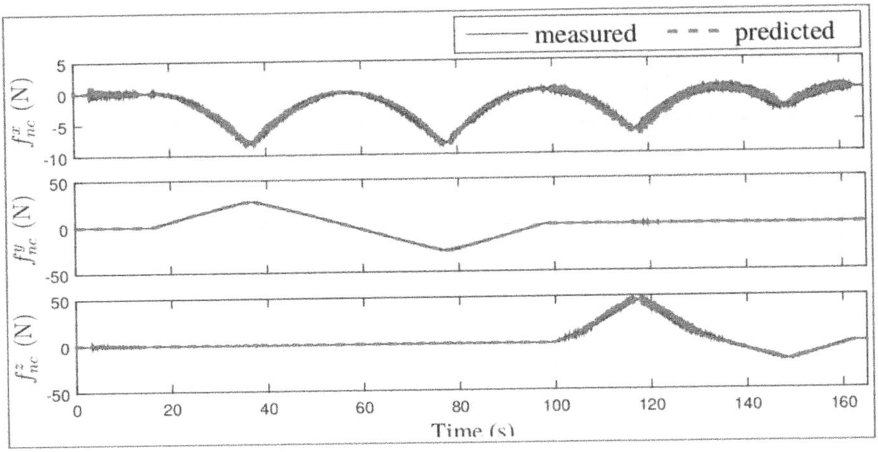

Figure 4.131. Non-contact forces (up) and torques (down) estimated by the RNNOB against the measured ones.

grinder on and moving the robot between random points in its workspace using various trapezoidal velocity profiles. The *rotational* trajectory was generated by rotating the sensor frame *{s}* and the compliant frame *{cv}* around each of their axis sequentially. The *manual* data was collected in 2 trials, each about three minutes long; similarly the *automatic* data was collected in 3 trials with an average time of 1.5 minutes each as well, and the *rotational* data was collected in two trials (in the frames *{s}* and *{cv}*) about 2 minutes long each. These datasets were then combined into one training dataset, where one trial of the automatic data and the rotational data around *{cv}* were separated to create a test dataset. The *manual* data was used entirely for training. The RNN model is trained for 20 epochs using an architecture with two hidden layers, with 15 and 10 Long Short-Term Memory (LSTM) units respectively, and a sequence length for the input layer of 15 time steps. Stochastic Gradient Descent was applied in the output layer with 0.01 learning rate to minimize the mean square error of the regression problem. A hyperbolic tangent sigmoid function was used as the activation function between the hidden layers and a linear activation function was used for the output layer. The efficiency of the RNNOB in estimating gravitational forces was evaluated by the rotational test around {cv} and is shown in Figure 4.131. Non-contact forces (up) and torques (down) estimated by the RNNOB against the measured ones.

Experimental evaluation

In order to validate the effectiveness of the proposed hybrid controller, wall grinding experiments are carried out on three different cases: (1) on flat surface; (2) on convex surface; (3) on concave surface. The curvature radius in the cases (2) and (3) is chosen such that the full contact between the disc and the surface is on its limits (e.g. if the radius of curvature is too big, the contact between the disc and the surface will be a disc, else the contact will be small segments at the contact points). The wall samples can be seen in Figure 4.132. In order to later evaluate the grinding quality, the initial state of the surface of each of the wall samples is evaluated using the coordinate-measuring machine (CMM). The CMM measures the surface geometry of the walls by sensing the discrete points of the mesh projected on their surface using the probe shown on the left side of Figure 4.132. Thus, the initial surface geometry of the walls is shown in Figure 4.132. The experimental grinding setup is shown in Figure 4.126. The KUKA lightweight arm is controlled by an external computer. The control framework is implemented on he computer using Robot Operating System (ROS) that communicates with the KUKA Fast Research interface (FRI) at a rate of 500 Hz. The six-axes ATI force-torque sensor is attached to the face of the six joint. Then, on the other end of the force-torque sensor, the grinding tool with a speed of 11000 rpm is attached. The tool rotates a disc with abrasive grains of 125 mm diameter. The overall end-effector (tool + disc + tool-frame) is around 7.3 Kg; its weight, inertia, Coriolis and centrifugal effects on the force-torque sensor readings are accurately estimated and compensated using the RNNOB previously presented. The force sensor publishes data at a rate of 1000 Hz and the RNNOB runs at 500 Hz. The force sensor output is filtered using a moving average window of 25 samples. The robot exerts on the wall the instantaneous normal force required for grinding. previously obtained and equal to 15 N. Along with exerting normal force on the wall, the disc moves on the wall with the identified desired velocity v norm = 0.015 m/s while it adapts automatically the surface based on the adaptive orientation controller.

In order to evaluate the grinding quality, the ground surface of the flat wall is evaluated in Figure 4.133. Figure 4.133a shows the superposition of the meshes measured before and after grinding. The two meshes are subtracted and the depth of the cut along the lines L1 and L2 (see Figure 4.133b) is plotted in Figure 4.133c. Figure 4.133c shows that the grinding depth varies between 1 and 2.67 mm along the ground path. The average depth of the cut under L1 and L2 is 1.79 and 1.86 mm respectively.

The grinding quality on the convex surface is evaluated as well in Figure 4.134. Figure 4.134a shows the superposition of the meshes measured before and after grinding. The depth of the cut can be obtained by calculating the Euclidean distances between the points of the meshes measured before and after grinding. Addition to that Figure 4.134c shows the depth of the cut evaluated under the curves C1 to C7 (see Figure 4.134b). The Figure shows that the grinding depth varies between 0 and 3.5 mm in the ground area. The average depth of the cut under each Curve Ci is also shown in Figure 4.134c and the overall average depth of the cut in the ground area is 1.64 mm.

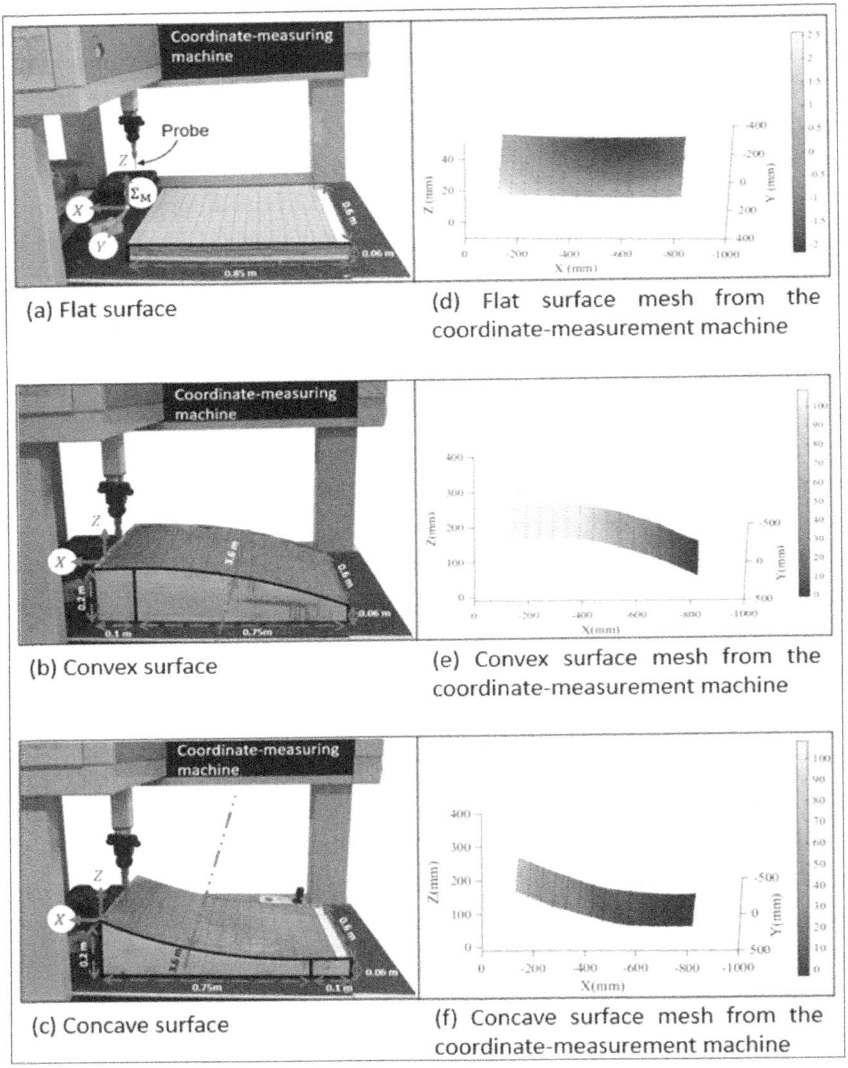

Figure 4.132. Different wall geometries. On the left side of the figure, the walls are placed on coordinate-measuring machine (CMM) to evaluate their surface before grinding. On the right side is the mesh obtained for each wall, it is measured in the machine frame.

Finally, the grinding quality on the concave surface is evaluated in Figure above. The superposition of the meshes before and after grinding is shown in Figure 4.135a. Similar to the previous case, the depth of the cut is obtained by calculating the Euclidean distances between the points of the meshes measured before and after grinding. Figure b shows the depth of the cut evaluated under the curves C2 to C5 (see Figure b). The Figure c shows that the grinding depth varies between 0 and 2.5 mm in the ground area. The average depth of the cut under each Curve

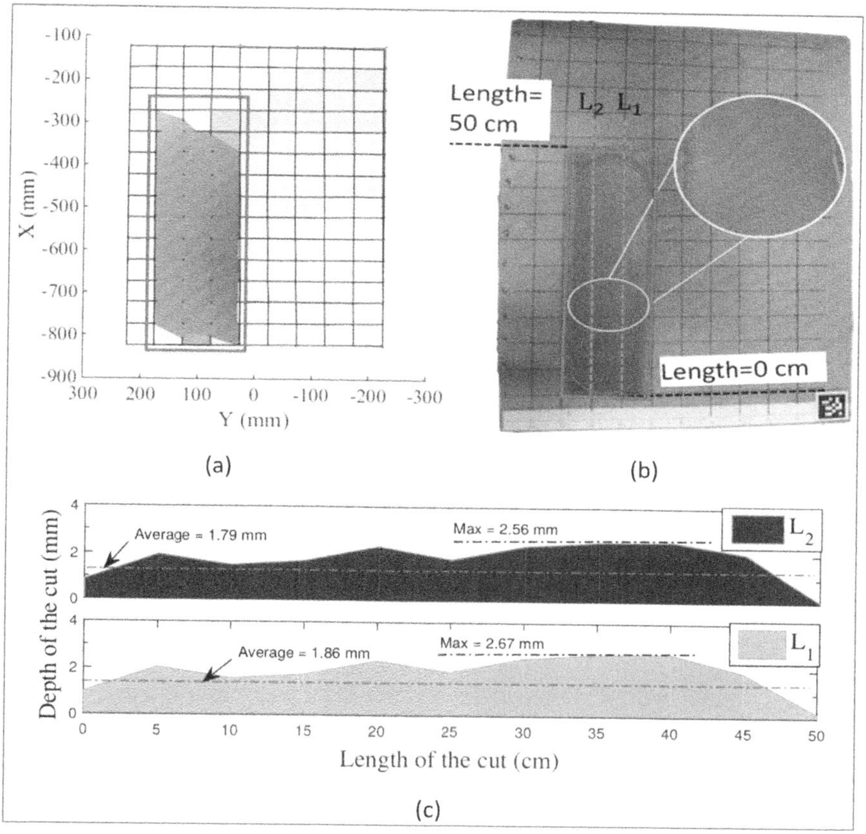

Figure 4.133. The surface state of the flat wall evaluated before and after grinding: (a) Top view of the superposed meshes describing the flat wall surface before and after grinding, (b) Flat wall after grinding, the depth of the cut is evaluated for 50cm under two lines (L1 and L2). (c) Grinding path depth evaluated under lines L1 and L2.

Ci is also shown in Figure 4.135c and the overall average depth of the cut in the ground area is 1.43 mm.

To sum up, although the geometries of the three surfaces were not known before hand, the grinding task was achieved with the disc exerting a normal force of 15 N on the wall surface while tracking its curvature. The experimental results showed good velocity, force and orientation tracking performances. Concerning the grinding quality, it was evaluated for the three test cases and good grinding was shown. The depth of the grinding cut in the flat wall case is around 1.8 mm (Figure 4.133c). For the convex wall case, the depth of the cut varies between 0 and 3.5mm (Figure 4.134c). And for the concave wall case the depth of the cut is around 2 mm (Figure 4.135c). The main reason why the depth of the cut in the tests is not constant is that it is not controlled directly, i.e., the depth of the cut should be around 2mm (the depth of the automatic grinding tests used for identifying the nominal

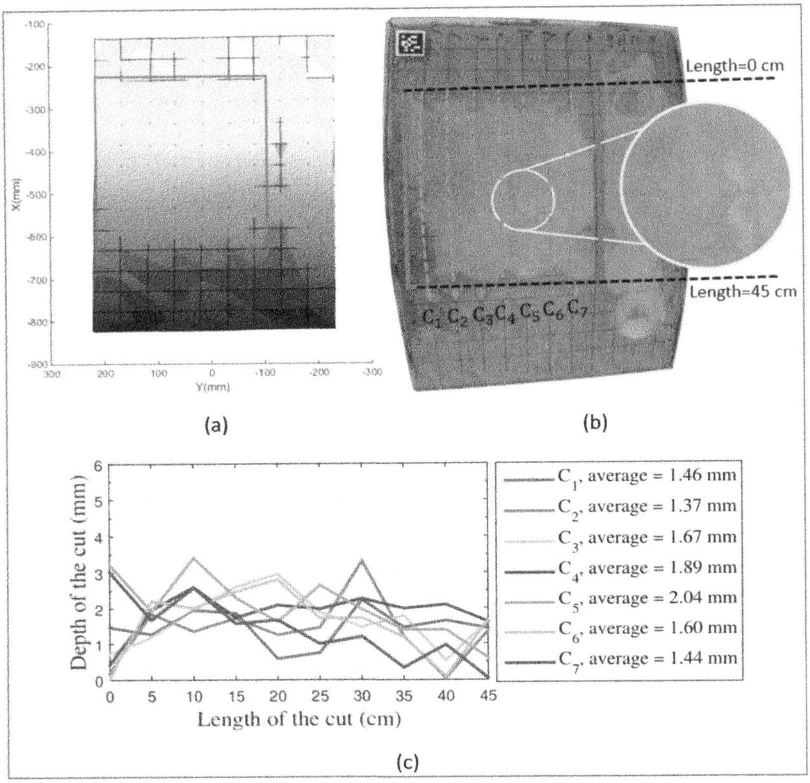

Figure 4.134. The surface state of the convex wall evaluated before and after grinding: (a) Top view of the superposed meshes describing the convex wall surface before and after grinding, (b) Convex wall after grinding, the depth of the cut is evaluated in a area with length of 45cm and under 7 curves (from C1 and C7). (c) Grinding path depth evaluated under the curves C1 to C7.

grinding lateral force) if the robot executes the identified normal force and lateral velocity on the wall. Addition to that, the walls surface was not smooth before grinding (it has some bumps). Moreover, the lateral grinding force model used is valid when the disc is in full contact with the wall. In case the radius of the curvature of the surface to be ground is small, the disc will not be in full contact with the wall, and then the lateral force model will have some error which in turn will lead to some variations in the depth of the cut.

Grinding test campaign

Because of lack of such studies and data in the public sphere a test campaign of thirty grinding tests was made in controlled conditions. The main goal is to understand with a first global practical vision the consequences of this construction materials machining operation on a robotic arm [1].

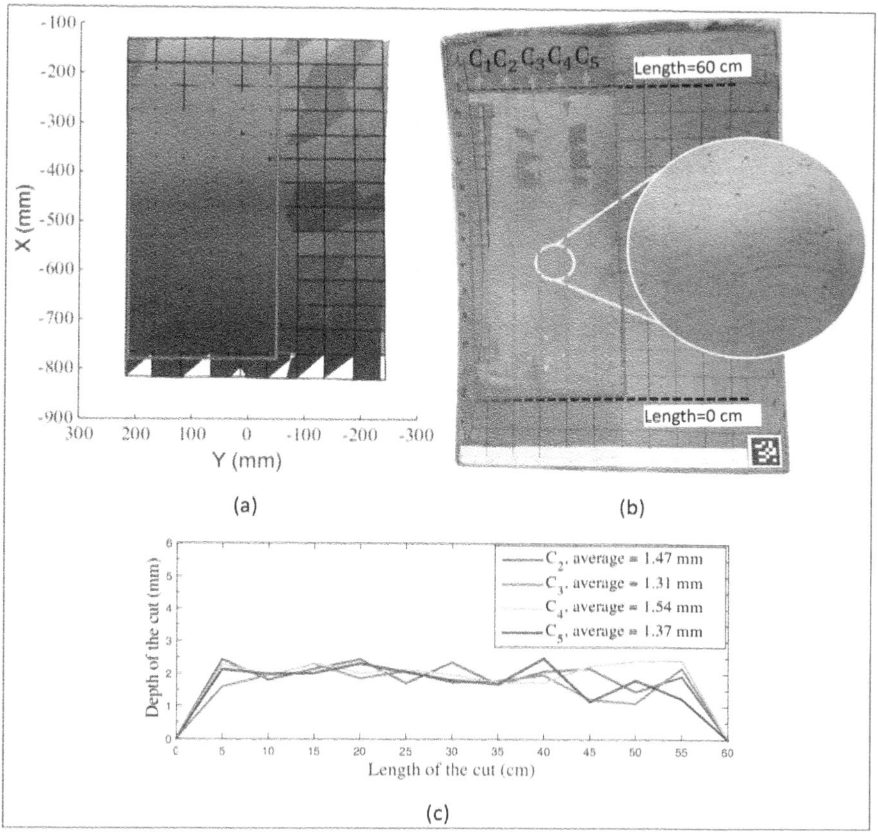

Figure 4.135. The surface state of the concave wall evaluated before and after grinding: (a) Top view of the superposed meshes describing the concave wall surface before and after grinding, (b) Concave wall after grinding, the depth of the cut is evaluated in a area with length of 60cm and under 5 curves (from C1 and C5). (c) Grinding path depth evaluated under the curves C1 to C5.

Tests care about the removal of a material layer considered as "contaminated by asbestos" from a material support with a grinding operation. This last disintegrates in dust the concerned layer by abrasion mechanism. It separates definitively the two materials sealed together. A cell test was prepared and a protocol established in order to record physical parameters and output physical variables issued of the tests done. A major part of the test set up and their course are explained in the proceeding paper published for IFToMM 2019 [1].

Parameters like feed speed, depth of cut, grinding wheel rotation speed and material machined were considered as of these constant and imposed. Physical output variables took care were in a first time the grinding forces and torques and in a second time the electrical power absorbed by the grinder.

Test made are:

- Tests I
  - Transverse grinding (figure a)
  - Constant feed-speed
  - Full resurfacing concrete layer machined
  - 9 tests were done
- Tests II
  - Transverse grinding (figure a)
  - Constant feed-speed
  - Full plaster layer machined
  - 6 tests were done
- Tests III
  - Transverse grinding (figure a)
  - Constant feed-speed
  - Intermediary layer (including resurfacing concrete, raw concrete and their border surface) machined
  - 7 tests were done
- Tests IV
  - Plunge grinding and transverse grinding (figure b)
  - Non-constant feed speed (acceleration after plunging)
  - Full resurfacing concrete layer machined
  - 4 tests were done
- Tests V
  - Plunge grinding and transverse grinding (figure b)
  - Non-constant feed speed
  - Full plaster layer machined
  - 4 tests were done

All needed parameters and all physical variables output for tests I and II were correctly recorded.

Considerations

- Two main construction material were studied: one "soft" (plaster) and one "hard" (resurfacing concrete).
- Material mechanical characteristics are unknown in the literature.
- Particular cases of construction material grinding are analyzed: recent materials without asbestos with adapted but particular tool. None are presented in literature.

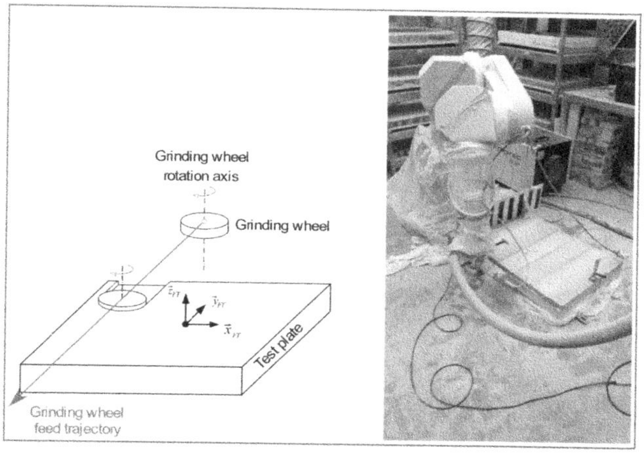

Figure 4.136. Transverse grinding operation scheme and Staübli RX130b robotic arm equipped with grinder set up in initial configuration to grind for vibration test picture.

- Measurement and machining devices used for the tests are the ones that where available at Sigma Clermont.

Data exploitation limitations

For tests IV and tests V (figure b) It was not possible to record easily the robotic arm position in space and time for tests due to the old communication interface of this machine. So, speed and accurate robotic arm segment positions are missing and not rebuildable from others data in our study. In this case it would not wise to exploit other data with this lack to analyse the tests IV and V data.

Another limitation appeared during the tests progress: Lack of knowledge about the real machined material layer. The layer has a many form errors due to material support layer form errors and due to its building technics used. As consequences the 3D volume of the studied layers are not perfect parallelepipeds. They have random forms errors making it difficult to characterize their real geometry and attribute them a precise geometric marker which locate them accurately in space. When the grinding operation only machine the contaminated material layer, and not the support material, no problem appears (figure c). However, the thickness of the contaminated layer has to be pretty significant in relations to the dimensions of the grinding tool. Otherwise a layer of the contaminated material plus a layer of the support material (interface materials layer) have to be removed (figure d).

In this case, we did not know the geometric characteristics of the frontier surface between both materials, in order to remove all the contaminated layer and the minimum volume of the support layer.

In fact, it necessary to remove the minimum of the support material to prevent from mechanical weakening of the whole concerned structured including this

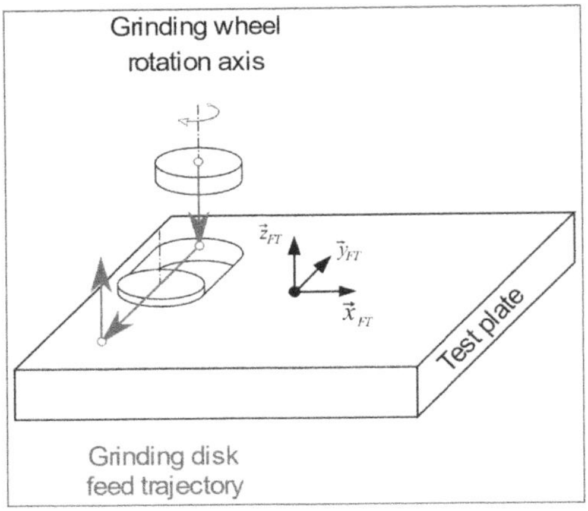

Figure 4.137. Plunge + transverse grinding operations scheme.

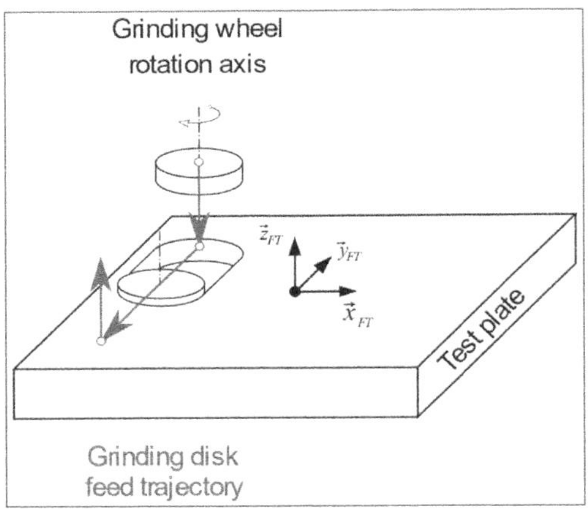

Figure 4.138. Grinding of contaminated layer.

material. At this point a new problematic appears: how can we determine the frontier surface? State of art seems leads at the uses of scanning devices designed for inspection of material layers after their construction in the building area. It would permit, without damaging the construction, to rebuild the geometry of the contaminated material layer about its material support.

In the grinding tests several successive transverse grinding operations were done until a support material surface appears free of every traces of contaminated material. But in these cases, the proportion of each machined material were unknown and finally the output physical variables could not be attributed correctly to define the machining operations. So, data of test III are not analyzable. Finally, the principal work of data treatment and analyses is concentrated in the tests I and II (Figure I).

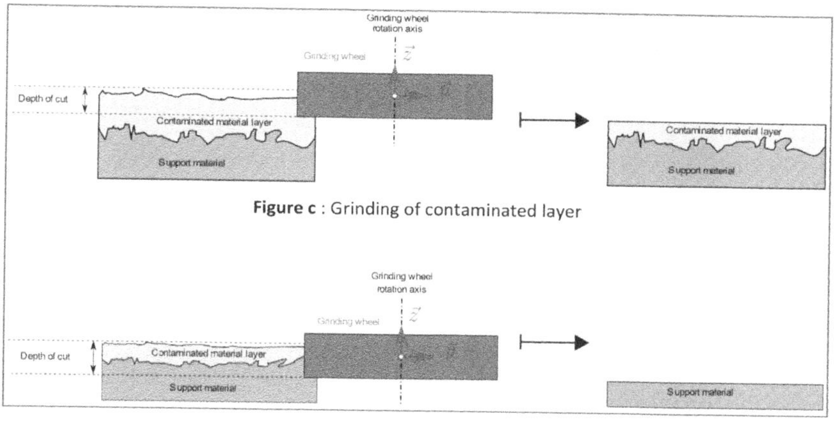

Figure 4.139. Grinding of layer of interface layer (contaminated material layer + support material layer).

Technical observations during tests

- Dust and chip production are very important and may participate in significant way to increase at the consumption electrical power when they are evacuated by the aspiration device because of rubbing phenomenon (with the grinding wheel and the dust protection cover).
- Dust protection cover does not adapt its orientation with efficiency to the surface of the layer machine when the grinder is attached to a robotic arm. The cover is too stiff and apparently lacks two rotations motions with respect to the grinder head for perfectly surrounding the grinding disk and creating an efficient depression chamber on the surface. The aspiration of the construction materials dust and chips (Numatic HZD900 [2], Power: 900W, Double filters of class H, different device than the proto V1) were not efficient enough and that is why a great) quantity of dust was disseminated in the whole test cell.
- A strong filtering device must be installed to remove the very fine particles from the cell test air and protect the operator's lungs.
- Moreover, each time dust protection cover had touched the grinding wheel because of an excessive deformation, the grinding operation was stopped.
- Mechanical vibrations, created during the grinding operations and propagated in all the test cell, felt by operators are very important in frequency and amplitude.
- The grinding wheel was damaged when grinding the steel armatures embedded into the concrete support. After this, an important decrease of its cutting performances was noticed although no observable damages were observed on the grinding wheel. We presume that a major metallurgy change in the bond material of the grinding wheel has happened.

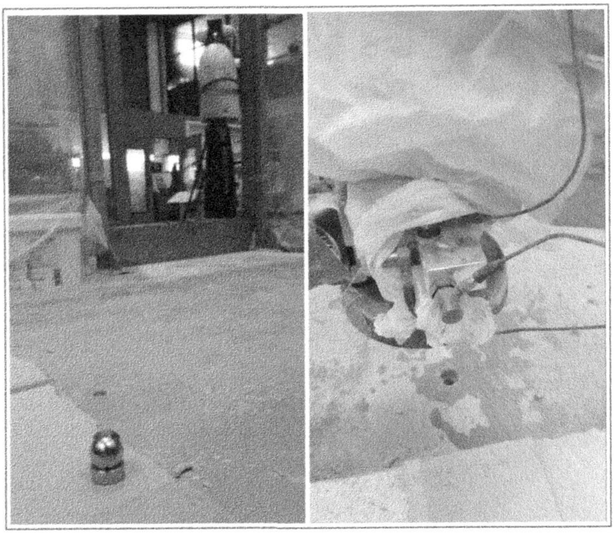

Figure 4.140. Measurement with laser tracker of one surface point from resurfacing concrete layer, Accelerometers glued on the terminal organ of the robot along its tree principal axis.

- The frontier surface position and 3D geometry between contaminated material layer and support material are unknown in spite of the care taken in the production of the test plates.
- A pinch angle of about one degree was observed between the robotic terminal segment and the grinding wheel due to the assembly of the grinder on the robotic arm.
- The assembly between the grinding wheel and the grinder spindle was not adapted for an automatic use. The constraints generated during the tests lead to a more important tightening constraint than manual use. In consequence, the grinder had to be dismounted from the robotic arm in order to change, with a lot of difficulties, the grinding wheels from the grinder.

Data treatment

Test I and II have shown that two parameters considered as constant were not: depth of cut (DOC) and rotation speed of the grinding wheel. For each studied test, they were computed with indirect data.

First, the lack of stiffness of the robotic arm coupled of the lack of relative position data between the terminal organ of the robot and the contaminated layer, completed by the lack of position loop control use for the robot, lead to a non-constant thickness of the material layer removed.

Before and after each grinding test, about twenty points of the considered layer surface were measured with a laser tracker. With an interpolation method, the

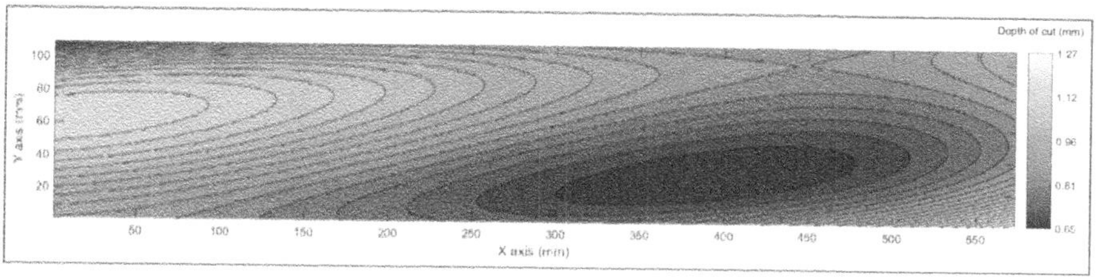

Figure 4.141. "Real DoC" computed along the tool trajectory for the RC grinding test with" imposed DoC (1 mm)" and feed-speed of 31.25 mm/s.

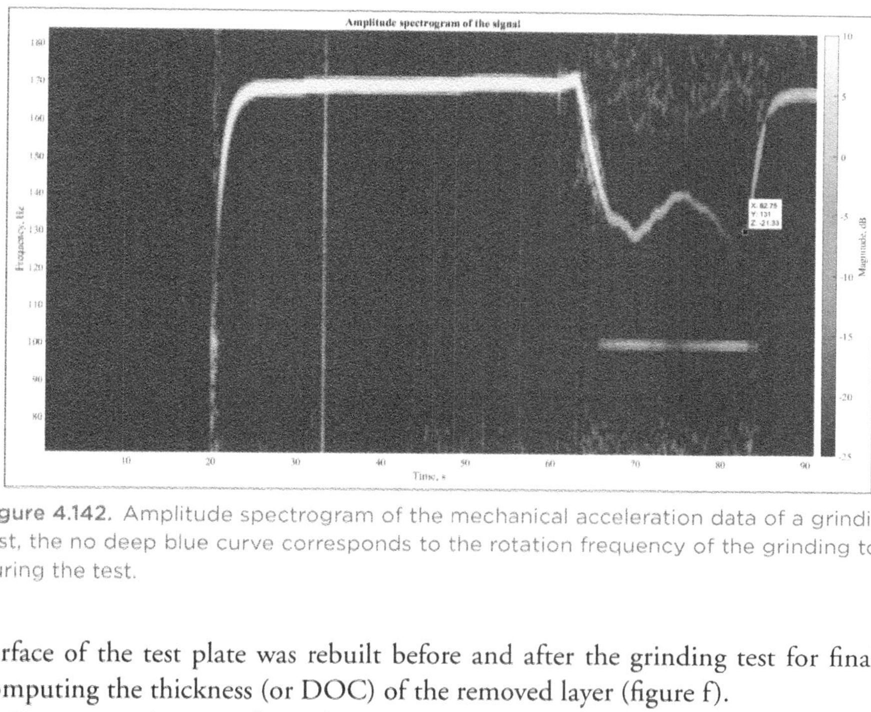

Figure 4.142. Amplitude spectrogram of the mechanical acceleration data of a grinding test, the no deep blue curve corresponds to the rotation frequency of the grinding tool during the test.

surface of the test plate was rebuilt before and after the grinding test for finally computing the thickness (or DOC) of the removed layer (figure f).

Due to an absence of speed rotation loop control implemented physically in the grinder, the grinding wheel rotation speed was not constant for each grinding test. After discussion with vibration tool specialist Dr. Vincent Gagnol, we have considered that the fundamental frequency measured by the accelerometers in the robot terminal organ was corresponding to the frequency of the grinding wheel rotation. Initially the accelerometers had to be used to characterize the vibration phenomena of the test assembly. Unfortunately, the used grinder had very poor assembly clearance and the setup of the vibration measurement chain was not precise enough (measured frequency range: 0-10 kHz), which made this complex study unfeasible. But the acceleration data gave access to the grinding wheel speed rotation with a Short Time Fourier Transformation of the measured raw acceleration data (figure g).

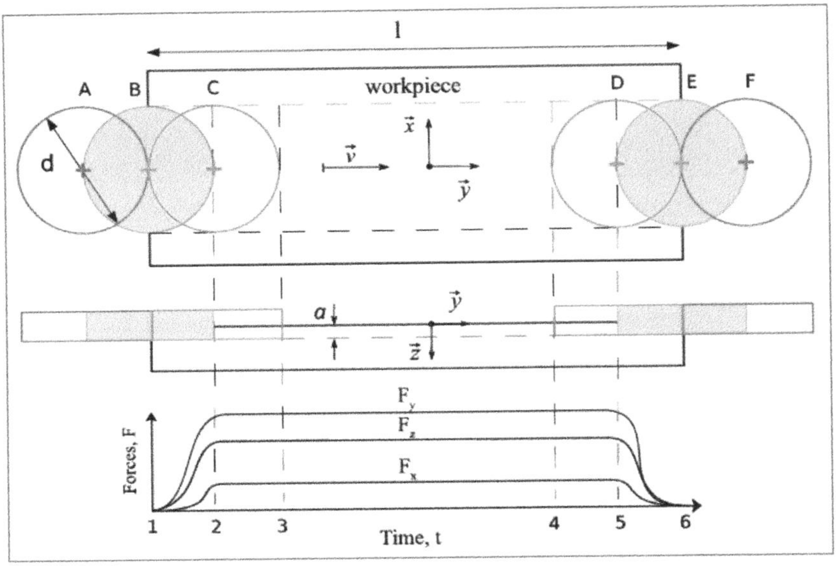

Figure 4.143. Transient cut-in and cut-out stages for vertical spindle surfacing during a Transverse Grinding Operation (TGO).

As the different measures had different temporal bases, they required a treatment for proper synchronization. Grinding torques were computed at the intersection of the grinding wheel axis and the machined material layer surface thanks to the torques measure by the force table in its reference marker. However, it appears that the grinding torques strongly depend on the arm pose, that was not recorded in these tests. So the recorded grinding torques were not analyzed because of the lacking data.

For each test of category I and II, the grinding forces, the electrical power absorbed by the grinder, the material flow rate and the grinder spindle rotation rate were computed from the raw measurements. The diagram made between both material grinding shown different behaviours for the grinding operation. Figure i reminds the different cut-in and cut-out phases of the operation.

Figure i, j, k present the curves parameters and the force grinding measured for a transverse grinding operation. Full resurfacing concrete layer is machined in this test with imposed parameters like DOF of 1 mm and feed speed (vt = 29.5 mm/s).

It makes sense to analyze the curves when the grinding disk is completely inside the test plate limit (figure h, configuration C) and omit the data when it is not. So, in the figures i, j, k, only the curves from 4,4s to 20,5s (stage from C limit to D limit or from time 3 to time 4 in the time graph of figure h).

The constant parameters normally expected for Material Flow Rate (MFR = Depth Of Cut/t in mm^3/ms) and Grinder Spindle Rotation Rate (rad/s) are

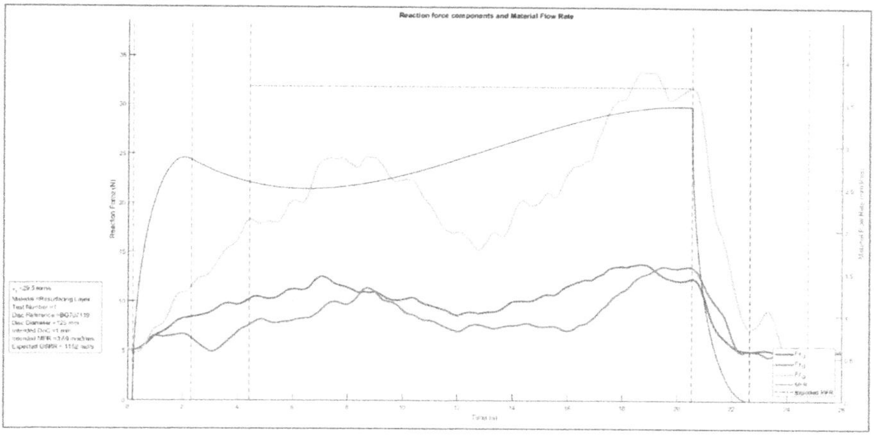

Figure 4.144. Grinding force and material flow rate for transverse grinding operation of resurfacing concrete layer. DOC targeted = 1 mm and feed speed targeted vt= 29.5 mm/s.

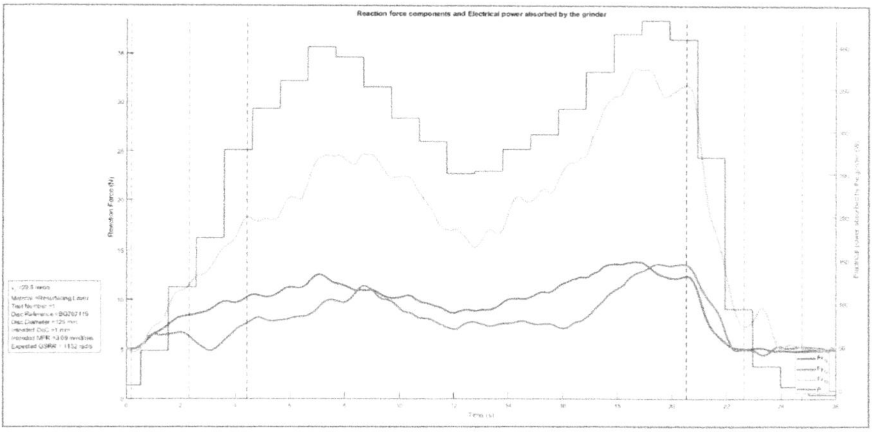

Figure 4.145. Grinding force and absorbed electrical power by the grinder for transverse grinding operation of resurfacing concrete layer. DOC targeted = 1 mm and feed speed targeted vt= 29.5 mm/s.

respectively displayed in figure j and k. As explained above and as shown in these last figures both of these parameters are non-constant and non-linear. The MFR increases from 2,5 to 3,5 mm^3/ms. The GSRR varies from 800 to 900 rad/s.

A general trend appears about the grinding force behavior. For the full resurfacing concrete material grinding tests the normal grinding force (Fzg) (Figure 4.144, 4.145, 4.146), dominates the tangential grinding forces (Fxg and Fyg). This trend is exactly inverse for plaster material (Figures 4.147, 4.148, 4.149) where the Fyg force against the toot advancement dominates. At the considered slow feed speed of 29.5 mm/s, the normal force Fzg is always under 35N for resurfacing concrete,

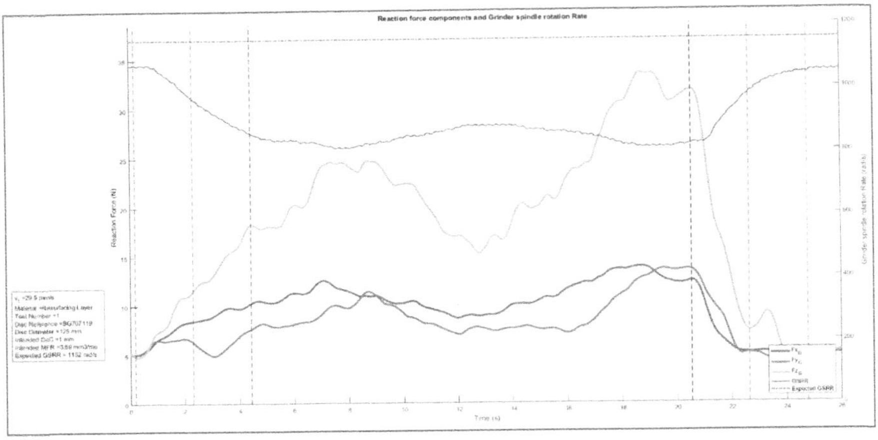

Figure 4.146. Grinding force and grinder spindle rotation rate for transverse grinding operation of resurfacing concrete layer. DOC targeted = 1 mm and feed speed targeted vt= 29.5 mm/s.

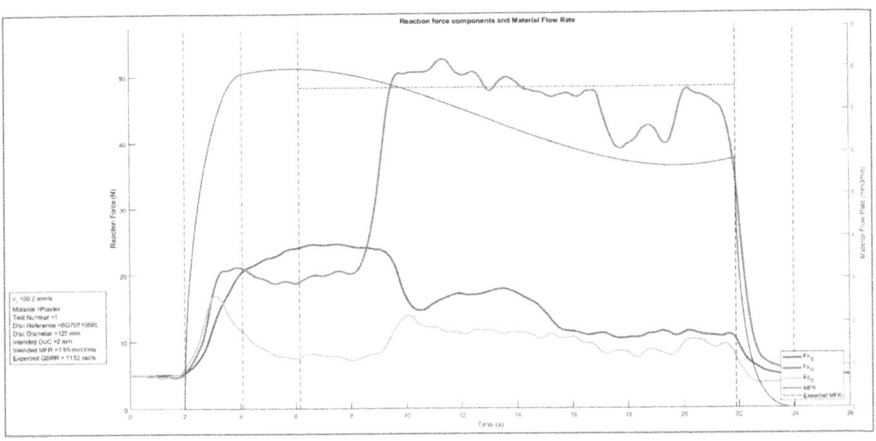

Figure 4.147. Grinding force and material flow rate for transverse grinding operation of plaster layer. DOC targeted = 1 mm and feed speed targeted vt= 29.5 mm/s.

whereas Fxg and Fyg are below 15N. For plaster and at the same feed speed, Fyg never exceeds 50N, Fxg is under 25N and Fzg under 15N. Obviously, these forces can vary a lot according to machining conditions, specifically feeding speed, that was limited on our robot. But these orders of magnitude remain interesting.

The electrical power consumed by the grinder follows the trend of the grinding normal force (Fzg) in figures j and m.

For only Fzg, a behavior asymmetry in time is observed and which find in the electrical power absorbed by the grinder. This asymmetry does not exist for the MFR (following an increase trend) and GSRR (symmetric behavior).

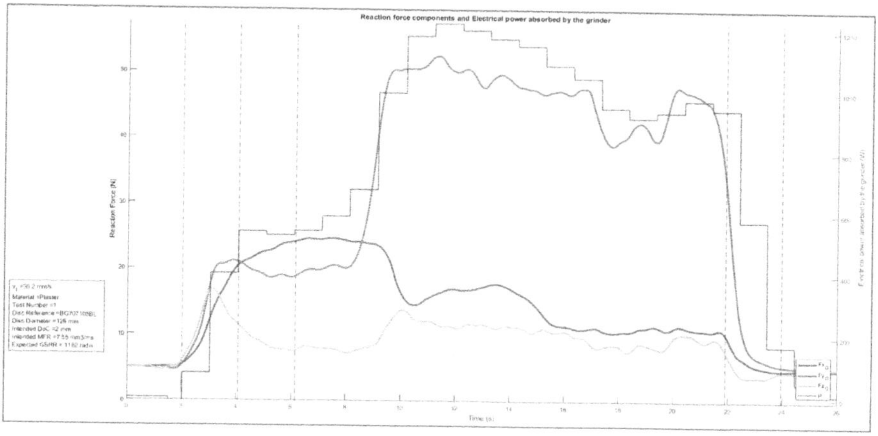

Figure 4.148. Grinding force and absorbed electrical power by the grinder for transverse grinding operation of plaster layer. DOC targeted = 1 mm and feed speed targeted vt= 29.5 mm/s.

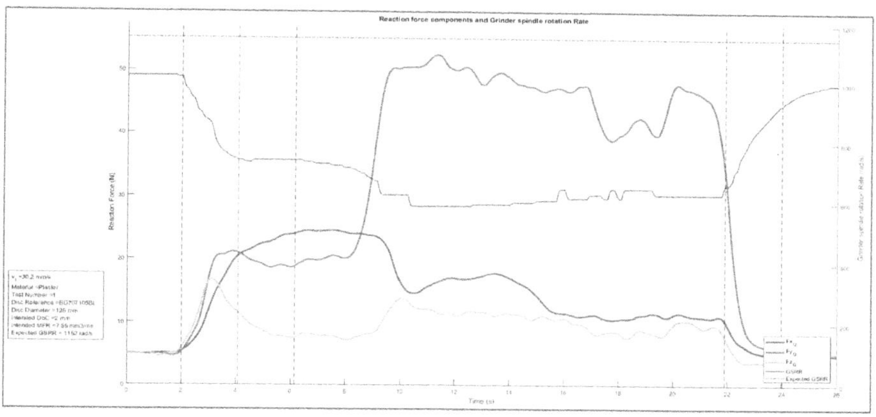

Figure 4.149. Grinding force and grinder spindle rotation rate for transverse grinding operation of plaster layer. DOC targeted = 1 mm and feed speed targeted vt= 29.5 mm/s.

The non-constant MFR shows non-constant thickness of the removed layer. We can assume the grinding torque asked on the grinding wheel by the machining operation is higher at the end of the machining operation due to the increase of material volume to remove and the GSSR low level (not controlled) for this test.

This example of test shows it is difficult to characterize the connection between the input parameters and the grinding force if the parameters are non-constant. According to the material, the dominating force components are not the same: the normal force Fgz dominates for resurfacing concrete, whereas the feeding force Fgy dominates for plaster.

Figure 4.150. Preliminary test on standard machine.

Two points are highlighted by our tests:

Firstly, to master efficiently the grinding force, the parameters (GSSR and DOC) have to be controlled. This is not simple and has to be considered for future work.

Secondly, a grinding force model (in relation with the parameters observed above) can be extracted with a realization of the experimental protocol designed here and following the different recommendations done above.

The model will be valid for one specific couple tool-material and cannot be general. However, it will give knowledge about what can happen to our prototype during grinding operation.

The model will require a non-linear regression with multiple parameters, based on our experimental results.

Tiles tool development and testing

A specific tool for tiles removal was developed and tested in parallel to main grinding tool integrated with the arm. Not to overlap with other ongoing testing activities on the integrated robotic system the tool was tested and interaction force evaluated on a independent simulated mock-up. The following section describes the solution developed for tiles removal and collect results of the different tests performed with the tool in the stand alone configuration.

The first goal of this activity was to develop and test a scaled model of a scarifier tool, broadly used in surface preparation on concrete. As first investigation a standard Blastrac 215 was tested on a tiles mock up.

Test performed with the support of Blastrac Italy shown the potential of the scarifier tool to beak tiles in a controlled manner on small collectable debris. As

Figure 4.151. 100 mm drum scarifier machine.

in previous similar cases TLABS developed a "scaled" version of the Blastrac tool using Blastract standard consumables.

The first version of this tool shown in the CAD view above was driven with a WurthEWS14-125 1400W unit. Initial trials performed with this motor shown the potential performance of the tool but the lack of control of a Wurth motor limited significantly overall tool reliability. At the same time the motor unit, forseen for manual operation, was not enough robust for continuous automated operation.

The tool was hence updated using the same driver/motor solution adopted for V2 robot arm: a Merkes MN3-350-30-320 3Nm@3000 rpm (920 W) AC brushless motor controlled in torque look with an Infranor Ethercat PAC-ET-230/11 driver. With this configuration, both torque and speed control are possible allowing full control of the tool performance. The following picture shows a view of the tool integrated with the final motor and some "traces" performed as first qualitative trials.

A second set of tests was then performed to define the interaction forces of the tool and the required torque at motor. Tool was positioned on a dedicated horizontal test surface with a lower layer of concrete and an upper layer of tiles. The tool own weight in test configuration is about 15 kg. Test were performed with different ballast on the tool and rotating speed to identify the minimum contact force required to keep stable the tool on the surface in working conditions. The following pictures shows the tool with 25,15 and 5 kg of added ballast (total mass is reported in each picture).

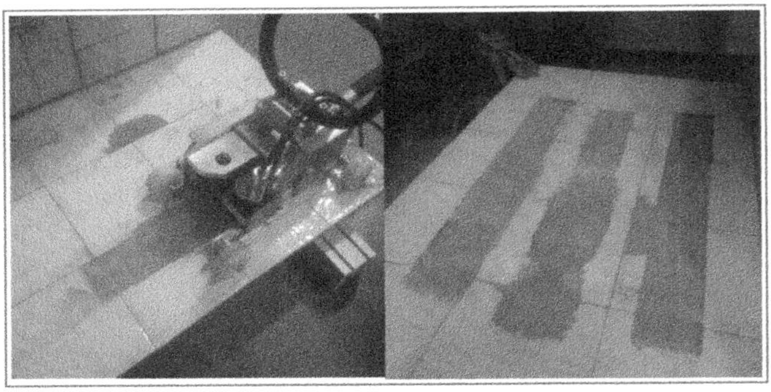

Figure 4.152. Tool in final configuration and qualitative tests.

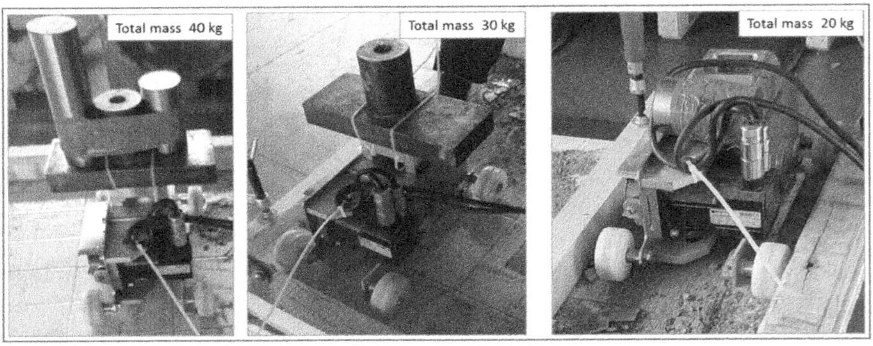

Figure 4.153. Different ballast configuration.

The following plots show the torque reading at tool motor shaft in different configuration tested:

The first two plot are torque reading for test performed with 15 kg ballast and motor set at 3000 and 2000 rpm. As expected torque increase with rotating speed as the energy fed to the tool increases.

The second plot compares torque reading at the same tool speed with different ballast mass (5kg vs 15 kg).

In this case, the two plots have roughly the same maximum torque values so ballast mass increase over the minimum set value of 5 kg has no impact on tool performance.

Tool developed for tiles removal as scaled version of a standard scarifier tool was tested in stand alone configuration with positive results. Motor and belt transmission sizing are in line with measured tool process torque (Merkes motor nominal torque is 3 Nm with peak torque of 10.5 Nm). Tool required contact force with the wall is estimated about 200 N where tool feed force is approximately 20:50 N.

The following last two pictures shows tool/robot flange forces as reference values for robot payload sizing. As discussed with tool laying on the floor an extra force of

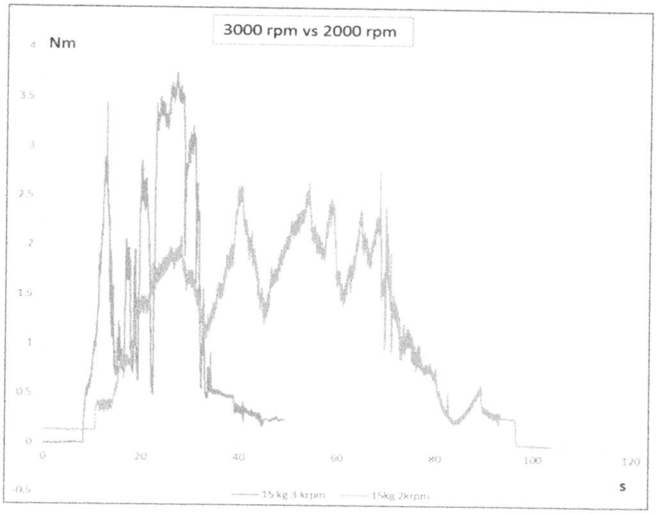

Figure 4.154. Tool speed test.

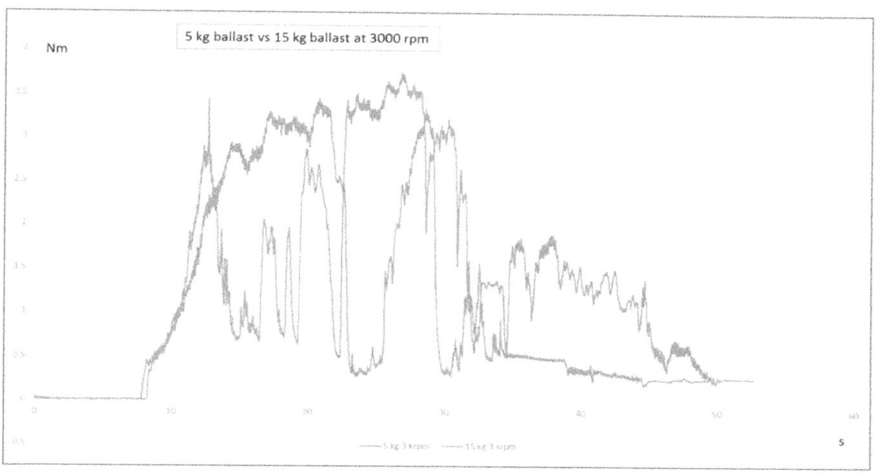

Figure 4.155. Tool ballast test.

50 N is recommended to reach a total contact force (dead weight+ added force) of 200 N. Tool feed force is 20:50 N typical. Total robot arm required process force is between 60 and 80 N while 150 N of lifting capability are required to handle the tool.

The same consideration for tool working on vertical wall gives different results: The robot arm must in this case support tool dead weight and push with the tool on the wall for 200 N. Robot payload in this case is in the range of 250:300 N. Version 2 prototype arm is not compatible with vertical wall tool processing but could be efficiently used on floor tiles removal.

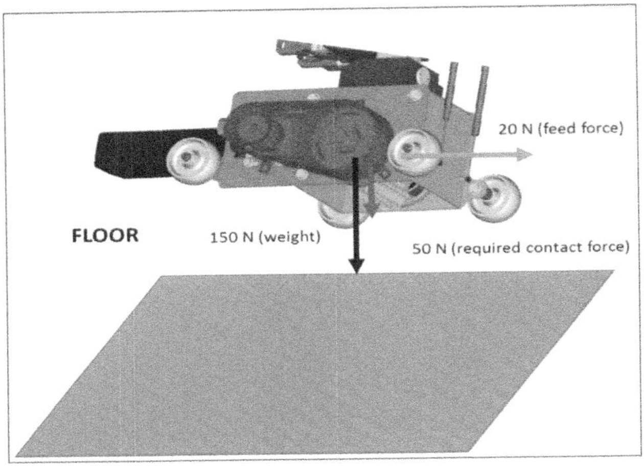

Figure 4.156. Vertical wall tool forces.

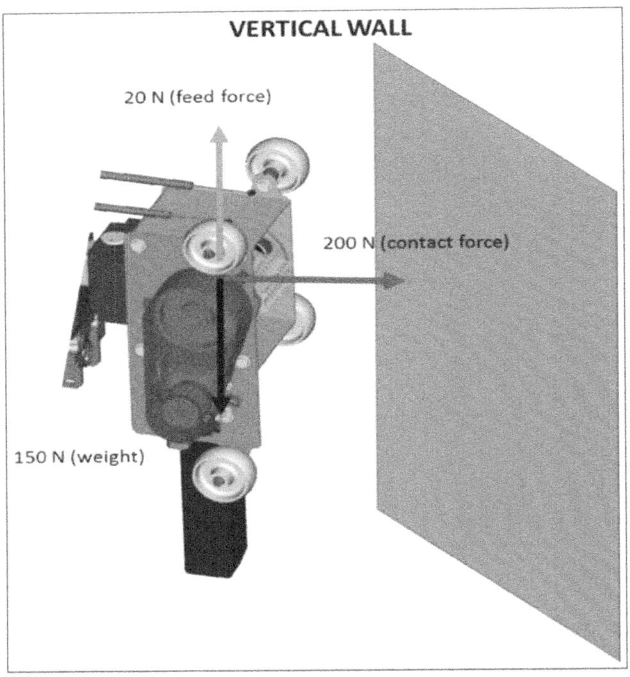

Figure 4.157. Floor tool forces.

Updates on grinding tool design

Grinding tool integration on the robotic arm was futher reviewed in the last period.

The first two prototypes built were both based on a sensorized fully compliant interface between the tool and the robot. The following picture took from the first periodic report shows the original concept.

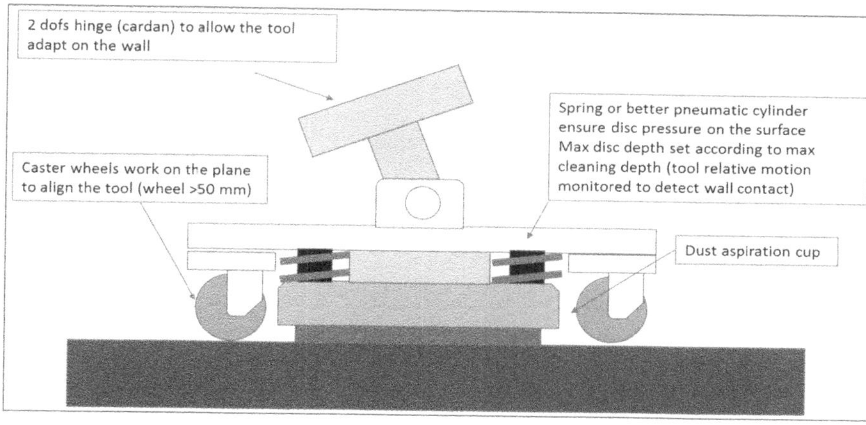

Figure 4.158. Original grinding tool concept.

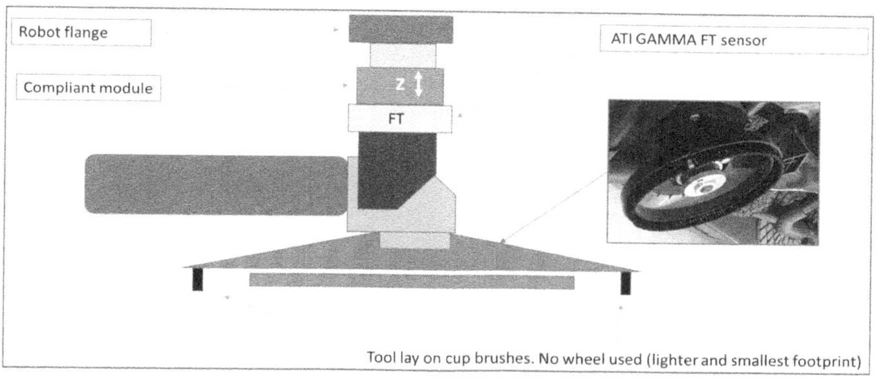

Figure 4.159. Updated Grinding Tool Concept.

This solution was implemented in two different release, first integrating caster wheels as in the sketch then with omni-wheel with a lighter and smaller tool footprint. A different, simpler approach was then developed and implemented on V2 prototype: The reference idea is to exploit, as human worker will do, the dust collection brush that surrounds the grinding disc as reference support and guide.

An active tool orientation control based on the FT sensor integrated on the tool can be used to ensure the even contact of the disc with the wall.

For smoother control, a compliant module is the integrated on the tool support so to compensate small wall/robot misalignment that may otherwise lead to a change in the grinding depth. The following picture shows a detail of the Z compliance module as integrated in the existing tool interface.

The next picture shows the new tool integrated in the V2 prototype. Tool control embed FT sensing and Indurad sensors.

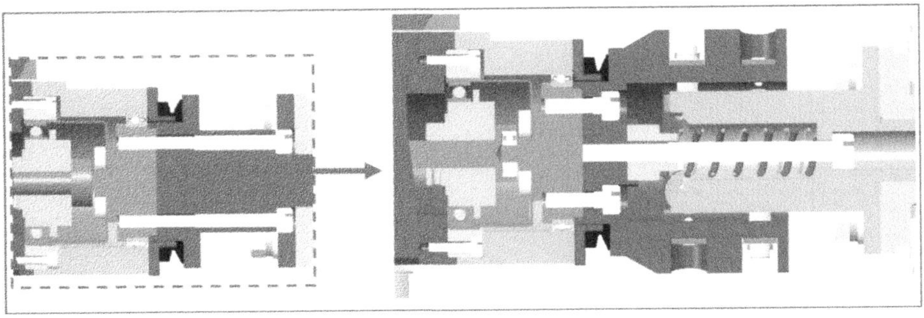

Figure 4.160. Z compliant module integration ref. design.

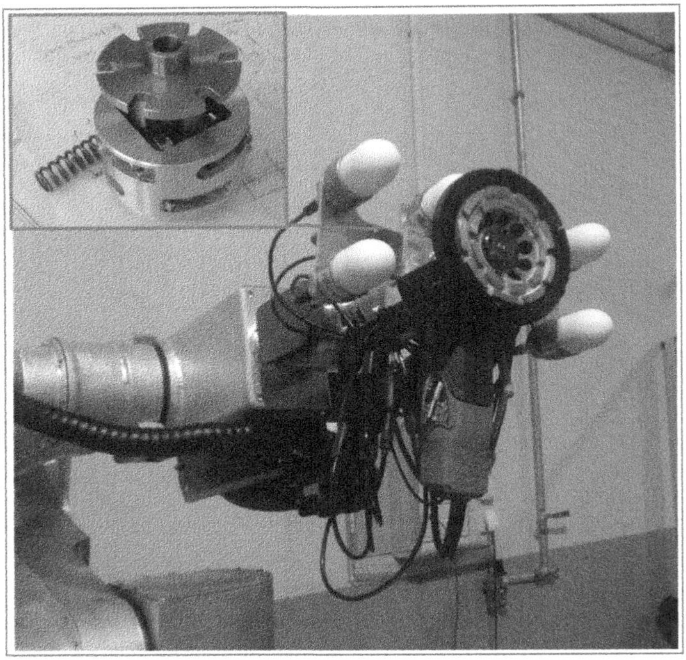

Figure 4.161. New tool integrated on the V2 robot arm.

The arm joints can be actuated independently, through the respective joint drive, or all together in a coordinated manner. In this latter case, a trajectory planner and controller is needed. For those cases where the arm moves freely in space without touching the wall, moving to a certain arm configuration or tool pose, the MoveIt! interface is used, which takes care of the calculation of the collision-free joint trajectory and its execution. For grinding, a pre-defined Cartesian path is used, which covers the wall area to be treated with an S-shaped path. In this case, both the pose of the tool and the contact pressure must be controlled within the grinding process. Hence, a simultaneous position and force control structure was implemented using

Inverse Differential Kinematics. Given the compliant property of the tool, the contact pressure can be controlled by slightly moving the robot wrist in the direction normal to the wall (without losing tool-wall contact). In this way, the error in the contact force is converted to a Cartesian speed command that is added to the one corresponding to the tool's position error. The Cartesian command is converted to a joint velocity command by the Jacobian pseudo-inverse method. In addition, use is made of the null space of the arm (it has 7 DOFs) to avoid singularities and reaching the joint limits. The joint velocity is sent to the robot, closing the loop.

DOI: 10.1561/9781680837155.ch5

Chapter 5

Results and Performances of Automated Asbestos Removal

By Roberto Guzmán, Francesco Becchi, Florian Dufour, Tobias Haschke, Miguel Moreno, Raul Molina and Tim Detert

5.1 Benchmarking of Subsystems

Deliverable D4.1
D4.1 Mobile platform V1 tested and ready for integration to the robotic system

Authors: Miguel Moreno, Roberto Guzmán

5.1.1 Deliverable in the Context of the Work Program

This deliverable report is an output from Task 4.5 "Testing of the Subsystem" to show the tests and validation procedures performed for the mobile platform of the first prototype (V1) of the Bots2Rec platform. This report contains the methods of validation and the result of the tests, beside this report pictures, videos of the robot behavior during tests have been recorded. Following to this deliverable is the integration of the mobile platform with the rest of the subsystems included in the robotic unit V1.

5.1.2 Methods and Work Carried Out

The purpose of this deliverable is to show the preliminary subsystem validation. The validation of the robot platform includes the validation of the traction system, power supply, charging system, associated safety circuits, localization, navigation sensors, the associated device drivers and interfaces to the central process control system.

The criteria to pass the validation tests is a correct behavior to fulfil the system requirements ("Specification of the technical Requirements for the V1 prototype" defined within task 3.2). These requirements are mainly related to the complete prototype, but some of them must be firstly used to validate the mobile platform, they are summarized in the following sections:

5.1.2.1 Geometry

- All units of the robotic system measurements shall comply with the following dimensions: Width <0.6 m x Length <0.8 m x Height <1.8 m with or without the dustsheet
- The robotic system shall be able to fit through an entrance of a site: 0.70 m x 1.80 m

5.1.2.2 Mass

- The overall mass of the robotic system respectively the mass pressure of each unit of the robotic system shall be <300 kg/m^2

5.1.2.2.1 Motion capability

- The robotic unit must be able to cross obstacles (level differences, thresholds, little excavations) up to 0,05m
- The robotic unit must be able to move on a slope <5% measured from the horizontal
- The robotic unit shall be able to cross one little step up to 0, 2m
- The robotic unit shall be able to move with a speed of 1 m/s
- Wheels and traction system must not damage or leave marks on the floor (or damage the plastic cover over the floor when it is present)

5.1.2.2.2 Sensors for environmental perception

- The robotic system must be able to detect and avoid (dynamic) obstacles including downstairs, upstairs, holes, persons, etc.
- The robotic system shall be able to create a local cost map of obstacles near the robotic mobile platforms

5.1.2.2.3 Navigation

- The robotic system shall allow environmental navigation
- The robotic system shall be able to navigate within a known environment

To validate the requirements regarding geometry and mass simple measurements have been done. To validate the other requirements more complex tests have been performed:

- Motion Capability: Test of the general mobility and maneuverability of the platform.
- Payload Capability: Motion of the platform with different payloads
- Plastic cover damage: Motion over the plastic surface with different payloads and trajectories
- Obstacle crossing: Crossing different obstacles (step and ditch)
- Slope climbing: move loaded platform over different slopes
- Navigation with laser sensor: Test of different mapping algorithms and robot navigation

5.1.3 Mobile Platform Tests

The mobile platform subsystems have been tested during the manufacturing processes and during the end phase of the production. The basic functionality of the complete platform including the correct performance of supply system, communications, safety stop, sensors, drives, the associated device drivers and interfaces to the process control system have been tested.

The general parameters of finished platform have been measured:

Length: 760 mm
Width: 595 mm
Height: 355 mm (without arm slider support)
Weight: 95 kg

The first two parameters meet the specification. The height and weight specifications depend on the complete robot but the platform measured values are under the expected results agreed in WP3. In a second phase higher level tests have been performed: Motion Capability, Payload Capability, Plastic cover damage, Obstacle crossing, Slope climbing and Navigation with laser sensor.

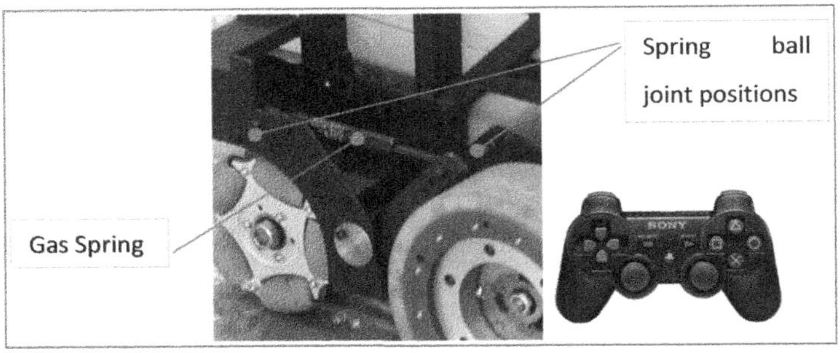

Figure 5.1. Spring location on the suspension system and manual controller used for tests.

5.1.4 Motion Capability

The platform has been tested using a manual controller for teleoperation. The general mobility and maneuverability has been tested by following different trajectories on a flat surface.

The robot has 6 wheels that transmit the weight of the system to the floor: It has 4 Omni-wheels at the corners and 2 traction wheels on the sides. Each of these driven wheels is mounted in a gas spring loaded suspension arm. The suspension system of the traction wheel allows to easily test different spring configurations by changing the spring position or the spring model. By this it is easy to change the way the robot weight is distributed between the wheels. As an example with a 220N spring, the suspension system adds a force of 26Kg to the traction wheels, while a spring of 300N adds 53 Kg.

The platform has been tested with no load (its own weight), with a payload of 110 kg (additional to the own weight) and with a payload of 200 kg using different spring configurations. As expected the platform needs a small spring to have a right grip with no load, while a strong spring causes the platform to tilt back or forward in case of strong acceleration or deceleration. On the contrary, if the platform is loaded with high weight, the grip needed in the traction wheels has to be bigger otherwise the wheels slide at high acceleration or deceleration. Based on these tests, a set of springs of 520 N has been empirically selected to move the platform with a payload of 200 Kg. Further analysis will be done with the complete robot to adjust this parameter.

A maximum speed of 1.7m/s has been measured during motion tests, which has been reduced by software to a comfortable range in a common house.

Figure 5.2. Platform loaded with 110 Kg, with 216 Kg, and with 286 Kg.

Figure 5.3. Tests over 37 microns plastic cover unloaded (left) and with 216 kg (middle) and resulting damage (right).

5.1.5 Payload Capability

The platform payload has been tested with different loads, equivalent to the weight of the arm, slider, aspiration unit and the rest of the robot components, and also with extra weight to have a safety factor. The mobility of the robot has been good and there has not been insufficient torque in any case. Some sliding on the traction wheel appeared in case of heavy load and high acceleration or deceleration.

The maximum load tested is 286 kg that considering the 95 kg of the platform results on a total weight of 381 Kg. Expected weight of the complete system is around 300 Kg so the test is considered passed.

5.1.5.1 Plastic cover damage

Moving the platform over the plastic cover has been tested with different payloads and trajectories. The results of previous tests on these plastics covers revealed the possible problems that the platform could challenge: Rotation of a single wheel on the spot was the most destructive behavior, consequently the test of this circumstance has been specifically targeted. The plastic covers used have been two Polyethylene films with different thickness: 37 microns and 125 microns.

There are no damages in the cover in general trajectories: Move straight, rotation on the spot 180° and rotation + translation 180°. Problems appear in repetitive trajectories:

Figure 5.4. Tests over 125 microns plastic cover with 216 Kg (left) and 286 kg (middle) and resulting damage (right).

- Rotation + Translation with turning radius equivalent to half distance between wheels: In this situation the center of rotation is one of the driven wheels (the wheel rotates on the spot). Damages appear around 1.5 complete turns of the robot (540°) at 37 microns plastic cover and 4 complete turns at 125 microns plastic cover.
- Rotation on the spot (turning radius zero): Several turns can cause the plastic to be pulled to center of the robot and then create folds in the film, which can cause problems in later movements over this area.

If the plastic cover is not flat, but partly folded by accident, the damages caused by movement of rotation + translation appear earlier than for a flat cover, as one of the driven wheels rotates nearly on the spot. In case of big folds (robot wheel rolls over several layers of plastic cover) damages could appear even in general trajectories.

The test is considered passed by the mobile platform subsystem, furthermore considering that usual plastic film is 125 microns thick and that trajectories will be set by the central process control system. However, for the complete robot further tests will have to be done in more realistic scenarios.

5.1.5.2 Obstacle crossing

The robot may have to surpass very different obstacles such as level differences, thresholds, little excavations, cables, and so on. Most of them unexpected until the system arrives to the working area. It is expected that obstacles are in the trajectory between working areas more often than inside working areas so they are passed in teleoperation mode. For a first test two kind of obstacles have been defined and tested: Step and Ditch

Perpendicular Step: Difference of height between two floor surfaces are usually found for example on a door threshold. 3 different steps have been tested: 10, 20 and 30 mm.

Figure 5.5. Mobile platform crossing different steps.

Figure 5.6. Mobile platform passing ditch (left) and mobile platform moving along a slope (right).

In all the tests it has been realized that it is much easier to climb the step backwards because the oscillating axle (that is in rear shaft) climbs the step easier. It is also important to arrive to the step with one wheel slightly before that the other so the first one climbs first. The results are:

- 10 mm: Crossed easily. Apparently it could be passed by the platform in automatic mode.
- 20 mm: Crossed most of times but sometimes operator has to move back and start the process of climbing again.
- 30 mm: It can be crossed but requires skilled operator and higher speed.

Ditch: It is a long, narrow open hole that is dug into the ground. The passing of 50 mm and 100 mm ditches have been tested.

The platform is able to pass 50mm ditches if the trajectory is not perpendicular to the ditch axis. Passing this obstacle is only desirable in teleoperation mode.

The platform is able to pass 100mm ditches only with a skilled operator at medium speed. At this speed if by error the platform attacks the ditch perpendicularly the platform can get blocked against the ditch with a high deceleration peak that could compromise the platform stability when the robot arm is mounted. Alternative options and presented in section 4.

Figure 5.7. Robot calculating trajectories in ROS RVIZ interface.

5.1.5.3 Slope climbing

The platform has been moved along a defined slope (see right in Figure 5.6). No difficulties were observed for a 5% slope, consequently the slope has been increased up to 13%, when the movement started to be less predictive because of eventual sliding of the wheels: In climbing in straight trajectory the robot is able to pass a 13% slope, but with additional rotation, one of the wheels starts sliding slowly and the operator has to correct commands to maintain desired trajectory.

5.1.5.4 Navigation with laser sensor

The platforms mechanical odometry parameters have been set up in the robot differential controller. The robot odometry has been tested according to Robotnik standard procedure. The procedure describes a set of trajectories starting in a position and ending on the same one and measuring the distance in the according ROS */odom* topic. The coordinate transformations of the laser range finders have been located according to mechanical manufacturing drawings. The positions of the lasers have been tested for correctness by comparing the measurements in the overlapping zones and testing the measured distances to walls.

The laser range finders have been converted to a single scan, referred to the ROS *base_link*, by the use of the ROS *laserscan_multi_merger* node (*ira_laser_tools*). Several ROS mapping algorithms have been tested succesfully in the robot:

- Karto
- GMapping
- HectorSlam

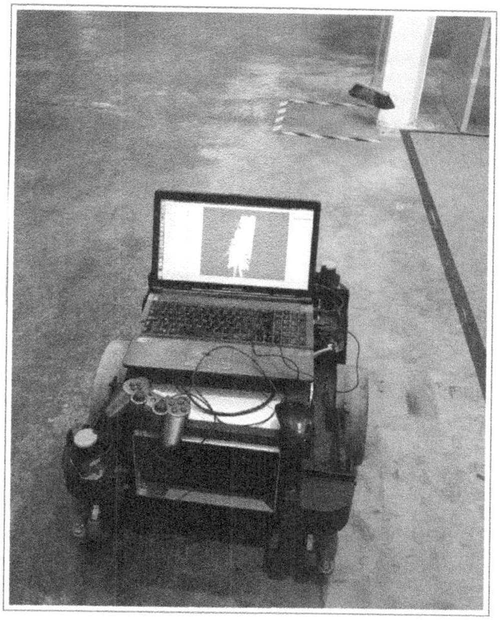

Figure 5.8. Robot AMCL localization inside Robotnik facilities.

In the Robotnik facilities and in a larger industrial area, the robot has been able to correctly map the environment. These tests have proven the ability of the robot to map 2D environments by using the range finders. The next test was to validate that the robot was able to successfully localize itself using the created maps. For this purpose the AMCL (Adaptive Monte Carlo Localization) algorithm was used.

AMCL localization has been configured and correctly tested in the two environments for about 1 hour on each environment without a single localization failure.

5.1.5.5 Outlook and future work to be carried out

The Mobile platform of prototype V1 is constructed, tested and ready for integration to the robotic system. It has passed the general mobility tests with the expected behavior and no problem is foreseen on this feature. Anyway the controller of the wheels will be modified to avoid high acceleration or deceleration commands.

Regarding the plastic cover test, the platform has proved to be capable of working in this environment. In the future it is necessary to keep in mind the exposed problem of the turning radius, and it will be considered in the trajectory control algorithms to avoid damage of the seclusion.

Regarding obstacle passing, the system is able to pass small steps and ditches but. Considering the great variety of obstacles that could appear before and after the autonomous asbestos removal work, it makes sense to prepare the environment

Figure 5.9. Final platform with all sensors and covers (left), platform with manual stabilizers (right).

before the robot enters. If there are big obstacles, simple ramps or bridges can be prepared to help the robot. As the platform has demonstrated ability to climb access ramps up to 13% they could help to pass steps much bigger than specification. Wide ditches could be passed by simple bridges, converting this big obstacle in a small step of the height of the bridge.

Navigation tests have been performed successfully and the work in real environment will define if they are valid for the operational process or if there are additional challenges due to dust interference.

However, these have been the first cycles of the design and we are expected to learn a lot of the integration process and the testing in the real environment before completing the design of version V2.

5.1.6 Summary and Conclusion

In this deliverable the validation of mobile platform for V1 prototype has been presented after performing several practical tests. We demonstrated that the platform has the necessary mobility and payload capability to carry the rest of the robot and to allow it to perform the tasks. Preliminary tests for the expected environment problems in the working areas have been performed: Different obstacles and the presence on plastic film that has not to be damaged. The capability of mapping and localization using the laser rangefinders in two different environments has been demonstrated so it is expected that they will be a good solution in the working area.

Following tasks are to integrate the mobile platform with the rest of subsystems and to perform the validation tests of the complete robot (Task 3.5: Assembly and validation tests of the robotic system).

Deliverable D4.3
D4.3 Mobile platform V2 tested and ready for integration to the robotic system

Authors: Roberto Guzmán, Raul Molina, Miguel Moreno

5.1.7 Deliverable in the Context of the Work Program

This deliverable report is an output from Task 4.5 "Testing of the Subsystem" to show the tests and validation procedures performed for the mobile platform of the first prototype (V2) of the Bots2Rec platform. This report contains the methods of validation and the result of the tests, beside this report pictures, videos of the robot behavior during tests have been recorded. Following to this deliverable is the integration of the mobile platform with the rest of the subsystems included in the robotic unit V2.

5.1.8 Methods and Work Carried Out

The purpose of this deliverable is to show the preliminary subsystem validation. The validation of the robot platform includes the validation of the traction system, power supply, charging system, associated safety circuits, localization, navigation sensors, the associated device drivers and interfaces to the central process control system.

The criteria to pass the validation tests is a correct behavior to fulfil the system requirements ("Specification of the technical Requirements for the V2 prototype" defined within task 3.2). These requirements are mainly related to the complete prototype, but some of them must be firstly used to validate the mobile platform, they are summarized in the following sections:

5.1.8.1 Geometry

- All units of the robotic system measurements shall comply with the following dimensions: Width <0.6 m x Length <0.8 m x Height <1.8 m with or without the dustsheet
- The robotic system shall be able to fit through an entrance of a site: 0.70 m x 1.80 m

5.1.8.2 Mass

- The overall mass of the robotic system respectively the mass pressure of each unit of the robotic system shall be <300 kg/m^2

5.1.8.3 Motion capability

- The robotic unit must be able to cross obstacles (level differences, thresholds, little excavations) up to 0,05m
- The robotic unit must be able to move on a slope <5% measured from the horizontal
- The robotic unit shall be able to cross one little step up to 0,02m
- The robotic unit shall be able to move with a speed of 1 m/s
- Wheels and traction system must not damage or leave marks on the floor (or damage the plastic cover over the floor when it is present)

5.1.8.4 Sensors for environmental perception

- The robotic system must be able to detect and avoid (dynamic) obstacles including downstairs, upstairs, holes, persons, etc.
- The robotic system shall be able to create a local cost map of obstacles near the robotic mobile platforms

5.1.8.5 Navigation

- The robotic system shall allow environmental navigation
- The robotic system shall be able to navigate within a known environment

To validate the requirements regarding geometry and mass simple measurements have been done. To validate the other requirements more complex tests have been performed:

- Motion Capability: Test of the general mobility and maneuverability of the platform.
- Payload Capability: Motion of the platform with different payloads
- Plastic cover damage: Motion over the plastic surface with different payloads and trajectories
- Obstacle crossing: Crossing different obstacles (step and ditch)
- Slope climbing: move loaded platform over different slopes
- Navigation with laser sensor: Test of different mapping algorithms and robot navigation

5.1.9 Mobile Platform Tests

The mobile platform subsystems have been tested during the manufacturing processes and during the end phase of the production. The basic functionality of the complete platform including the correct performance of supply system, communications, safety stop, sensors, drives, the associated device drivers and interfaces to the process control system have been tested.

The general parameters of finished platform have been measured:

Length: 800 mm
Width: 600 mm
Height: 480 mm (without arm slider support)
Weight: 110 kg

The first two parameters meet the specification. The height and weight specifications depend on the complete robot but the platform measured values are under the expected results agreed in WP3. In a second phase higher level tests have been performed: Motion Capability, Payload Capability, Obstacle crossing, Slope climbing and Navigation with laser sensor.

5.1.9.1 Motion Capability

The platform has been tested using a manual controller for teleoperation. The general mobility and maneuverability has been tested by following different trajectories on a flat surface.

The robot has 4 mecanum wheels that transmit the weight of the system to the floor. The rear axis is mounted on a damped pivoting joint to guarantee contact of the traction wheels with the ground.

The platform has been tested with no load (its own weight), with a payload of 215Kg (additional to the own weight) and with a payload of 300 kg. The platform mechanics do not need any adjustment to operate with different payloads.

A maximum speed of 2.5 m/s has been measured during motion tests, which has been reduced by software to a comfortable range in a common house.

5.1.9.2 Payload Capability

The platform payload has been tested with different loads, equivalent to the weight of the arm, slider, aspiration unit and the rest of the robot components, and also with extra weight to have a safety factor. The mobility of the robot has been good and there has not been insufficient torque in any case. There was no slippage detected in any of the payload nor acceleration / deceleration configurations.

The maximum load tested is 300 kg that considering the 110 kg of the platform results on a total weight of 410 Kg. Expected weight of each of the 2 V2 platforms is expected to be <300 Kg so the test is considered passed.

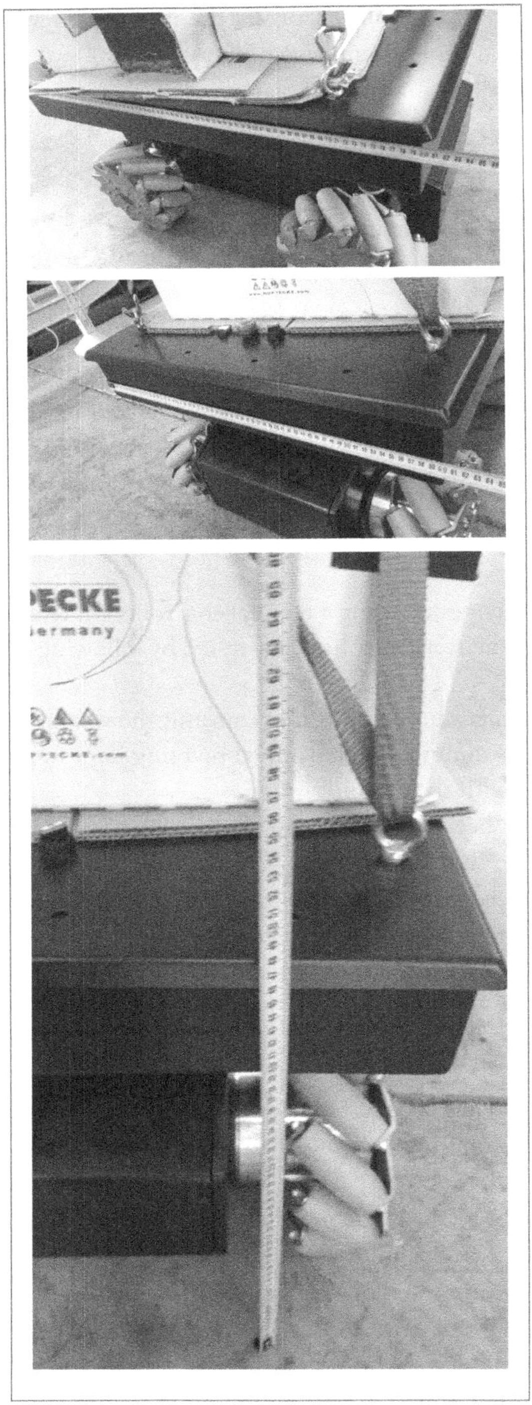

Figure 5.10. Robot dimensions L x W x H (top cover is not used in the complete robot).

Figure 5.11. Robot platform and manual controller used for tests.

Figure 5.12. Platform loaded with 215 Kg.

5.1.9.3 Obstacle crossing

The robot may have to surpass very different obstacles such as level differences, thresholds, little excavations, cables, and so on. Most of them unexpected until the system arrives to the working area. It is expected that obstacles are in the trajectory between working areas more often than inside working areas so they are passed in teleoperation mode. For a first test two kind of obstacles have been defined and tested: Step and Ditch

In all the tests it has been realized that it is much easier to climb the step backwards because the oscillating axle (that is in rear shaft) climbs the step easier. It is also important to arrive to the step with one wheel slightly before that the other so the first one climbs first. The results are:

- 10 mm: Crossed easily. Apparently it could be passed by the platform in automatic mode.

Figure 5.13. Mobile platform crossing different steps.

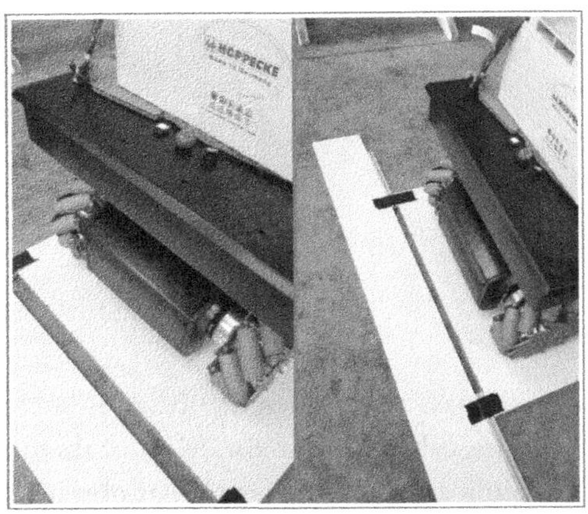

Figure 5.14. Mobile platform passing ditch of 40 mm (left) and 20 mm (right).

- 20 mm: Crossed easily. Apparently it could be passed by the platform in automatic mode.
- 30 mm: It can be crossed but due to low acceleration configuration in the drivers it cannot be done at an extremely low speed.

Ditch: It is a long, narrow open hole that is dug into the ground. The passing of 20 mm, 30 mm and 40 mm ditches have been tested.

Figure 5.15. Mobile platform climbing with no payload (left) and climbing 27% slope with 215 Kg (right).

The platform is able to pass 40 mm ditches independently of the angle of the platform with the ditch. This test could be repeated in larger distances, the robot may be able to pass ditches up to 100 mm.

5.1.9.4 Slope climbing

The platform has been moved along a defined slope (see Figure 5.15).

The robot was able to climb up to 15°, which about 27% slope loaded with 215 Kg payload. At this percentage, the current configuration of the drivers was in the limit and once the platform was stopped, it was not possible to start again upwards. This made this test stop at this 27%, but it was clear from the traction performance that the platform could climb higher slopes. The control of the platform sideways works also fine while climbing. Further tests will be done once the servoamps have been reprogrammed to know the real limit.

5.1.9.5 Navigation with laser sensor

The platforms mechanical odometry parameters have been set up in the robot differential controller. The robot odometry has been tested according to Robotnik standard procedure. The procedure describes a set of trajectories starting in a position and ending on the same one and measuring the distance in the according ROS/*odom* topic. The coordinate transformations of the laser range finders have been located according to mechanical manufacturing drawings. The positions of the lasers have been tested for correctness by comparing the measurements in the overlapping zones and testing the measured distances to walls.

The laser range finders have been converted to a single scan, referred to the ROS *base_link*, by the use of the ROS *laserscan_multi_merger* node (*ira_laser_tools*).

In this robot the default slam_gmapping configuration has been installed.

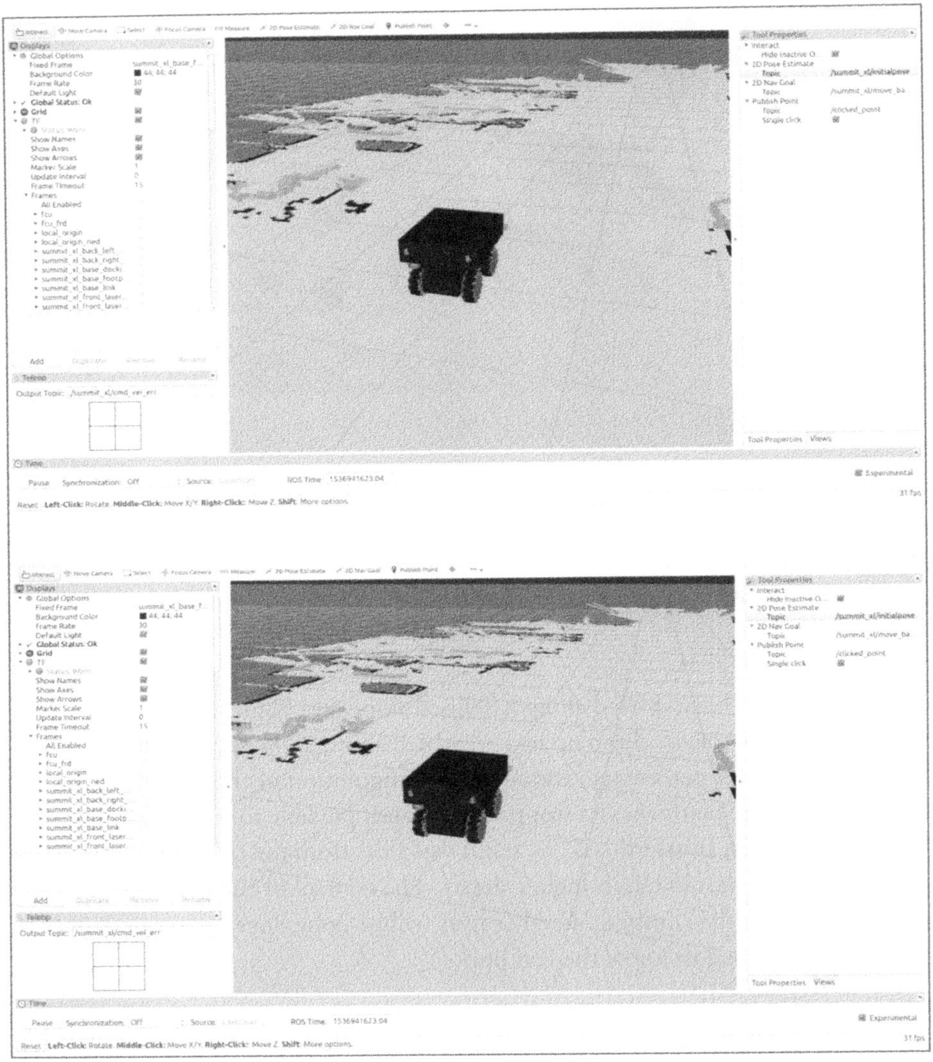

Figure 5.16. Robot navigation shown in ROS RVIZ interface.

In the Robotnik facilities the robot has been able to correctly map the environment. These tests have proven the ability of the robot to map 2D environments by using the range finders. The next test was to validate that the robot was able to successfully localize itself using the created maps. For this purpose the AMCL (Adaptive Monte Carlo Localization) algorithm was used.

AMCL localization has been configured and correctly tested in the two environments for about 1 hour in the environment without a single localization failure.

Figure 5.17. Robot AMCL localization inside Robotnik facilities.

5.1.9.6 Outlook and future work to be carried out

The Mobile platform of prototype V2 is constructed, tested and ready for integration to the robotic system. It has passed the general mobility tests with the expected behavior and no problem is foreseen on this feature.

A minor adjustment will be done in the wheel drivers before shippment to get a configuration that is comfortable to operate with high payloads.

The system is able to pass small steps and ditches. As stated for the previous robot version, considering the great variety of obstacles that could appear before and after the autonomous asbestos removal work, it makes sense to prepare the environment before the robot enters. If there are big obstacles, simple ramps or bridges can be prepared to help the robot. As the platform has demonstrated ability to climb access ramps up to 27% with the max payload they could help to pass steps much bigger than specification. Wide ditches could be passed by simple bridges, converting this big obstacle in a small step of the height of the bridge.

Navigation tests have been performed successfully and the platform is ready to mount 2 additional radar sensors in the other 2 corners, the robot will be able to localize even in the worse dust conditions.

5.1.10 Summary and Conclusion

In this deliverable the validation of mobile platform for V2 prototype has been presented after performing several practical tests. We demonstrated that the platform has the necessary mobility and payload capabilities to carry the rest of the robot and to allow it to perform the tasks. Preliminary tests for the expected environment problems in the working areas have been performed: different obstacles, ramps, ditches. The capability of mapping and localization using the laser rangefinders in our facilities has been demonstrated so it is expected that they will be a good solution in the working area.

Following tasks are to integrate the mobile platform with the rest of subsystems and to perform the validation tests of the complete robot (Task 3.5: Assembly and validation tests of the robotic system).

Deliverable 5.1
D5.1 Robot Arm V1 tested and ready for integration to the robotic system

Authors: Francesco Becchi

5.1.11 Deliverable in the Context of the Work Program

This deliverable is an output from *WP 5: Customization of the robotic arm: Design, control and implementation* and in particular from *Task 5.5 Testing of the sub system.* The deliverable summarizes the outcomes of the test procedure implemented on the V1 robotic arm before its shipment to the integration site.

5.1.12 Methods and Work Carried Out

With reference to the Robotic System V1 architecture this report focuses to the set of subsystems integrated with the robotic arm.

More specific the modules that will be covered in this test report are

- Robotic Arm
- Aspiration
- Supply cable automated reel
- Arm grinder tool for wall and ceiling

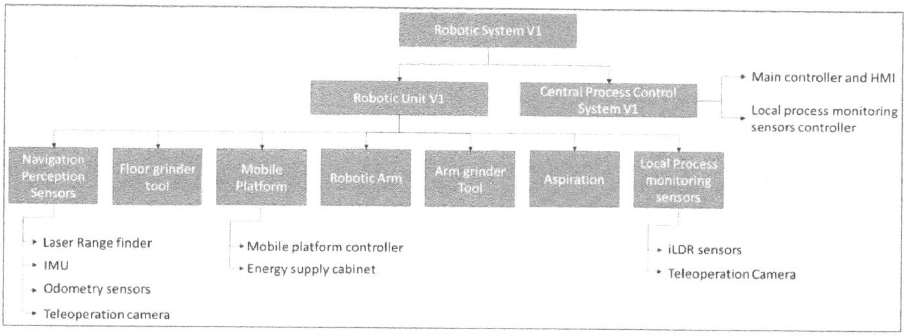

Figure 5.18. General system architecture.

In line with TLabs Quality Control system FAT (Final Acceptance Test), the modules undergo a test procedure aimed at the incremental verification and validation of all relevant subsystems.

The main structure of the test sequence covers the different topics resumed in the list reported below:

1.0 *Arm integration*
1.1 *Robot arm power supply*
1.2 *Robot arm electrical interfaces*
1.3 *Emergency loop ok*
1.4 *Motor calibration*
1.5 *Ethercat network configuration ok- motors*
1.6 *Ethercat network configuration ok- IOS*
1.7 *NUC/Beckhoff communication*
1.8 *Tool*
1.9 *Dust extractor*
1.10 *Dust extractor aspiration valve*
1.11 *Dust extractor top valve*
1.12 *Cable reel actuation*
1.13 *Cable reel encoder*
2.1 *Arm test sequence*

The following section will detail the results of the test performed.

5.1.13 Robotic Arm Test Report

5.1.13.1 Arm integration

The scope of this first test is to check that the robotic arm and the main subsystems are correctly integrated according to the reference CAD model. The reference cad model considered for this purpose is the T000120A001.asm (available in the SharePoint platform for data exchange). The following table summarizes the tests done and results achieved:

§	Short Description	ok	nok	NA	Remarks
1.0.1	Robot arm is correctly integrated on the vertical slider and all actuators and sensors are correctly connected to cabinet B2	x			No relevant issues found in the assembly. No relevant review/update on reference drawings
1.0.2	Cabinet B1 is integrated	x			
1.0.3	Cabinet B2 is integrated	x			
1.0.4	Tool is correctly integrated and connected to B1	x			Connection of tool moved to B2 for better internal space management.
1.0.5	Dust extractor is correctly integrated and connected to B1	x			Electrical connection moved to B2 for better internal space management.
1.0.6	Cable reel is correctly integrated and connected to B1	x			
1.0.7	FT sensor integrated and connected to its controller			x	Dummy sensor integrated. Sensor integration will be performed at a later stage
1.0.8	iLDR integrated and connected to ILDR NUC			x	Support frames integrated on tool support. iLDR integration postponed
1.0.9	Global cables routing compliant with robot arm motion	x			
1.0.10	Aspiration pipe integrated	x			

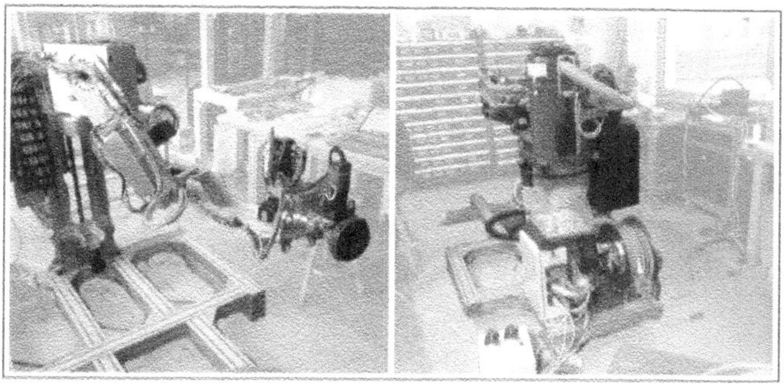

Figure 5.19. Robotic arm V1 prototype after complete integration .

As visible in the above table all robotic arm subsystems are correctly installed. Only the Force Torque (FT) sensor and the iLDR radar system are not yet integrated in the robot at this stage. The following pictures show front and rear view of the system fully integrated. A dummy frame is used in this stage to reproduce the platform interfaces and allow the correct positioning of the robotic arm ancillaries.

5.1.13.2 Robotic arm power supply

With this test, it is checked if the robotic arm power supply is correctly managed and is ready for system integration. In line with the robotic arm architecture the power supply is divided over a 230VAC and a 48VDC line, the latter provided in the integrated configuration from the platform battery pack. In this stage 48VDC are provide with a 20 A power supply.

5.1.13.3 Robotic arm power electrical interfaces

The goal of this step is to check that the connection between the robotic arm and platform are compliant with the defined interfaces. The interface was successfully verified on pre-integration meeting held in TLABS on 17/02/2017 with Eurecat.

5.1.13.4 Emergency loop ok

The scope of the test is to check that the emergency loop is correctly implemented and working. The functionality of the emergency buttons located and integrated at the robotic arm cabinet and the functionality of the button integrated in the junction box at the end of the cable reel is correct. The functionality of the bypass connector that can be used replacing the junction box (with no bypass connector in place the emergency loop is open) was also tested.

5.1.13.5 Motor calibration

The motor calibration test ensures that driver and motor are correctly connected and tuned. Referencing to the home position and jog control can be activated by the motor driver configuration interface. The following table summarizes the test performed (all tests were positive):

§	Short	Description	ok	nok	NA	Remarks
01.04.2001	motor 1 calibration	driver and motor are correctly connected and tuned.Homing and jog control can be activated by Copley configuration interface	x			arm yaw
01.04.2002	motor 2 calibration	same as above	x			arm shoulder pitch
01.04.2003	motor 3 calibration	same as above	x			elbow pitch
01.04.2004	motor 4 calibration	same as above	x			upper arm roll
01.04.2005	motor 5 calibration	same as above	x			wrist pitch
01.04.2006	motor 6 calibration	same as above	x			wrist roll
01.04.2007	motor 7 calibration	same as above	x			slider

5.1.13.6 Ethercat network configuration ok: motors network

The scope of this test is to verify the correct configuration of the Copley motor drivers in the Ethercat network. All motors are correctly mapped on the Beckhoff PC via Ethercat protocol. The test was passed.

5.1.13.7 Ethercat network configuration ok: I/O network

The scope of this test is to verify the correct configuration of the Beckhoff Input Output (I/O) modules connected in the Ethercat network. All inputs and outputs are correctly mapped on the Beckhoff PC via Ethercat protocol.

5.1.13.8 NUC/Beckhoff communication

The communication between Beckhoff and NUC was tested establishing the ADS communication protocol. The NUC implementation is ongoing through an update of the operating system from Ubuntu to “Real Time Preempt”. Consequently, the communication between NUC and Beckhoff will be updated in the system integration and testing phase.

5.1.13.9 Tool

The grinding tool is integrated on the robotic arm and connected to robotic arm cabinet. The tool can be activated and deactivated from Beckhoff output signals.

5.1.13.10 Dust extractor

The dust extractor is integrated on the slider structure and connected to robotic arm cabinet. Aspirator can be activated and deactivated from Beckhoff output signals.

5.1.13.11 Dust extractor aspiration valve and top valve

The dust extractor aspiration and top valve are each actuated with a linear electrical actuator, controlled via two Beckhoff outputs. The position of the actuator is detected with two proximity sensors at the two end of stroke. The actuator activation in the two directions and sensor feedback are correctly managed from the Beckhoff IOs.

5.1.13.12 Cable reel actuation

The cable reel motor is integrated on the slider structure and connected to the robotic arm cabinet. The reel can be rotated correctly in the two directions (wind/unwind) using Beckhoff output signals. If an overload is applied to the cable the friction transmission slides (preload can be tuned with the belt tension). When the motor is off, it can be actuated passively pulling on the cable (gearbox is back drivable).

5.1.13.13 Cable reel encoder

The cable reel encoder is connected to the Ethercat bus. The encoder is correctly mapped on the Beckhoff. A rotation of the reel is correctly detected. The encoder is connected to the reel with a timing belt, independent from the motor transmission.

5.1.13.14 Arm test sequence

A test displacement sequence of the arm for all joints was tested using a simplified user interface running on the Beckhoff embedded PC. The referencing procedure for the home position for all robot joint is programmed, tested and documented in a dedicated guideline document (T000120M00_quick start guide).

Figure 5.20. Shipment of the Robotic Arm V1 after complete test and integration.

5.1.13.15 Dust extractor cleaning sequence

The cleaning sequence of the dust extractor was tested successfully as follows:

1. The dust extractor is activated;
2. The aspiration valve is closed;
3. The aspirator top valve is cyclically opened and closed;
4. The aspiration valve is opened;
5. The dust extractor is deactivated.

The entire signals related to the sequence were correctly actuated and read with the Beckhoff interface.

5.1.14 Outlook and Future Work to be Carried Out

The robotic arm prototype V1 was integrated and successfully tested. The Force Torque sensor and iLDR integration together with further testing is planned in the integration activity for the full robotic system. At least two more tools will be integrated and tested with the same arm during the ongoing testing and development. The tool development is currently ongoing and tool integration and test will follow.

5.1.15 Summary and Conclusion

The robot integration of the V1 robot arm prototype was successfully validated following a detailed set of tests discussed in this report. No relevant issues were found at this stage and the robot was consequently shipped for platform integration:

Deliverable 5.2
D5.2 Robot Arm V2 tested and ready for integration to the robotic system

Authors: F. Becchi

5.1.16 Deliverable in the Context of the Work Program

This deliverable is an output from WP 5: Customization of the robotic arm: Design, control and implementation and in particular from Task 5.5 Testing of the sub system.

The deliverable resumes the outcomes of the test procedure implemented on the V2 robotic arm before its shipment to the integration site.

5.1.17 Methods and Work Carried Out

With reference to the Robotic System V2 architecture, this report focus to the set of subsystem integrated with the robotic arm.

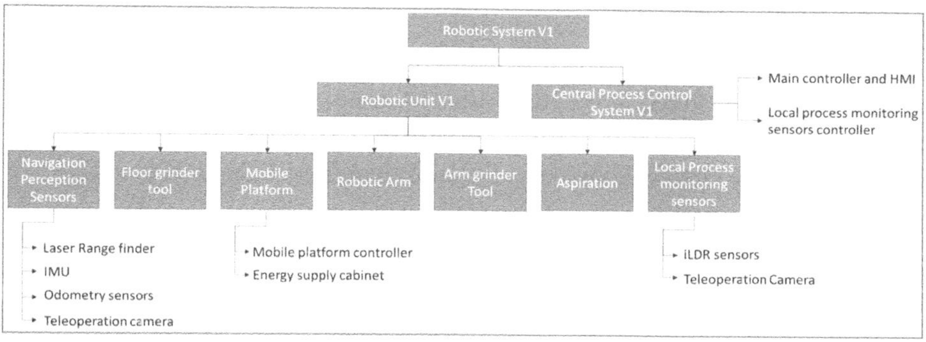

Figure 5.21. General system architecture.

More specifically the modules that will be covered in this test report are

- o Robotic Arm

In line with TLABS QC system FAT (final acceptance test) is managed through a test procedure aimed at the incremental verification and validation of all relevant subsystems.

The main structure of the test sequence covers the different topics resumed in the list below reported:

1.0 arm integration
1.1 robot arm power supply
1.2 robot arm electrical interfaces
1.3 emergency loop ok
1.4 motor calibration
1.5 ethercat network configuration ok- motors
1.6 ethercat network configuration ok- IOS
1.7 tool
1.8 arm calibration check
1.9 arm test sequence

The following section will details the results of the test performed.

5.1.18 Robot Arm Test Report

5.1.18.1 Arm Integration

Scope of this first test is to check that robot arm and main subsystems are correctly integrated according to the reference CAD model.

The reference CAD model considered for this purpose is the T000120A100.asm shown below for reference. The following table summarizes tests done and results achieved.

As visible in the above table all robot arm subsystems are correctly installed. Tool is currently not included in the integration. A review on the "reference" tool was performed and a new tool is currently in construction.

The following pictures show some views of the system being integrated.

5.1.19 Outlook and Future Work to be Carried Out

Within the next weeks, the following test will be performed in order to ensure, that the robotic arm is ready for the final integration at Eurecat facilities.

Table 5.1. Tests and results.

§	Short	Description	ok	nok	NA	Ref.	Remarks
1.0	arm integration	robot arm and main subsystems are correctly integrated according to reference CAD model	x				
1.0.1	robot arm is correctly assembled		x				
1.0.2	robot arm is integrated on robot cabinet		x				
1.0.3	tool is integrated on the robot			x			tool integration is postponed with respect to arm integration
1.0.4	cabinet is integrated with the motor drivers, Beckhoff CPU and other auxiliary components		x				
1.0.5	robot arm is electrically connected to robot cabinet		x				

Test schedule:

1. Robot arm power supply
2. Robot arm power electrical interfaces
3. Emergency loop ok
4. Motor calibration
5. Ethercat network configuration ok- motors network
6. Ethercat network configuration ok- I/O network
7. Beckhoff arm calibration test
8. Tool
9. Arm test sequence

Figure 5.22. Arm assembly.

Figure 5.23. Arm assembly.

This deliverable will be updated accordingly as soon as the tests are finalized.

5.1.20 Summary and Conclusion

The robotic arm for the V2 prototype was manufactured and assembled. First tests have been performed with satisfying results. Before the robots will be shipped to the Eurecat facilities for the final integration, further tests will be done according to the above listed schedule. After completion, this deliverable will be updated and finalized.

Figure 5.24. Arm integration.

ANNEX: Full Table To Be Completed

§	Short	Description	ok	nok	NA	Ref.	Remarks
1.0	arm integration	robot arm and main subsystems are correctly integrated according to reference CAD model					
1.0.1	robot arm is correctly assembled		x				

§	Short	Description	ok	nok	NA	Ref.	Remarks
1.0.2	robot arm is integrated on robot cabinet		x				
1.0.3	tool is integrated on the robot		x				tool integration is postponed with respect to arm integration
1.0.4	cabinet is integrated with the motor drivers, bechoff CPU and other auxiliary components						
1.0.5	robot arm is electrically connected to robot cabinet						
1.1	robot arm power supply	the robot arm power supply is correctly managed and is ready for system integration					
1.2	robot arm electrical interfaces	connection btw robot arm and platform are compliant with the defined interface					

§	Short	Description	ok	nok	NA	Ref.	Remarks
1.3	emergency loop ok	the emergency loop is correctly implemented and working					
1.4	motor calibration						
1.5	ethercat network configuration ok- motors	all motors are correctly mapped of the beckhoff PC via ethercat protocol					
1.6	ethercat network configuration ok- IOS	all IOS are correctly mapped of the beckhoff					
1.7	tool	tool can be activated b dedicated beckhoff output signal					
1.8	arm calibration check	beckhoff sw manage robo calibration.					
1.9	arm test sequence	simple test sequence is implemented o beckhoff. Moto are moved on b on and the low level configuration contro stops motion in case of need.					

5.2 Benchmarking Results of the Entire System

Deliverable 1.3
Document with benchmarking directions, specifying detailed performance objectives and metrics

Authors: Florian Dufour, Tim Detert

5.2.1 Deliverable in the Context of the Work Program

To judge the economic viability and to enhance the number of buildings cleaned, it must be guaranteed that the robotic process and system can achieve the best performances in terms of quality and in terms of time saving.

Performance objectives and metrics were developed for this deliverable to benchmark the operational process, the system, and its integration to an intervention on an asbestos contaminated site (as specified in task 1.1) under real life conditions, as referred to in task 1.5.

The task 1.3 is performed parallel to the development of the operational process (task 1.4) to be able to identify important factors for the automated process.

5.2.2 Methods and Work Carried Out

Before being able to define any objectives, we had to define a typical process for the robot, as a basis for measurements and improvements.

The first step was to analyze and document the process, from the arrival of the transport of the robot on site to the end of the intervention, with the robot back to the transport. In parallel to this, we analyzed and documented a typical day for a human asbestos removal worker, as a comparison base.

Based on this, we will perform a precise time analysis of the overall process and its sub-tasks, as main indicator for the performances assessment. Additional to this time analysis, we will define the performance objectives needed to ensure a safe process in accordance with regulations. Particular attention will be paid to the dust levels monitoring and the way to manage wastes.

Finally, the objectives for the degree of autonomy for the process control and the usability of the user interface will be defined.

5.2.3 Benchmarking Directions, Specifying Detailed Performance Objectives and Metrics

5.2.3.1 Robotic process

As a first step for the performance objectives and metrics, we analyzed and documented a general process, taking in account all the stages of the project, from the arrival of the robot on site, passing by its routing to the working area, the removal action, till the way back to the transport means.

This process is divided in sub-tasks, to define on which ones we can focus and which we can't impact. The documentation of the process was done as a maximum detailed one, with different cases and way we can deal with the tasks.

We start by the most basic level, as following:

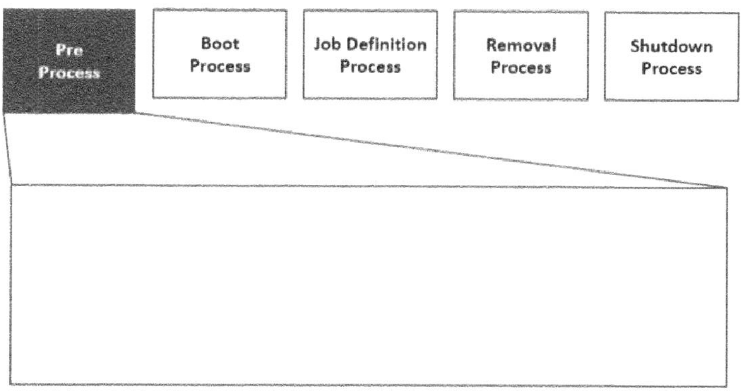

Then, we got deeper and deeper in the process, to finish at a detailed level like following:

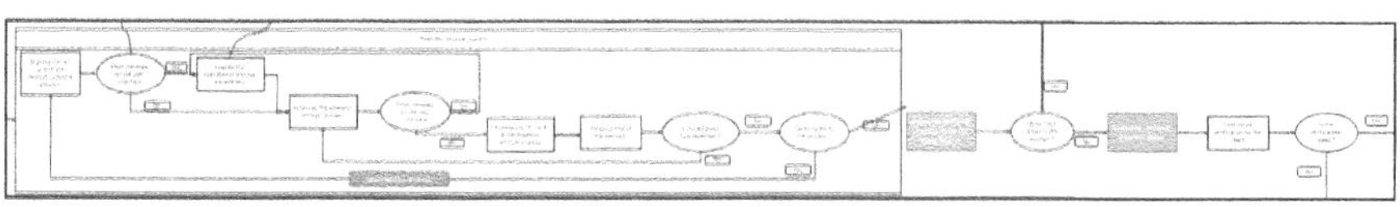

5.2.3.2 Overall timing estimation

After documentation of the process, we summarized the different main actions in a simpler process:

Thanks to this, we were able to estimate what would impact, in terms of timing, the process, and to compare it with a human. For example, we know that a worker needs 15 to 20 min to dress before getting into the secluded area, oppositely to the robot, which would be able to go inside faster. However, a worker can go faster than the robot to the right floor, so we will have to be as efficient as possible to bring the robot there, to lose as less time as possible.

In parallel, we will create a table that will allow us to compare the robot with a human worker more precisely during the different stages. First, we will perform identical tests for the main process tasks in the Bourg les Valence testing facility with both, human workers and the robotic solution. The idea is to perform a typical process, grinding 3 walls for example, in the exactly same conditions. This will take into account the time to get inside the seclusion, the time to process and the time to clean and get out. Thanks to this, we will be able to compare, action by action, the difference between the robot and a human, and so, on what topics we should improve the robot.

When the robot will be more developed and that we will be able to proceed tests on site, we will do again the comparison test, to be as close as possible to the final application context.

5.2.3.3 Tasks automation

After the documentation of the process and the time estimation of each task, we had a look to the different tasks that might be automated, from the most obvious one, like removing the contaminated material, to more complex ones, like moving from a secluded area to another.

Afterward, we prioritized them, from the most important to least. This step is currently under investigation and will include the user feedback form asbestos removal experts. The purpose is to be able to concentrate and to automate what is the most important, then go deeper and deeper into the automation. This is the preliminary prioritized table at the moment:

Task	Priority	Remarks
Removing asbestos	1.1	Assisted teleoperation for grinding process at the wall/ceiling - autonomous distance /tool-force control Positioning of the platform relative to the wall could be tele-operated
Autonomy for the grinding process at the surface	1.2	Autonomous grinding process at the wall/ceiling with autonomous feed and distance /tool-force control

Task	Priority	Remarks
Autonomous platform positioning	1.3	Autonomous maneuvering of the mobile platform to progress the working area at one wall or ceiling and move relative to the specified area
Moving inside a secluded room autonomously	1.4	Autonomous maneuvering of the mobile platform to change to a different working area
Moving inside the seclusion	2	Being able to move from one room to another inside a secluded area
Managing wastes	3	Handling, dropping and storing the waste containers. The alternative is the tele-operated dropping of the waste containers.
Moving in between secluded area	4	Decontamination subject linked, the robot leaves the contaminated site temporarily
Cleaning the confinement	5	Vacuum cleaning protection sheets and non-protected surfaces
Moving everywhere	6	Crossing obstacles like stairs, steps, gaps. This capability can be applied in the above mentioned autonomous tasks. For now they would be interrupted and need manual teleoperation in tricky situations.
Booting	7	Getting ready to process the materials. This includes the whole set-up and ramp up of the robot and process and would include full self-diagnostics and fully autonomous mapping and avigation based on existing task plan.
Dust measuring while processing	8	Existing but not officially used in the legislation

5.2.3.4 Benchmarking tables for the robot

As a benchmark of our solution to asbestos removal, we will produce at least 2 different tables.

One to compare our solution with other possible solutions for every main task of the robot. The purpose of this is to be able to enlighten and measure the efficiency of possible improvement for our robot, and the following ones, for any other fields. We will compare the solutions through the main topics that are the timeline and the cost, but we will also take into account the special constraints linked to the task (dust production, geometry, decontamination...).

On another hand, we will develop a table to be able to measure the success of our project. We will list the main goals the robot has to reach, on every tasks, and rate it, with a comparison with a human when possible, to be able to scale the success of the overall project. The main goals will come from the initial description of the tasks, and from the different topics that appeared during the development.

Those two tables will allow us to measure the success, and to compare the different solutions that might be used to improve the robot.

5.2.4 Outlook and Future Work to be Carried Out

The overall process and different steps of the robot intervention were analyzed and defined on different levels. We got deeper and deeper, from an overview process to a more detailed one.

We started the time analysis of the process, and defined performance objectives and metrics based on it. We specified the degree of autonomy we are aiming to, in order to process the asbestos removal as efficiently as we can.

We will now develop the robot, and build the tables that will allow us to benchmark our robot, following the main thematic described before.

5.2.5 Summary and Conclusion

The benchmarking of the system will be based on the overall process and not only consider the technical implementations to make sure all user needs and the full process are considered.

In this deliverable, we have analysed and described the overall process including the lower level details to benchmark the full process according to it.

Afterward, we made a first estimation of the time linked to the actions, and planed tests to make it more precisely. Then, we defined the level of automation of each task, and their priority.

We have now everything to test our robot, benchmark it through table we will create and develop, and measure all part of the process.

Thanks to those different points, we will be able to develop a robot very close to the definitive use case.

Deliverable 1.4
Final report for the demonstration and benchmark of process and system

Authors: Florian Dufour, Tobias Haschke

5.2.6 Deliverable in the Context of the Work Program

After developing the operational process for the overall intervention, the final robotic system have to be demonstrated in an operational environment. It was initially planned to do so on a real world rehabilitation site provided by BYCN. The complete workflow including the mapping of contaminated material, the containment of the site, the removal of asbestos and the disposal of waste will be executed (Milestone 9: Robotic system and operational process validated on a real life rehabilitation site). The conventional and automated removal of asbestos contamination will be benchmarked according to the performance objectives and metrics from task 1.3 (D 1.9-1.12 and D 1.13) and support the further dissemination in task 7.3 and project follow up strategy and business plan in task 7.4.

5.2.7 Methods and Work Carried Out

As the development of the robotic unit is behind schedule due to the unforeseen complexity of some essential challenges, the initial plan to test the overall process on a real site is not worth the time and money at this point in time. We will instead demonstrate the robots capabilities in scenarios close to reality within RWTH facilities, to show the actual performance and limits. Before running the real benchmark with the fully assembled robot, we will take the process written previously, and see what we can already benchmark according to the requirements and the objectives determinate in "Di3.5 – Specification of the technical Requirements for the V2 prototype". To do so, we will decompose task by task the process, determine if the robot meets the requirements and explain those results when it needs to be done.

5.2.8 Benchmark

5.2.8.1 Locating asbestos and confining the area

The first two parts of the process are the location of the asbestos and the confinement of the contaminated area.

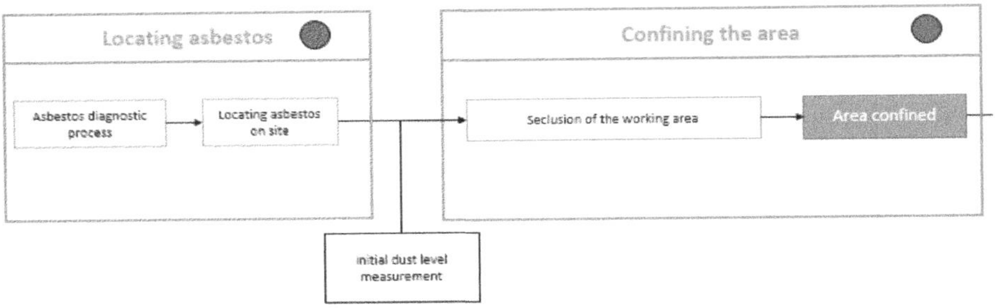

Those two first steps are out of the scope of the project, and will be anyway done the same way for a classic human intervention or a robotic one. Therefore, we don′t need to benchmark those steps.

5.2.8.2 Bringing the robot to the confined area

The first real part of the robot process is bringing on site then inside the confined area the robotic unit. This part of the process will be benchmark from two different perspectives; the first will be according to the requirement, to confirm that the robot fits where it should. The second benchmark will be a time one, as the overall time for the removal of asbestos start there. We will compare this part with the time human needs to get ready to go inside the processing zone, for every session.

5.2.8.2.1 Technical benchmark

The first sub part is to bring the robot out of the transport vehicle.

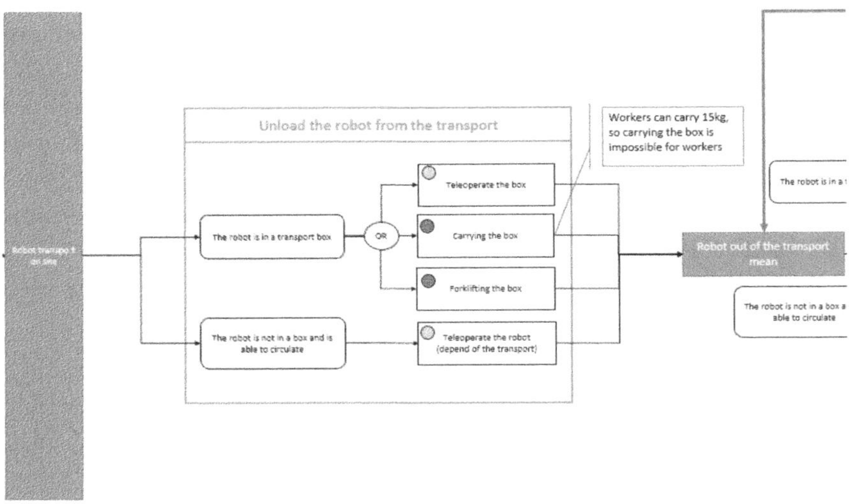

6.1.8 Transport

ID	V1	V2	Requirement description	Source
	W	W	The robotic system shall be transportable by a single operator with a pallet truck and a van	BBNE/BYCN
	W	W	A connection (e.g. hooks, handles) for transportation shall be provided at the robotic unit and at the supply unit	BBNE/BYCN
	W	W	The robotic unit shall be transportable in two pieces (as few as pieces as possible)	BBNE/BYCN
	W	W	The robotic system will be transported in a sealed safety (containment of asbestos contamination according to regulations) transport container/bag from site to site	BBNE/BYCN

To do so easily, the main aspect is the weight of the robot. This weight was defined by the acceptable load for a housing building floor slab and in respect to the typical handling tools at hand on normal construction sites.

ID	V1	V2	Requirement description	Source
	M	M	The overall mass of the robotic system (including all units, cables, computers, etc.) resp. the mass pressure of each unit of the robotic system shall be <300 kg/m^2	BBNE/BYCN

The weight of the robot couldn't be measured exactly yet, as a suitable crane or a suitable scale wasn´t available during the last benchmarking period. The result will be handed in immediately in January as a lift truck with integrated scale is already ordered. Provided that the weight is as specified, the robotic system must be unloaded by means of a forklift truck or the classic integrated tailboard of the truck.

Truck with Tailboard[1]

Truck loading with forklifter[2]

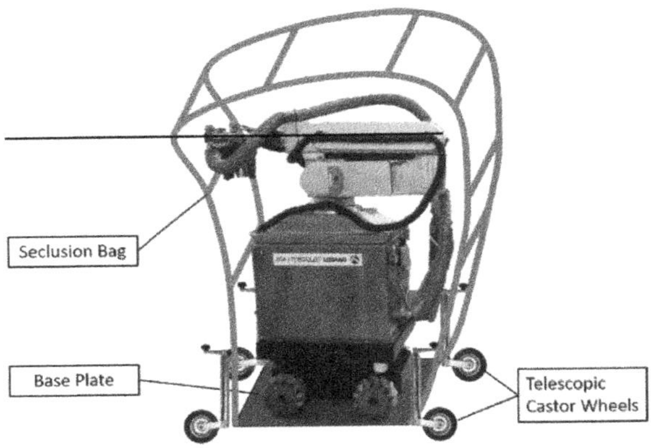

As the robot will be contaminated with asbestos, a two chamber system is targeted. The normal handling process is done within typical wooden transport boxes with a standardized EU pallet as the base. This enables the handling with hand-guided lift trucks or forklifts. After bringing these transport boxes close to the building entrance, the secluded robot is unloaded from it. Due to the missing real site tests a functional secluded robot garage is defined in concept but not manufactured (see figure on the left). Depending on the progress during the final benchmarking in January, a first concept of this secluded mobile robot garage may be installed. In principle, it is a solid plate with castor wheels, where a plastic cover can be attached airtight. After opening the seclusion, the robot is ready to be teleoperated directly, to be well positioned in the secluded area to be reconstructed.

1. Source: https://www.fahrzeugsysteme.de/wp-content/uploads/Schoon-Koffer-mit-Ladebordwand.jpg
2. Source Eurotransport: https://imgr1.eurotransport.de/Gabelstapler-Beladen-entladen-Lkw-Fotolia-bigMobile-d8666afc-56723.jpg

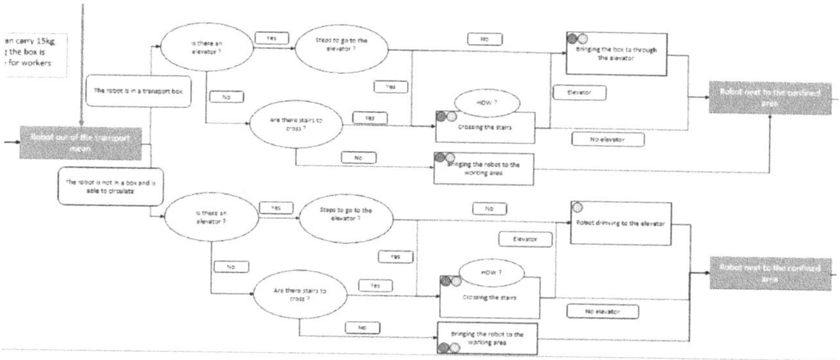

The other geometric requirements have to be validated as well, to be sure the robot is conform for the movements within the building. In detail this is the movement from the location form the transport boxes into the secluded area.

Therefore, it has to be able to fit into an elevator or able to cross stairs. To enable these movements the dimension of the robot are as important as the robot weight.

ID	V1	V2	Requirement description	Source
	M	M	All units of the robotic system (robotic unit, supply unit) measurements shall comply with the following dimensions: – Width <0.6 m – Length <0.8 m – Height <1.95 m, Wish: 1.90 or even 1.80 with or without the dustsheet	BBNE/BYCN
	M	M	The assembled robotic system shall be able to fit through an entrance of a site: 0.70 m by 1.95 m, Wish: 1.90 or even 1.80	BBNE/BYCN

The benchmarking showed that the geometric requirements are met and the robot is able to be moved within elevators and to pass narrow doors. Both situations are pictured in the summing process at the end of this deliverable. Regarding the challenge to change levels within building without elevator, the option of moving the robot across stairs is considered as well. Special tools like the DOMINO AUTOMATIC form Zonzini[3] (see figure to the right) are capable to handle this job.

3. Source: https://www.bartels-shop.com

They are able to perform these moving tasks for up to 400kg. From this perspective, there is no limit for moving the robot over stairs as well. On the other hand, buildings and especially staircases may have complex shapes, which do limit the movability due to tight corners or narrow ceilings that may collide with the robots peripherals. On the first hand, buildings with elevators are generally larger and presumably much more economical to implement autonomous asbestos removal. Out of this, the benchmarking is marked as successful to this state.

The last part of the bringing process is the unloading of the robot from the seclusion box.

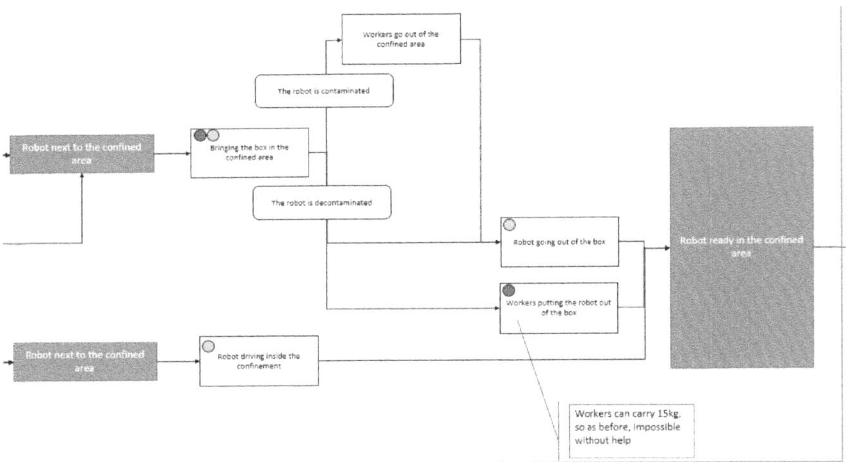

There is no specific requirement about this topic, except the fact that the robot has to be tele operable from outside the seclusion.

5.2.8.2.2 Classic human intervention benchmark

This all bringing the robot part of the process has to be compared to the time that human needs to get on site and ready to process. The classic schedule of intervention is the following:

- Workers need 30 minutes to get ready, inside the zone. They need the same time to clean themselves and undress after each session.
- A session lasts maximum 2 hours, taking into account the preparation before and after getting in the zone.
- They can't not work more than 6 hours per day inside the seclusion (still taking into account the time to get ready and getting out of the zone)

However, the robot won't move from the zone until the removal is complete, and can work more than 6 hours. Therefore, the comparison is hard to do task by task, but a real on site comparison will allow to acquire the data needed to compare both intervention. First measurements are targeted to be done in January to get deeper insights.

5.2.8.3 Removing the contaminated materials

The removing process starts by the booting process. As it has no comparison with human intervention, it will be taken for benchmark with the removing process itself. The requirements for the operational process are the following:

6.1.4.3 Operation

ID	V1	V2	Requirement description	Source
	M	M	The robotic system shall be able to execute a set of pre-planned asbestos-removal-tasks	BBNE/BYCN
	W	W	The operation of the robotic system shall only be done by trained users, but with a quick and easy training	BBNE/BYCN
	W	W	The verification of the asbestos-removal-task shall be done autonomously	BBNE/BYCN
	M	M	The robotic system shall be able to fulfill tasks by remote control	BBNE/BYCN
	–	M	The robotic system shall be able to fulfill tasks semi-autonomously	BBNE/BYCN

ID	V1	V2	Requirement description	Source
	–	W	The robotic system shall be able to fulfill tasks autonomously	BBNE/BYCN

The more we will raise the level of automation, the more efficient the robot will become. Therefore, we will have to benchmark again those values when the robotic unit will have a higher level of automation.

5.2.8.4 Checking process and moving to another zone

The checking process will be separated in different parts for the robotic unit intervention. Indeed, the first checking for the worker to know if the robot did the job correctly is different than the legal one. The first checking has to be made thanks to cameras by the worker, to check that the robot did correctly the removal. However, there are then different checking that have to be done legally, by a professional. It will then be the same as a human intervention and so, out of the benchmark. The robotic control will be taken into the overall removal process timing, as it is a task that doesn't exist for a worker intervention, but that is critical for the robot. Indeed, having to come back to a task that was already done by the robot makes it lose all the advantages and efficiency of the robotic removal.

When the secluded area will have been correctly cleaned from asbestos, the robot will have to reach another secluded area, which can be on another floor. It will then need to go back to the seclusion box, the box will have to be decontaminated, and the bringing the robot on the next secluded area will start again. To have a good overview and benchmark of the robot compare to human intervention, at least two different comparisons will have to be done. The first is purely on the removal of asbestos: is the robot faster than a human to remove a determined area of asbestos.The other one will be more general, taking into account all the other part of the process, that doesn't exist for human, but does for the robot. This will allow us to have a better idea of the target of the robot, where it is suitable to use it, and where a classic human intervention is more suitable.

5.2.9 Illustrated benchmark of the process

To show as clearly as possible the whole process and its benchmark, we simulated it within RWTH facilities

Both robots are in their transport boxes

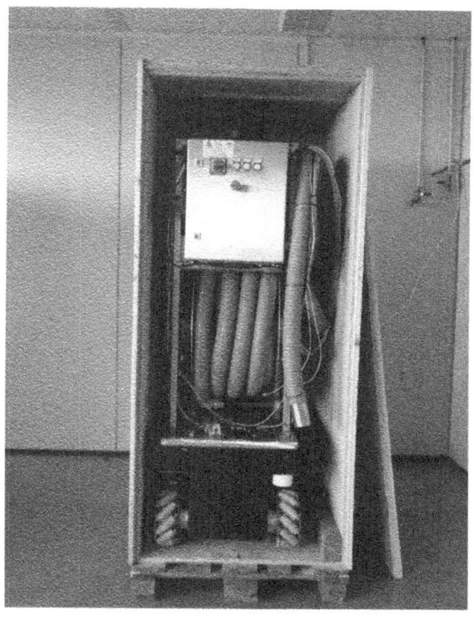

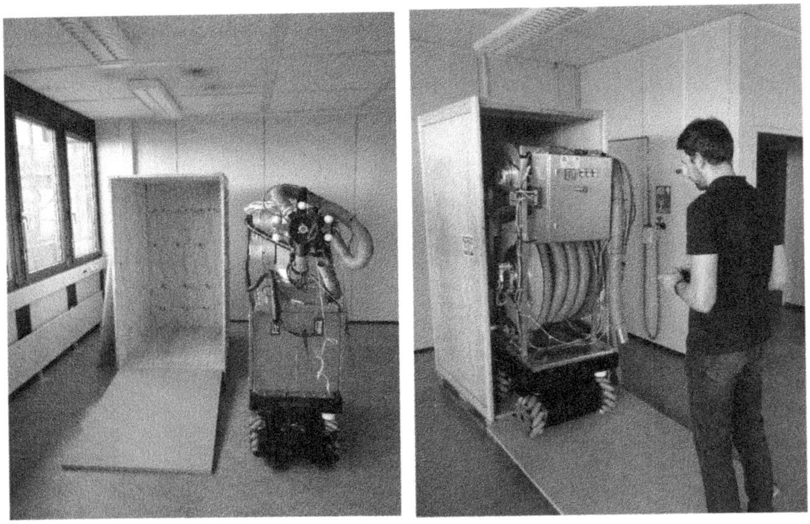

Robots are going out of the transport boxes

For this part we considered that the robots are still in a confined "box" that we don't currently have, but that will be necessary as soon as the robot will be contaminated by asbestos.

This seclusion on the robot will be removed into the secluded area where the robots will do the removal.

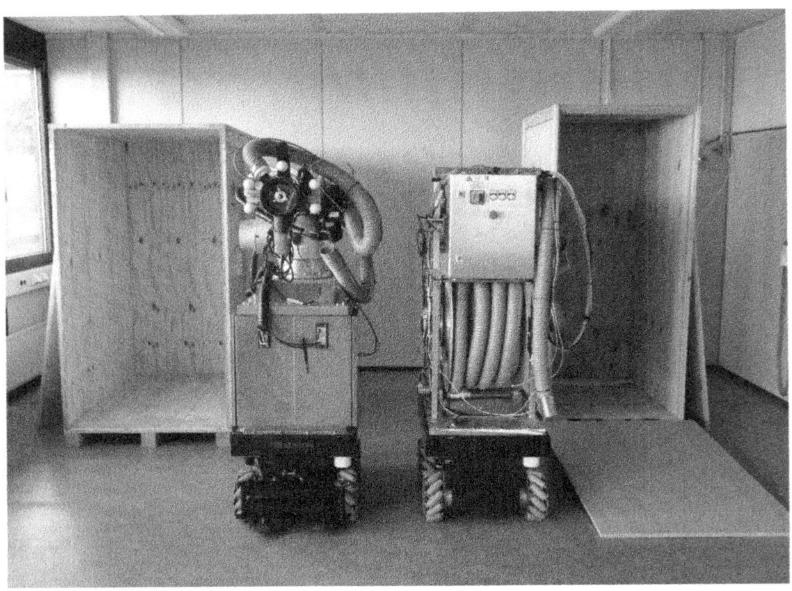

Both robots are out of their transport boxes (and still secluded).

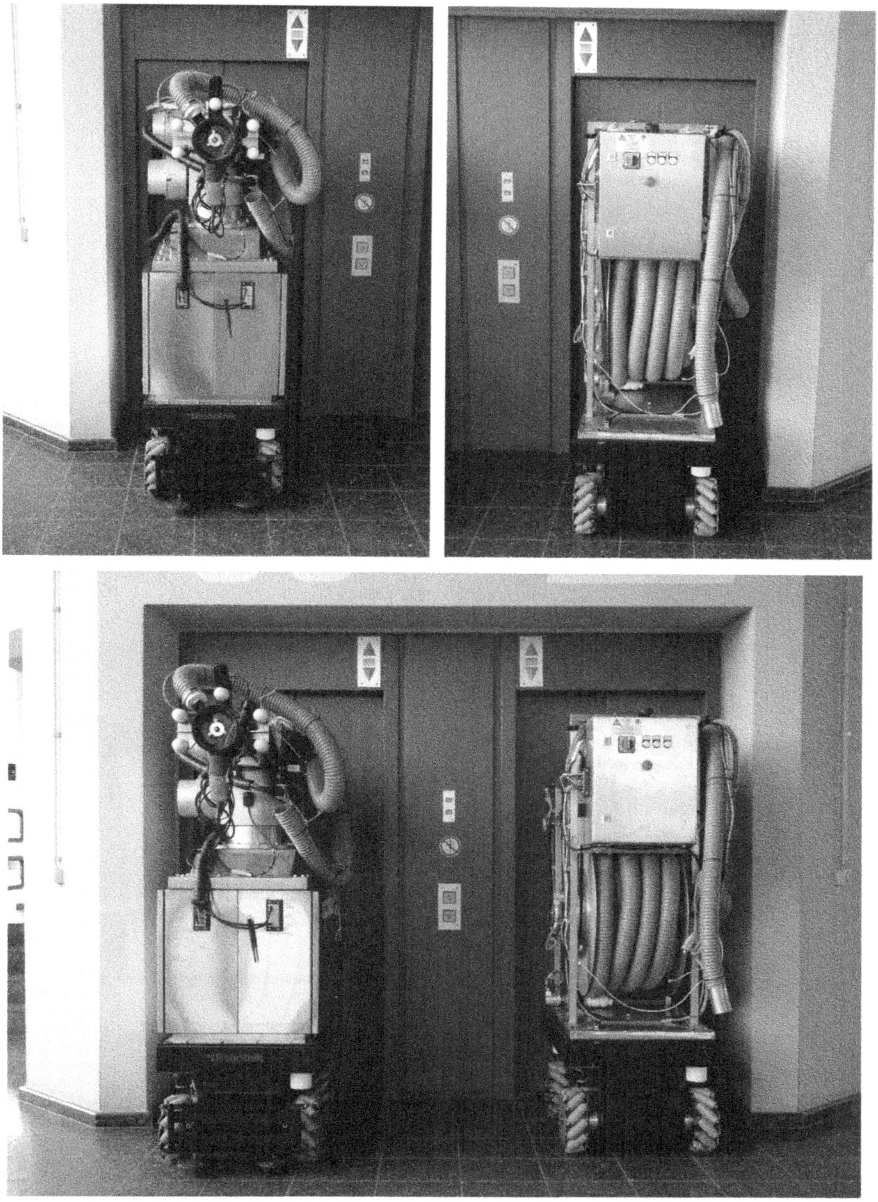

Both robots are brought on the removal site by elevator, or any other carrying means necessary

As required, the robots are fitting in a normal size elevator.

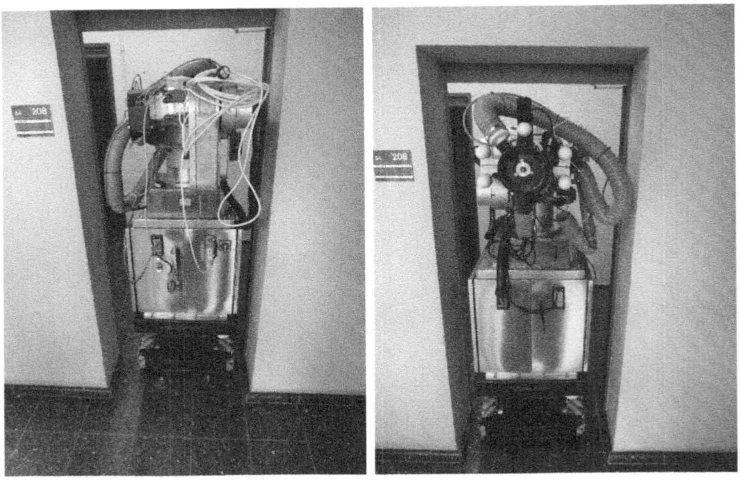

The robots are fitting into normal sized door without restriction, as stated in the requirements

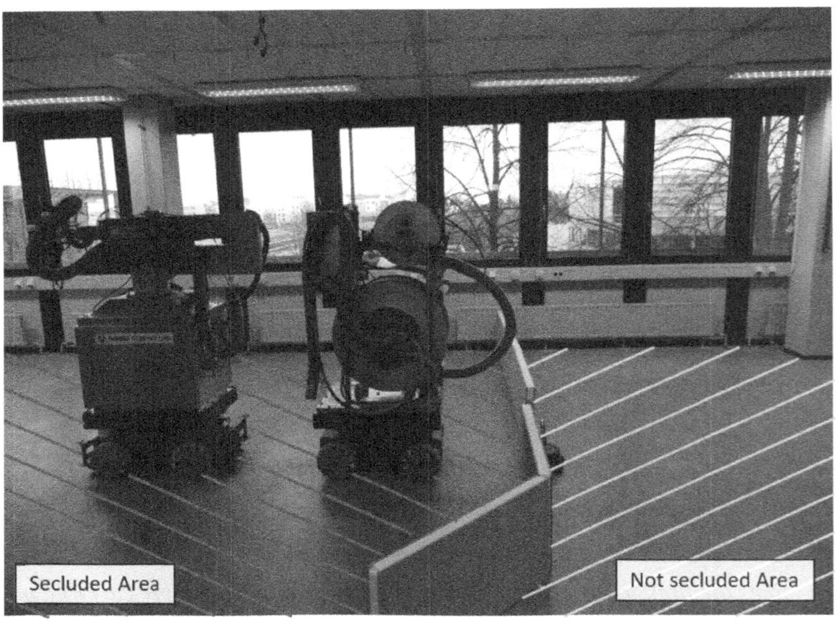

For the simulation, the secluded area is represented by the light panels on the floor.

The robots will go to the secluded area that will then be closes behind them. The seclusion around the robot will then be removed from the robots and stored into a box that will be put out of the zone. The robots are then ready to be connected, booted and started.

The robotic tandem is connected

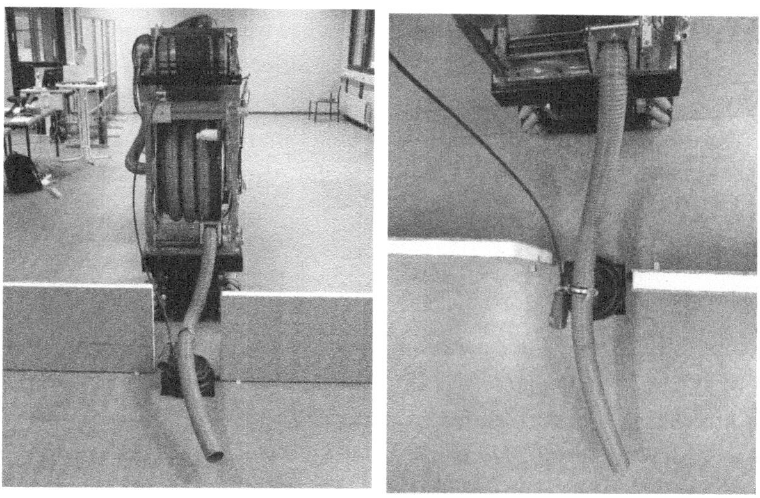

The rear unit is then connected to the aspiration nit and energy supply.

From this point in time the robot can be booted and calibrated. After this the normal removal procedure can take place as displayed in the following picture by targeting the next wall for refurbishment.

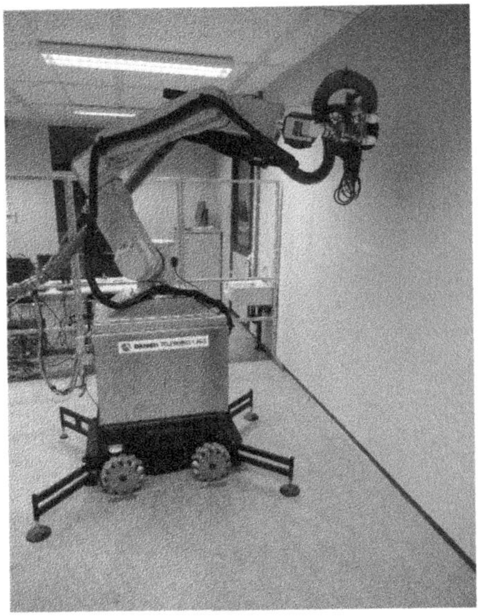

5.2.10 Summary and Conclusion

In this deliverable, we took back the overall removing process, took it task by task and compared it with the requirement to see what we could benchmark. We simulated an onsite intervention within RWTH facilities. Thanks to this, we were able to measure some characteristics of the robots to see if it fits the requirements. It does, on most of the points, and additional benchmarking tests in January are going to answer the still unanswered questions. So far, the realistic usage of the robotic system looks promising.

Task 1.4 Development and validation of the operational process

To develop and validate the operational process, we built a hangar on a technical base of Bouygues TP, one of the Bouygues subsidiaries. Inside this testing site, we built a concrete box with different materials on the walls, floor and ceiling. It allowed us to test the grinding on the robot on a close to reality case. As the integration of the V2 and its debugging took longer than expected, and that there was no easy access to needed debugging tools on the testing site, we decide to bring the V2 to Aachen to go on with the tests in simplified conditions. All the steps of the process already working were tested and validated there. This is detailed in the task 1.5 below.

Task 1.5 Demonstration and benchmark of the process and system

After developing the operational process for the overall intervention, the final robotic system had to be demonstrated in an operational environment. It was initially planned to do so on a real world rehabilitation site provided by BYCN. However, as said before, we decided to do so within RWTH facilites. Indeed, as the development of the robotic unit was behind schedule due to the unforeseen complexity of some essential challenges, the initial plan to test the overall process on a real site was not worth the time and money at this point in time.

In parallel, and before running the real benchmark with the fully assembled robot, we took the process written previously, and saw what we could already benchmark according to the requirements and the objectives determinate in "Di3.5 – Specification of the technical Requirements for the V2 prototype". To do so, we decomposed task by task the process, determine if the robot was meeting the requirements and explain those results when it needed to be done. We first checked the transportation requirements, concerning the dimensions and weigh of the robot, and provided different solutions to achieve that transportation, like a forklift for example.

We then did the same for the transportation of the robot from the bottom of the construction site to the secluded area. This aspect is conditioned by steps and elevators, so we check the dimensions and confronted it with the requirements. As for the transportation, we provided a possible solution to move the secluded robot (e.g. with the Domino shown left).

To show as clearly as possible the whole process and its benchmark, we simulated it within RWTH facilities, which is documented in the following figures and explanations.

When the robot arrives on site it is stored within the shown wooden transport boxes. Both are able to be moved by manual or drivable fork lifters. With those devices the boxes are brought near to the sites entry. From there the robot is moved into the seclusion. The figures are showing that the robotic system

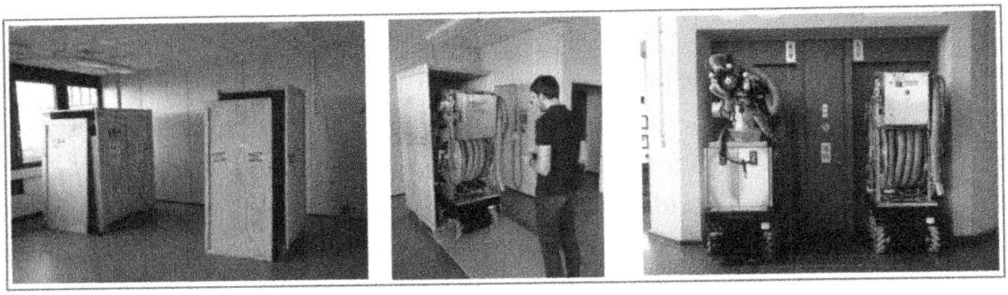

Figure 5.25. Bring robot to secluded area.

does fit into typical elevators for persons as they can be found in the targeted building.

After reaching the target level for the removal task the booting process can start. Therefore the currently independent two robots need to be coupled to the tandem configuration. First the rear unit is connected to the power supply of the site across the safety module and for sure to the central aspiration system. Those aspiration systems are often of the same size as one robot and can be moved with the robotic system from level to level. After connecting to the supplies offered by the site, the front and rear unit need to share power, data, the safety circuit and the aspiration pipe. The connections are assisted by hooks and elastic stripes for having a highly flexible tandem configuration. The booting process from two individual units to one tandem is shown in the following figure.

As already mentioned the raised complexity of the robotic system caused the consortium to simplify the benchmarking conditions. Therefore the information gained during the process from real construction sites was used and transferred into as close to reality simplified scenarios that could be performed in the available laboratory in Aachen.

Out of this it was possible to demonstrate corridors, corners, small and large rooms to benchmark the different developed methods. The mobile walls are build of wood and Styrofoam to have standalone small walls to create multiple scenarios. The thickness of the wooden plates is defined to also block the radar signals form the iSDRc to be able to demonstrate the implemented radar localization.

Having the tandem coupled and a flat with tight entrances to the first rooms in the hall it becomes obvious the autonomous planning into those rooms is highly challenging due to the physically irreducible inaccuracy of the localization approaches. Therefore in critical situation like those mentioned ones the operator needs to take over by teleoperating the robot into those areas based on the camera input and the sensor signals. A demonstration of a teleoperated entrance around the tightest accepted corner is shown in the following figure.

Within the room for the removal task the automatic movements by artificial planning can be active again, when the room is of a certain size. The flexibility of

Figure 5.26. Tandem coupling procedure.

the tandem and the different strategies are explained in the following paragraphs of the benchmarking.

After this first example that the needed robot for the task is already on the edges of the available space, an analysis of door width is made in the following figure. While a width of 1m for corridors or walls are tight but still possible to be handled automatically the desired width of 80cm doors is already too small for automatic movements, having an inaccuracy of +/-5cm in mind coming in by the localization process. On the other hand the major geometric restriction in the width is the necessary aspiration pipe on the left of the aspiration reel, that is forwarded to the

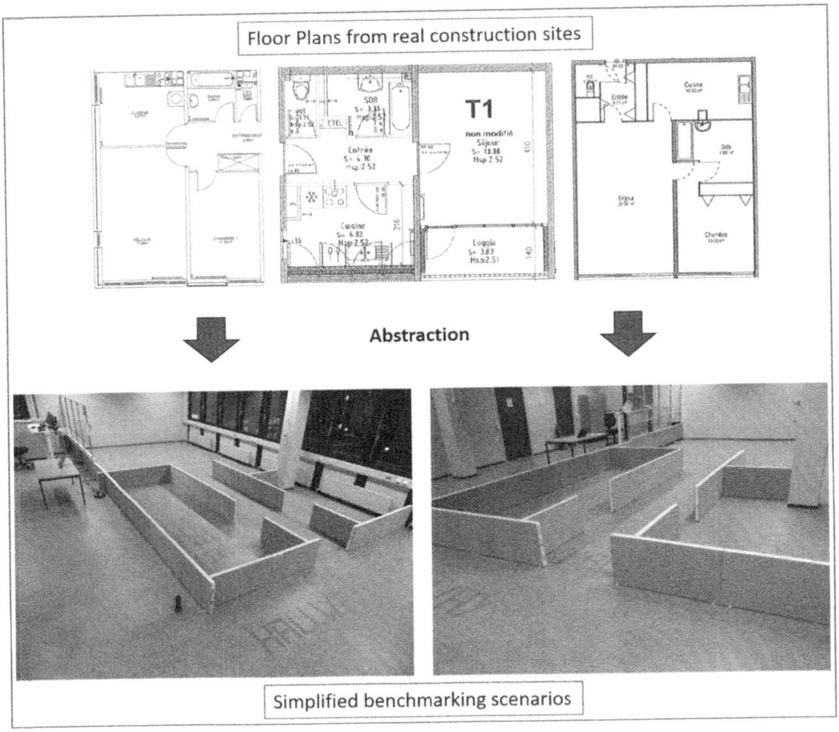

Figure 5.27. Benchmarking simplification.

front unit. This piping is adding +10cm of extra space to the sides which can be reduced easily with an alternative aspiration concept without changing the mobile base itself. The figure below illustrates those circumstance.

As mentioned above the tandem is one of the major complexity drivers of the system that weren´t foreseen during the time of the proposal. As the configuration planning of the redundant arm was planned, the configuration planning of the mobile bases towards each other was not present and raised during the project. The following figures are showing to different mobile base configurations for the removal job within targeted rooms.

One configuration is that the support unit is positioned directly behind the removal unit with a turn of 90 degrees. With a certain distance for the arm workspace the required distance to the wall is 2,90m and the needed width of the system is 1,50m. It is obvious that this configuration may won´t fit in every room that is going to be processed. Therefore another configuration is used as shown in this figure.

This configuration is in side by side mode where the removal unit is positioned in front of the wall and the support unit placed parallel to the site. The orientation

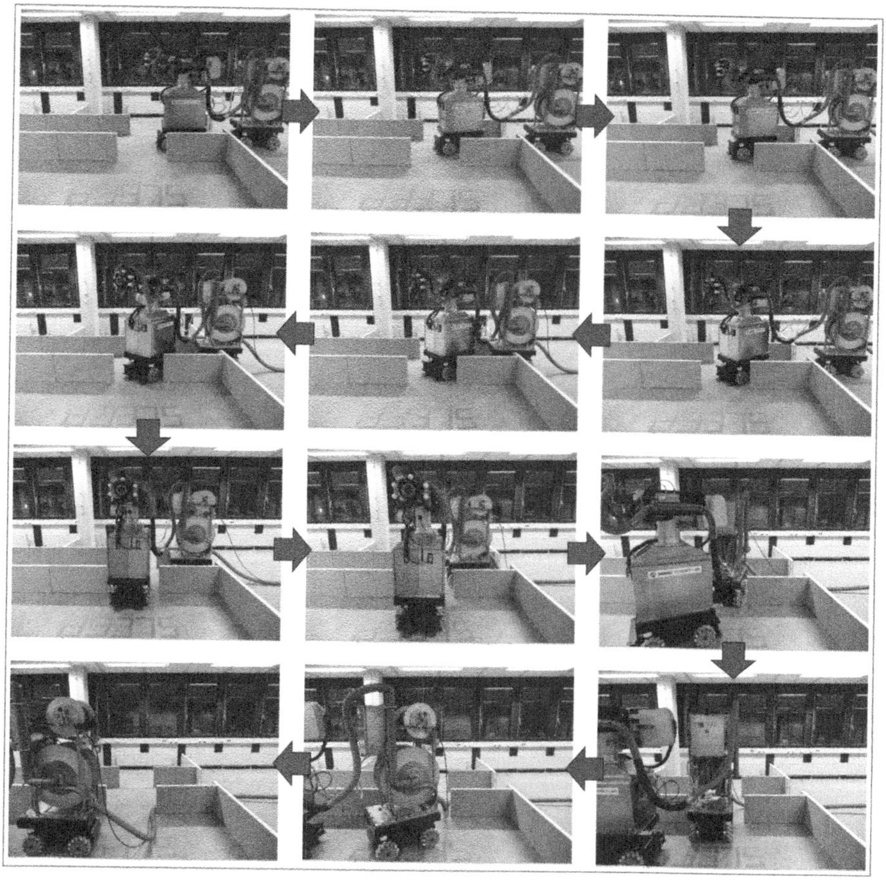

Figure 5.28. Teleoperated Room Entrance.

of the support unit is not fixed due to the flexible coupling system between front and rear unit. This freedom of movability for the support unit is given. Taking those configurations into account the targeted floorplans are analyzed on the reachability of the walls within them by those tandem restrictions given by the robot itself. Therefore the different configurations were designed in the scale of the floorplan at hand and tested which walls are in reach and which walls are too close to each other that no mobile base configuration can be placed inside.

The figure above is one example of this reachability analysis based on the tandem configurations. After the analysis it becomes clear that especially the twisty parts of the flat near the entrance are not fully covered by the robotic system. While 14 of 24 walls can be processed by 100% also 6 walls cannot be processed at all. In total numbers this seems high but the related surfaces are comparable small so that the benefit of the system is still given.

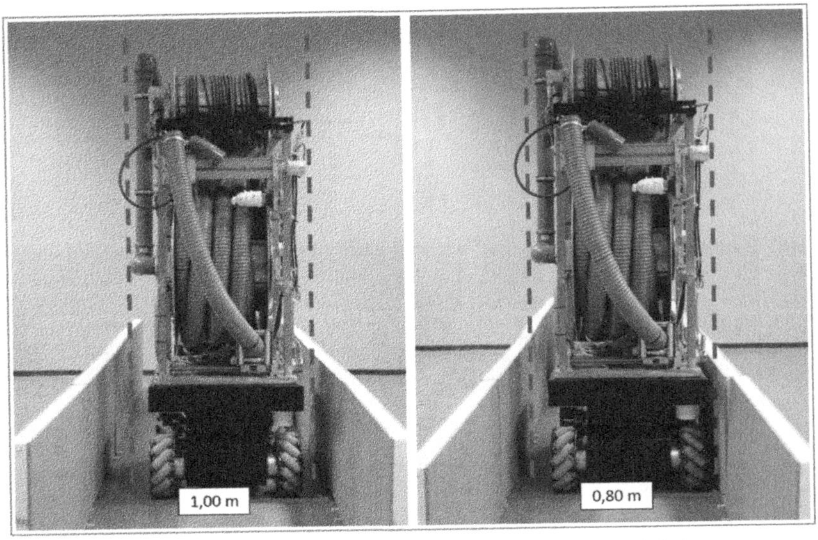

Figure 5.29. Showing the width conditions of the mobile bases.

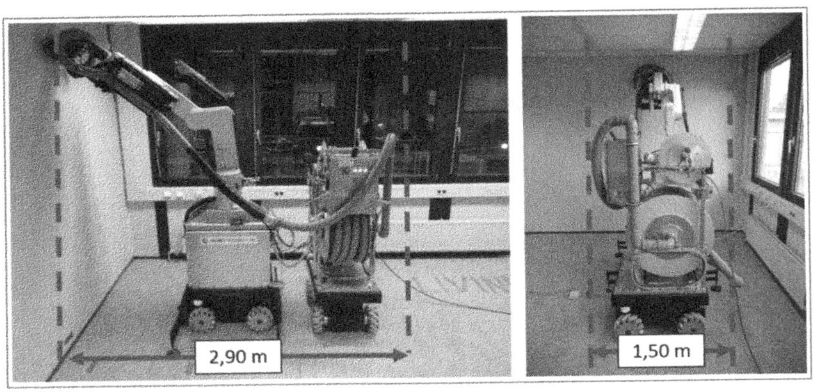

Figure 5.30. Tandem Configuration Behind.

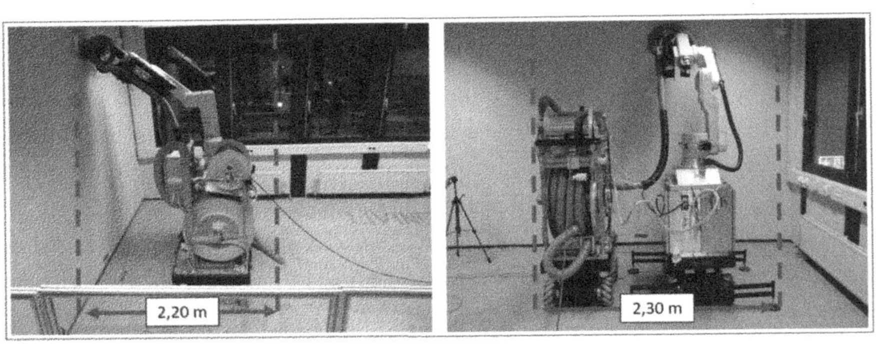

Figure 5.31. Tandem Configuration side by side.

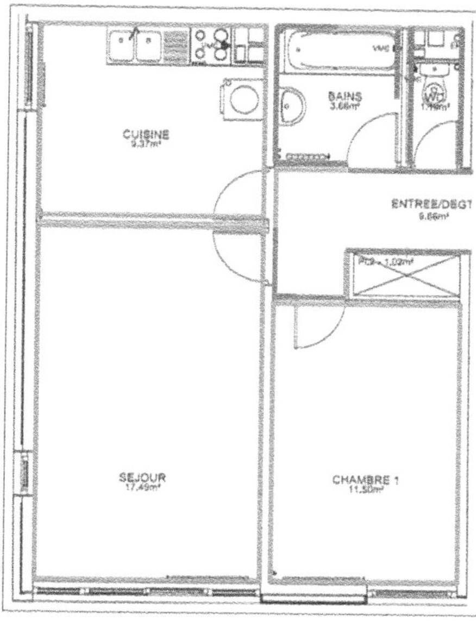

Status	Quantity
Total walls	24
Full removable walls	14
Partly removable walls	4
Non removable walls	6

Figure 5.32. Benchmarking - Wall coverage in real flat I.

A quite similar result can be found in the next flat taken from the old projects of Bouygues removed from asbestos. Within here the main living rooms are all covered by 100% but the small rooms as bath, kitchen and toilet may not be covered totally. But also here the majority of the surfaces are able to be processed and only a comparable small surface is left unprocessed due to the twisty floorplan conditions around.

Beside those examples of comparable small flats the number of removable walls is raising rapidly for bigger flats with wider conditions. In the third example shown below the only real bottle neck is tight small corridor to the left rooms and the directly connected toilet. Every other wall and room can be processed completely which underlines the flexibility of the system and the well-tailored design towards the target use case with only little special cases not covered automatically by the system.

Beside the tandem configuration and the resulting reachability of walls within flats the next level is the coverage per wall by the robotic arm. This is going to be explained in the following paragraphs with figures from the prototype and measurements in the benchmarking environment in Aachen.

As a first point, the lowest reachable removal point in front of the mobile base is found on the wall. The necessary arm configuration is shown in the left figure while the measurement of the removal level is shown on the right by 20cm. the limiting factor in this case is the sensor system around the grinding disk. This circumstance is explained later more explicitly.

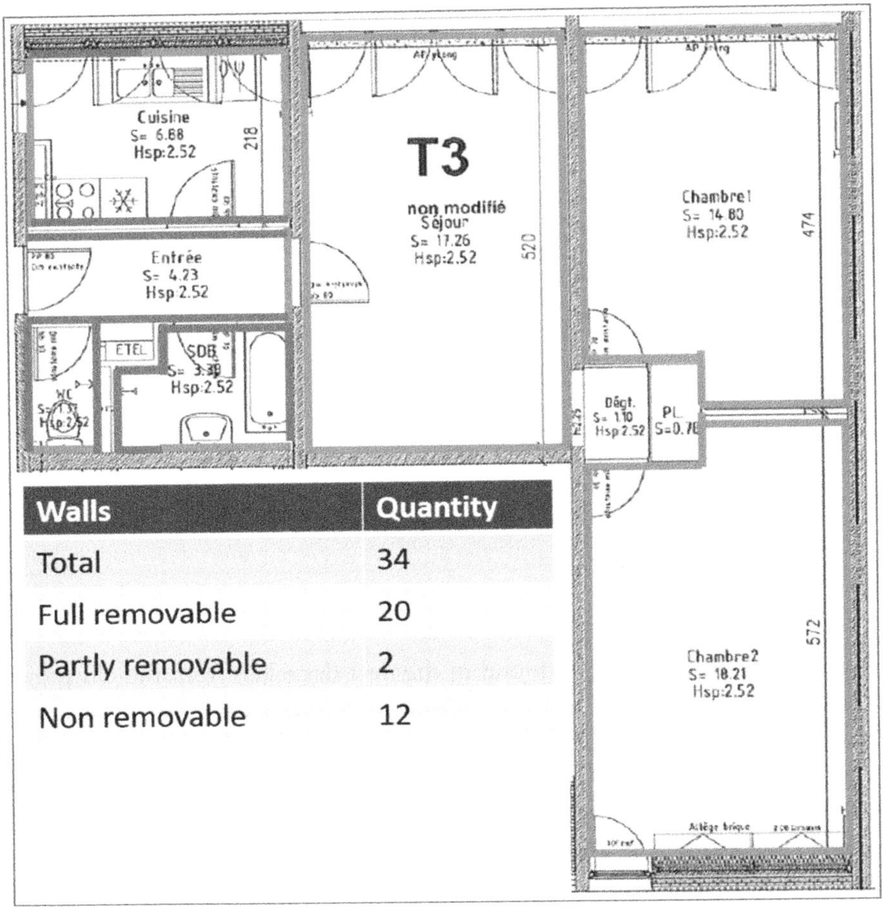

Walls	Quantity
Total	34
Full removable	20
Partly removable	2
Non removable	12

Figure 5.33. Benchmarking – Wall coverage in real flat II.

But not only the sensor system is limiting this configuration. As shown in the following figure the configuration of the arm is close to a collision with the cabinet mounted on the mobile base. The problematic elements are the mounted motors for the axes 5 and 6 and the cable guiding frame on the side of link 4. This is a typical engineering challenge as the cable guidance and the motors weren´t part of the arm synthesis made and were added later during the engineering process of the arm. Very important at this stage is that the reachability is still given but not as flexible as planning into this tight condition is quite too risky for the robot itself. The upcoming explanations will show the strengthens of the robotic arm in to ceiling area of rooms and give new insights into this challenge.

Having the mentioned collision challenge of link 4 and the cabinet in mind the distance between base and wall is enlarged a bit and the grinding disk starting position is placed 20cm higher than the minimal reaching point on the wall. From

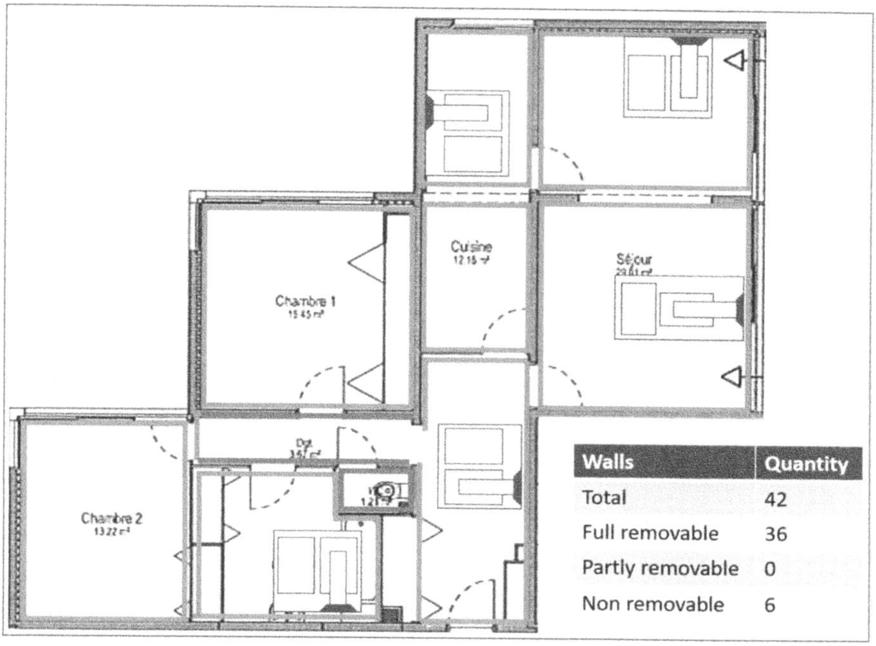

Figure 5.34. Benchmarking - Wall coverage in real flat III.

Figure 5.35. Minimal Removal Height.

here the robot is able to perform a straight trajectory fully up to the ceiling with no changes in the mobile base position and collision problems. The entire motion is going from 0,35m up to 2,75m of removable height represented by the grinding disk position.

The full motion of this trajectory can be seen in this sequence of figures.

Having in mind the collision challenge of link 4 and the cabinet and this information that the arm is able to reach heights of 2,75m vertical, a solution can be derived easily. As the cabinet on the mobile base is a standard purchased part, the dimensions were fixed. With a custom solution that is reduced by 20cm in height

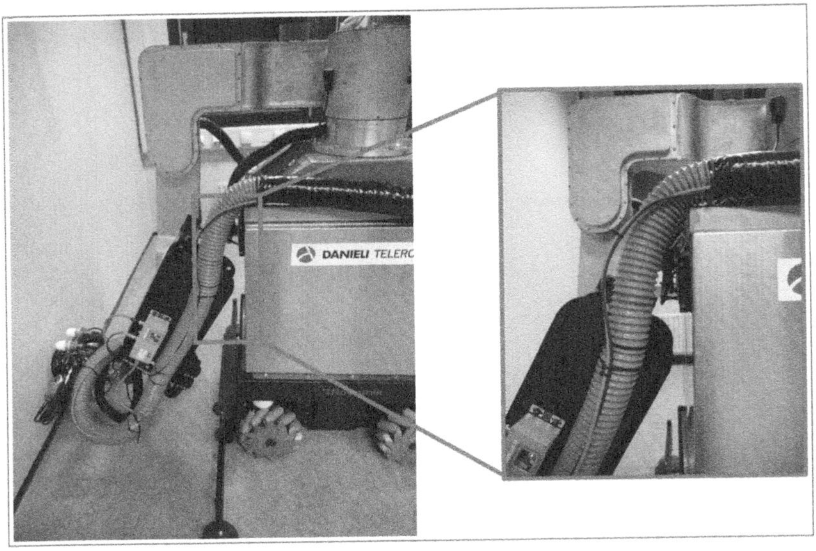

Figure 5.36. Minimal removal height explanation.

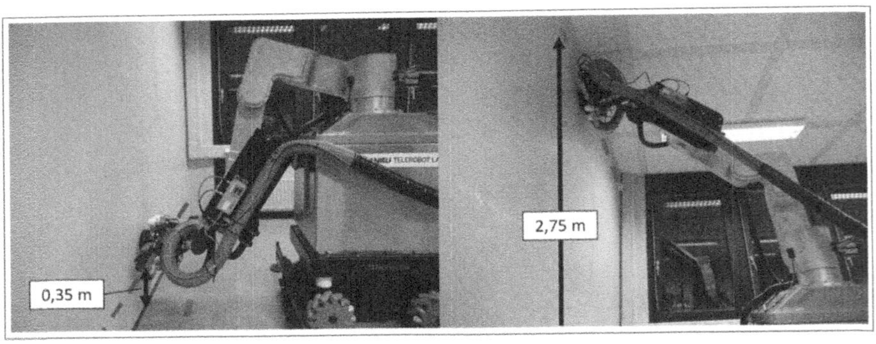

Figure 5.37. Length of maximal vertical wall trajectory.

the reachability of the system will cover an entire wall only restricted by the positioning sensors around the tool. All in all, this is for sure a challenge for the robot at hand but the derived structure is capable of performing the desired task and can be easily adapted to meet the requirements even more.

Based on the already mentioned restrictions given by the sensor system around the tool, the wall coverage can be discussed in the following closing paragraph of the benchmarking. As the iLDR frame and the camera for supervising the task are placed around the disk, a frame around each wall is created, that is impossible to be processed automatically. The following figure demonstrates this situation and the options for orientation control of the tool along the wall edges.

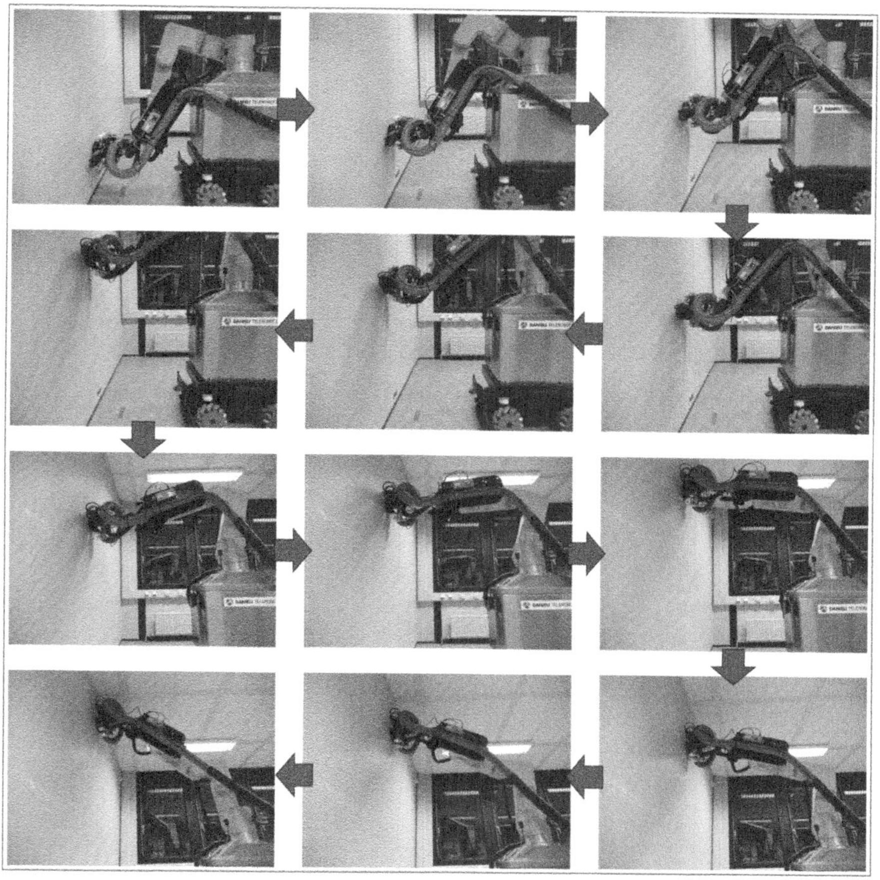

Figure 5.38. Maximal continuous vertical wall trajectory.

From this analysis, the size of a wall and the removable size of it can be compared. During the benchmarking, it was possible to show that the wall coverage for normal walls with the current sensor frame is above 70% and goes up 80% for huge walls (blue curve). Reducing the iLDR frame to its technical minimum, the wall coverage values rise to 90% (orange curve).

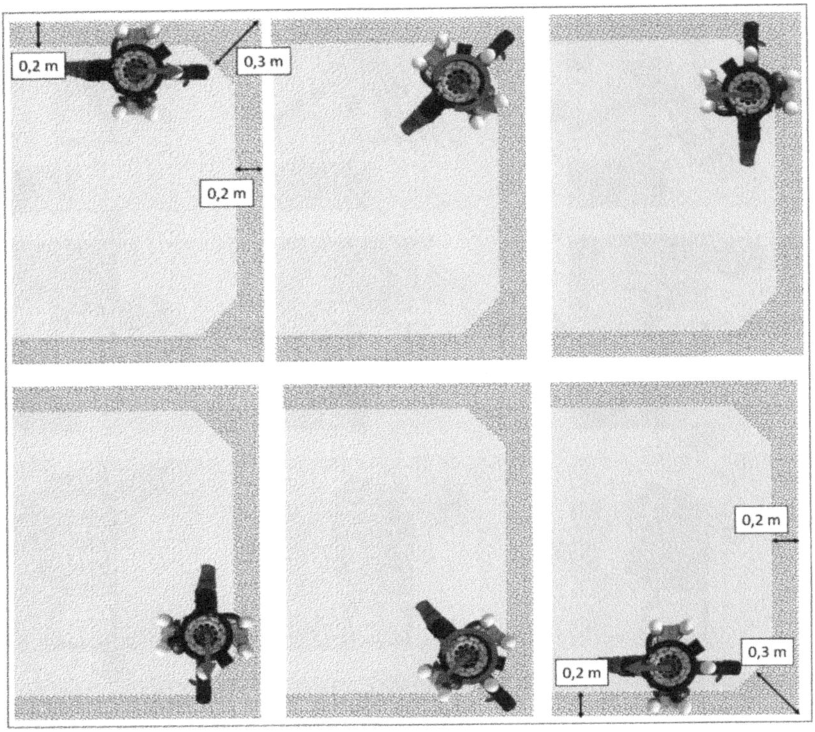

Figure 5.39. Wall coverage floor.

DOI: 10.1561/9781680837155.ch6

Chapter 6

Summary and Conclusion

By Tobias Haschke and Francesco Becchi

Deliverable 7.1
D7.1 Project follow up business plan and future agreements

Authors: F.Becchi, T. Haschke

6.1 Deliverable in the Context of the Work Program

This deliverable is the main outcome from WP 7: Dissemination and Exploitation as result of the Task 7.4 (Project follow-up strategy and business plan). The goal of the task was to structure and further develop a reference plan to support the project follow-up. Market survey investigation were performed at project early stage and a later stage to check the maturity level of the reference scenarios drafted in project original proposal. All considerations were based on the current development status of the robot and adapted accordingly.

6.2 Reference Exploitation Scenarios

Main target of the system developed in Bots2ReC project is to address autonomous asbestos removal.

This main target was selected for two main reasons:

- Based on existing data on past consumption in Europe, the market of asbestos removal for the first five to six bigger European countries is estimated exceeding several billion Euros deriving from this a potential application of several thousand robots
- Introduction of robotic asbestos removal could reduce in a single step both risks for building occupants and for workers involved in the task paving the way to a wider and faster site rehabilitation policy with beneficial environmental and public health returns
- BYCN, involved as potential "end user" owns the application and field expertise needed to write detailed project requirement for the successful operational deployment of the system.

6.3 Summary on Market Survey

During the exploitation and dissemination process the consortium focused on the asbestos removal market first but also had a glance at the wider market, since the sub-components of the full-application were already promising.

6.3.1 Reference Market

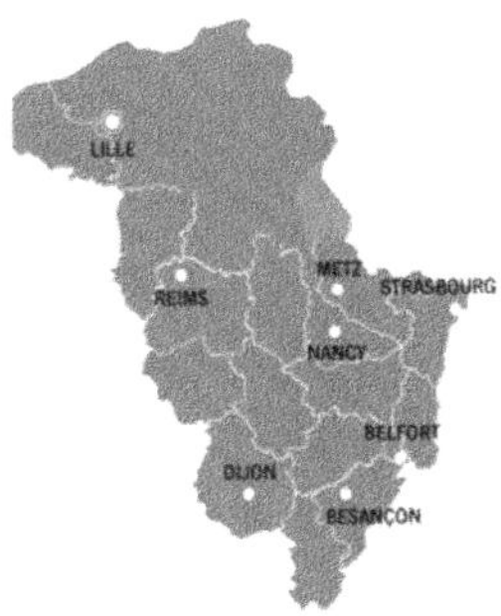

Quantity by situation	
Floor	19026 m²
Wall	16317 m²
Ceiling	5442 m²

Quantity by materials	
Plaster	20460 m²
Tiles	2084 m²
Plastic floor	18241 m²

From the expertise of BYCN some insights into the real asbestos removal market were given, as they still continue the asbestos refurbishment based on manual work without automation. Against this knowledge the set requirements were benchmarked and validated. The results are visible in the Deliverable 1.3 for V1 and V2 prototype. The market positions are the cities were the 20 investigated refurbishment projects were performed.

Beside the validated requirements the reference market survey showed that there is one competitor close to the automation solution of Bots2ReC. The AS Protek company has developed a XY-Portal system that can be moved manually from wall segment to wall segment and runs an automated pre-defined removal operation. This removal is also done with a diamond grinding disk but it is not using an aspiration system. The process is running based on a water cycle which washes away the produced dust during the grinding operatoin. The water is saved in a tank outside the building and asbestos is separated with a special filtering technique. The seclusion of the building is still required and the water filtering technique is very expensive. Additionally the water based grinding solution is more aggressive to the wall structure than manual grinding operations. The result in comparison to the targeted solution was, that a well-controlled grinding operation with an air based aspiration system can challenge this product due to the not needed costs for manual workers beside the robot and the filtering technique.

6.3.2 Wider market

As briefly introduced in the original market exploitation strategy included in the project proposal the development of a dexterous autonomous robotic platform specifically designed for harsh environment deployment that could be of interest in a wider application field.

The growing market of "construction robotics" is continuously identifying tasks where a similar platform could be of interest. BYCN already assembled and used in previous research programmes a big moving platform integrated with a standard Kuka robot, to test in site autonomous drilling and cutting tasks.

During the last four years, the use of robots in the construction industry has been showcased by several H2020 funded projects. The most relevant are:

- **HEPHAESTUS** (https://www.hephaestus-project.eu/the-project/)-2017-2020
- **BADGER** RoBot for Autonomous unDerGround trenchless opERations, mapping and navigation (https://www.badger-robotics.eu/)-2017-2019
- **HINDCON** (http://www.hindcon3d.com/)2016-2019

Harsh environment robotics application often require mobile robotic platform to increase the range of operation. TLABS already built and used a similar system, in this case with a dedicated tethered vehicle rather than a wheeled platform, to perform decontamination and surface sampling inside a 100 m tall nuclear power plant chimney (chimney now successfully dismantled after nuclear decommissioning). In the last period, thanks to the integration of Telerobot Labs in the DANIELI group TLABS started the deployment of different industrial robot in steel plants (mainly for steel sampling or handling tasks) and is now further fostering robotic integration in this field to overcome robot fixed installation constraint. A mobile robot with the mobility of a redundant arm and the payload of more than 20kg as developed during Bots2ReC can be a game changing technologie in this sector as well and is currently under investigation by TLABS.

6.4 Final Prototype Status and Perspective

The original project ambition was to develop a multi robot platform suitable for remote operation in harsh environment, provided with a specific set of tools tailored to surface rehabilitation and supported with a dedicated SW environment for robot in situ deployment and operation. The original concept of the robotic platform evolved through a first prototype testing to a final configuration, integrated and tested in the dedicated mock-up built in BYCN facilities. Even if the original goal was ambitious, the second prototype built integrates all most relevant function originally considered.

The high potential of this kind of platform leaves always a relevant space for further development, test and validation that will go beyond the scope of the project: a deep platform validation would be necessary to understand the gap from this second prototype to something considered as a "product".

Only as example for TLAB direct experience in FP6 ROBOTCUB project no-one in the consortium managed to use all the iCub platform degrees of freedom for a long time (early iCub platform started using legs to walk in the last few years while the robot was built in 2007!).

As more specific consideration: the robotic platform, as integrated unit of moving base and articulated redundant robot arm, went already through deep re-engineering in the transition from the first and second prototype. The SW infrastructure built to perform all the required function, from surface identification and classification to path planning and surface cleaning control, reached a level of significant complexity that would require, to be consolidated at an industrial level, still a deep involvement of the research partners of the consortium.

6.5 Follow up Alternatives Considered

This section summarize the different follow up perspective that were discussed within the consortium.

6.5.1 Consortium company endorser

The original follow up strategy was based on the idea that the end-user involved in the project, or a cooperation between the two robotics companies in the consortium, would directly foster validation and testing of the platform to move the robot on its target application. Looking back to the business plan sketched in the proposal the task considered, and potentially under control from the identified partners, were mostly related to certification, site trials and market promotion.

As briefly summarized in the previous section, the robotic platform can be considered in line with an industrialization perspective for building replicas and scaling costs. On the other hand the complex SW infrastructure developed to answer the target application requirements, whether being validated in terms of achievable performances, is still the output of a research project.

Even if exchanges with Asbestos removal stakeholders involved in the IAB confirmed the interest on the application, the consortium feeling is that a straight migration of the complete application to the originally targeted application is far from the immediate reach.

6.5.2 External company endorser

To complement the original exploitation vision presented in the original proposal a second alternative, preliminarily investigated with IAB members is below presented.

Several different player stated that Bots2Rec robotic platform, including the SW infrastructure developed, could have a straightforward application in other "neighbour" fields in construction industry where, for instance, while environmental pollution is less critical, operation in close interaction with human worker would require a safety related system review.

The impact and cost of a partial design review of the full system, with the goal to reduce as far as possible its infrastructure complexity, cannot be assessed at this stage as no formal expression of interest was sent to the consortium even if this follow up strategy for a neighbour application and hence market carry over looks promising.

Two different options to support this application migration were considered (of course with full respect of Bots2Rec consortium IPR ownership):

In case of relevant R+D activity need part of the consortium could structure with the new "end user" a research project, targeting a suitable funding scheme and in respect of the previous IPR. Alternatively, the end user could involve as "service provider" each member for its skills and competence to directly support the new end user (in this case "customer") for project review.

6.5.3 Bots2Rec technologies exploitation and follow-up

Beside the real application of asbestos removal a lot of standalone technologies were developed during this project which are all related to construction robotics and therefore with a huge generic potential. Especially during the IAB Meetings the interest of the external experts was already raised by subcomponents of the system and not always by the full application, as the full path to automation seemed to be long as mentioned above. The often asked and mentioned sub-technologies developed are listed below:

- The robot itself was often raised as very interesting for other tasks on construction sites
- The different developed removal tools for robotic use in removal tasks for construction industry
  - Grinding tools (3 different ones)
  - Scarifying tool
- Generation of semantic maps to enable task planning for construction sites without available BIM
- Task planning for robot fleets in highly restricted areas
- Radar SLAM for mapping and localization in rough environments
- Hybrid radar localization to be able to localize robots in pre-generated maps based on lidar sources
- Multi-modal control of robotized grinding
- Stability optimal path planning for removal tasks
- Mobile robot tandem following solution for narrow spaces
- Cable and aspiration pipe handling for mobile robots in tethered mode

Each element is analyzed and presented as a standalone technology on the final IAB Meeting to get feedback on the value for each different solution.

6.5.4 Start-up company

The creation of a start-up company based on the Bots2Rec platform was finally discarded in the original proposal and is here confirmed as non-viable solution in this case. Some further consideration to support this assumptions made in the consortium meeting are below reported as further comment.

- A "robot based" start-up company does not look to have enough potential and would be also partially in overlap with the mission of the two robotic partners of the consortium;
- A "service provider" start-up company could not afford, as discussed, the investment to put in place a working and validated system in asbestos removal market;
- A technology baser company has no sense as each industria partner involved that took part of the platform development has already its own commercial strategy in place.

6.6 Innovations Already Launched to Market

The three automation related companies of the consortium, Indurad, Robotnik and Telerobot Labs, have already got a significant impact on their products and their features according to the developments and findings in Bots2ReC. The individual new products and new features are shown here:

- Indurad has optimized and improved their 77GHz radar platform iLDR/iSDR to allow higher measurement rates, to reduce power requirements and to improve overall performance and usability on small robots. This includes a new ROS interface and adaptible new variants for signalprocessing, improving 2D scans.

  Additional, based on the experiments and experiences in the project Indurad is about to start a new product line of very small radars operating at 122GHz. Indurad assumes that it is ready to release it to the market during 2020.
- Robotnik already integrated in their product portfolio the wheeled base developed named *RB2 Base* for heavy loads in industrial indoor transport scenarios. In addition based on the Bots2ReC V2 Protoype Robotnik developed and launched the new Summit Steel that they sell as a general purpose platform for industrial and rough environments, and is the base of Robotniks *RB-Kairos mobile manipulator.*

- TLAB exported the same mechanical transmission tested in the V2 robot arm joint on other custom built robotic arm including a clean room heavy duty (500 kg) 4 DOF manipulator.

6.7 Future Agreements

Since the further development of the Bots2ReC idea and the robot contained in it depends strongly on the available financing and cooperation possibilities, the consortium has decided to be particularly flexible. In this way, it should be possible to react to different possibilities for further development. The idea behind this is as follows: The original consortium will continue to exist as a group of legal entities and will offer its expertise for the continuation of the robotic system in all subsequent projects. However, the aim is not to achieve a solution in which all partners are allocated fixed man-months, similar to the previous funding basis. Rather, there should be the possibility of weighting the competences of the consortium according to their own interests.

This shall be illustrated by an example:

If a future investor is interested in implementing the technology of automated sanding of walls and ceilings on an industrial scale, the present prototype will focus on increasing the robustness of the current control system and the further implementation of new handling features. A further development of the mobile platform or major changes to the robotic arm would be of secondary importance and would only require low capacities. A major revision of the radar sensors is equally unlikely. For this reason, it is obvious to pay the industrial partners involved for necessary adaptations through classical subcontracts without having to define work packages and person-months.

This example clearly illustrates how much all the consortium's expertise is also required in each subsequent project, but to varying degrees depending on the focus. For this reason, given the circumstances, a clear contractual agreement is impossible at this stage. However, expressions of interest from all partners have already been obtained for the proposed format. In a final Industrial Advisory Board meeting, the results of the project will be presented and potential partners will be invited for a continuation.

6.8 Summary and Conclusion

In summary, it can be stated that not all objectives of the planned project could be fully achieved. This was particularly due to the greater complexity of the task and the detailed processes than originally predicted. Nevertheless, major steps could be taken in the planned direction, so that in the end a fully integrated robot with a high degree of flexible application possibilities was developed. The focus is clearly on harsh environments, so that in addition to the original plan of automated asbestos removal, other industries and applications are also in the interest of project continuation. Overall, the consortium has unanimously decided to continue the efforts and is currently reviewing the interests and funding opportunities of existing interested parties from outside the consortium.

6.8.1 Work package 7: Dissemination and Exploitation

The dissemination and exploitation activities were implemented all along the project duration starting from the original assumptions reported in the *"Description of Action Part A"*.

The roadmap to CE mark defined in the second periodic report is here complemented with a first risk assessment table developed on the V2 prototype. Full built to print documentation of the robotic systems as well as detailed electric diagrams were collected in the project repository. The attached D7.1 structure and summarize the conclusions of the different discussions and contact held within the consortium and with external potential "end users". Details are given in the sections below.

Task 7.1 Roadmap to CE mark

As discussed in previous RP a first complete risk assessment (RA) on the integrated system was developed. Reference regulatory considered for the lack of a dedicated standard is Machine 2006/42/EC (annex 1 of guideline ed. 2.1 07/17). Risk classification is the based on ISO 14121-2 2007 A.5. As the final prototype is not in the state to be released as a product and some optimisations are already in mind, the way to the CE mark is clear but wasn´t followed along within RP3 as it was too early due to the system state.

Task 7.2 Accompanying market survey

As a construction expert, Bouygues Construction had to provide a market survey to confirm the target of the project. This survey was done on a large amount of projects within Bouygues Bâtiment Nord Est. It allowed us the see that all the rehabilitation projects were concerned by asbestos. However, in most cases nowadays, the asbestos is just hidden during the rework, as the removal is mostly expensive and complex to be deal with when people are still leaving in the surrounding flats. Therefore the robot would be a perfect solution for such interventions. Thanks to the market survey, we were able to verify that the materials targeted by the project were the main ones. We showed that plaster, tiles and plastic flooring were the main contaminated material. The whole description of the market survey was explained in the second periodic report.

(ref. annex 1 guideline 2006/42/EC ed. 2.1 07/17)			ISO 14121-2 2007 A.5									
Hazard		Failure	SS1	PS1	RS1	Risk	Preventing measures	SS2	PS2	RS2	Risk	Remarks
Control systems												
Safety and reliability of control systems	Incorrect or inappropriate functionality-OPERATION scenario	wrong/unexpected device output/motion can lead to unexpected uncontrolled device displacement	50	80	130	medium	system operation in nominal condition is foreseen only inside confined space that provide intrinsic safety barrier	30	80	110	low	access control of robot confined operational space is out of the scope of the project. The issue was addressed and integrated in the test setup implemented at Boyugues.
	Incorrect or inappropriate functionality-MAINTENANCE/TRANSPORTATION scenario		100	100	200	high	system in maintenance mode is operational with reduced functionality near operator	30	60	90	low	some functions of the system can be (and must be) disabled directly from robot HW cabinet. Battery operation, used to move the robot from storage to operation site, cannot be used to move the robot arm or the tool (reducing related risks)

(ref. annex 1 guideline 2006/42/EC ed. 2.1 07/17)			ISO 14121-2 2007 A.5									
Hazard		Failure	SS1	PS1	RS1	Risk	Preventing measures	SS2	PS2	RS2	Risk	Remarks
	Incorrect measurement – OPERATION scenario	wrong position measurement can lead to unexpected uncontrolled device displacement	60	80	140	medium	Process monitoring based on CCTV monitoring of system operation is foreseen in the reference scenario. In	30	40	70	negligible	robot arm calibration is ensured by resolver position check at robot start-up in case of lost of calibration the robot arm use is not possible
	Incorrect measurement – MAINTE-NANCE/TRANSPORTATION scenario		60	100	160	high	system in maintenance mode is operational with reduced functionality near operator	30	60	90	low	some functions of the system can be (and must be) disabled directly from robot HW cabinet. Battery operation, used to move the robot from storage to operation site, cannot be used to move the robot arm or the tool (reducing related risks)

(ref. annex 1 guideline 2006/42/EC ed. 2.1 07/17)			ISO 14121-2 2007 A.5									
Hazard		Failure	SS1	PS1	RS1	Risk	Preventing measures	SS2	PS2	RS2	Risk	Remarks
	Erroneous data transfer	distributed control architecture may lead to communication errors btw actuators and controls with consequent unwanted/uncontrolled system behaviour	100	80	180	high	ETHERCAT communication protocol used for main CPU/local unit communication manages on a deterministic way messages over bus reducing risks related to data loss. Continuous diagnostics functions on ECAT modules, drives, actuators, sensors.	30	40	70	negligible	System operation is foreseen with direct supervision by tvcc. In case of maintenance, qualification and test task monitoring with user supervision near emergency button located on main locomotion cabinet is required.

(ref. annex 1 guideline 2006/42/EC ed. 2.1 07/17)			ISO 14121-2 2007 A.5									
Hazard		Failure	SS1	PS1	RS1	Risk	Preventing measures	SS2	PS2	RS2	Risk	Remarks
	Loss or deterioration of function	system wearing can lead to loss or deterioration of function risk on injuries	95	70	165	high	product documentation must clearly define system periodic maintenance and systematic functional check needed to ensure correct system reliability and performances. system operation in nominal condition is foreseen only inside confined area that provide intrinsic safety barrier.	80	40	120	medium	
Multiple control positions	simultaneous control	risk of unexpected motion with consequent worker injuries	95	70	165	high	simultaneous control of the device from different interfaces is excluded by a local/remote switch positioned on the control module and connected directly to local PLC 80	40	120	medium		

(ref. annex 1 guideline 2006/42/EC ed. 2.1 07/17)			ISO 14121-2 2007 A.5									
Hazard		Failure	SS1	PS1	RS1	Risk	Preventing measures	SS2	PS2	RS2	Risk	Remarks
Failure of the power supply	loss of system functionality	in case of mains power failure robot operation stops	70	80	150	medium	In case of lost of power the system can be recovered and moved to a "storage" or "recovery" position as the robotic platform are both with local batteries	70	40	110	low	for operational safety reason the local power supply cannot be used to move the robotic arm. This could be in principle feasible to manage site power failures adding a 24Vto 230VAC power converter to be used to move for limited displacement robot joint in order to reposition the robot arm in case of need.

(ref. annex 1 guideline 2006/42/EC ed. 2.1 07/17)			ISO 14121-2 2007 A.5									
Hazard		Failure	SS1	PS1	RS1	Risk	Preventing measures	SS2	PS2	RS2	Risk	Remarks
Protection against mechanical hazards												
Stability	Gravity	gravity effect could create unexpected device displacement. Consequences could be relevant	100	70	170	high	failsafe brakes are installed on locomotion traction modules preventing vehicle free motion in case of power failure.Dedicated SW control robot configurations to avoid loss of stability	30	70	100	low	In case of functional tests, commissioning or qualification trials added safety precaution are recommended as boundary condition: string safety sling/ winch or physical barrier to limit loss of stability are recommended

(ref. annex 1 guideline 2006/42/EC ed. 2.1 07/17)			ISO 14121-2 2007 A.5									
Hazard		Failure	SS1	PS1	RS1	Risk	Preventing measures	SS2	PS2	RS2	Risk	Remarks
Moving parts	Moving parts	exposed actuated moving parts (fast spinning tool parts)could lead to personnel injury	90	70	160	high	moving part must be correctly marked, dangerous areas should be also clearly identified.Immediate stop on HMI and emergency button on control room module. Correct personnel position with respect to the device must be specified in the product documentation. Speed limitation in maintenance mode is implemented in SW HM (this is a default set to limit risk related to unplanned unsupervisioned function activation in maintenance). correct PPE must always be correctly dressed	50	50	100	low	safety shoes, safety glasses and safety gloves are required. Suitable working suit for mechanical maintenance is also recommended. Acoustic signal active when system is working to be considered

(ref. annex 1 guideline 2006/42/EC ed. 2.1 07/17)			ISO 14121-2 2007 A.5									
Hazard		Failure	SS1	PS1	RS1	Risk	Preventing measures	SS2	PS2	RS2	Risk	Remarks
Moving transmission parts	Torsion, shear and tensile force	actuators used may give high actuation forces	80	80	160	high	moving part must be correctly marked, dangerous areas should be also clearly identified. Immediate stop on HMI and emergency button on main locomotion module. Correct personnel position with respect to the device must be specified in the product documentation. correct PPE must always be correctly dressed	50	40	90	low	

(ref. annex 1 guideline 2006/42/EC ed. 2.1 07/17)			ISO 14121-2 2007 A.5									
Hazard		Failure	SS1	PS1	RS1	Risk	Preventing measures	SS2	PS2	RS2	Risk	Remarks
Moving parts involved in the process	Moving and positioning of devices	positioning of devices may trap/create injury to personnel in the surrounding	100	80	180	high	site preparation and device installation/positioning should be performed following clearly stated rules and standard sequence by trained personnel. Immediate stop on HMI and emergency button on main locomotion module.	70	40	110	low	

(ref. annex 1 guideline 2006/42/EC ed. 2.1 07/17)			ISO 14121-2 2007 A.5									
Hazard		Failure	SS1	PS1	RS1	Risk	Preventing measures	SS2	PS2	RS2	Risk	Remarks
Uncontrolled movements	Attentional failure	system process error	100	80	180	high	system operation starts from a job planned from the operator acknowledgement. Risk assessment consider as standard foreseen/intended use the machine operation in a confined area/room acting as main barrier for risk reduction . Correct use of the system is dependent only from user control and supervision as the system cannot check directly the correctness of its installation and compliance of operational setup with safety prescription. As further risk reduction the higher safety relevant operational steps (were e.g. high speed moving discs are activated) in the system are NOT the first step of the operational cycle were the first step is the vehicle positioning inside work area (task performed under user visual monitoring)	40	40	80	negligible	

(ref. annex 1 guideline 2006/42/EC ed. 2.1 07/17)			ISO 14121-2 2007 A.5									
Hazard		Failure	SS1	PS1	RS1	Risk	Preventing measures	SS2	PS2	RS2	Risk	Remarks
Risks due to other hazards												
Electricity	24VDC power supply	isolation failure	70	40	110	low	24VDC batteries are located inside the robotic platform in a fixed steel cabinet that avoid unintentional access to batteries					
	230VAC power supply		70	100	170	high	the 230VAC powered line are integrated in an industrial electrical cabinet properly grounded and provided with differential safety switches. Access to 230VAC cabinet is possible only in maintenance for trained and authorized personnel	70	30	100	low	
	Electrical connection	Direct Electrocution	100	80	180	high	All power connections are implemented with standard industrial rugged connectors.	95	20	115	low	routine inspection of the system includes also inspection of all connectors/cables that link the different modules.

(ref. annex 1 guideline 2006/42/EC ed. 2.1 07/17)			ISO 14121-2 2007 A.5									
Hazard		Failure	SS1	PS1	RS1	Risk	Preventing measures	SS2	PS2	RS2	Risk	Remarks
	Indirect contact with live parts	contact with the casing of the equipment	100	80	180	high	Ground line must always be properly connected before starting autonomous operation with external power supply	95	20	115	low	
Unwanted static	electricity high sped tooling and dust moving inside aspiration duct may create local accumulation of static electricity	risk oh shock	100	80	180	high	Ground line must always be properly connected before starting autonomous operation with external power supply	95	20	115	low	conductive aspiration hoses could be considered for future implementation
Extreme temperatures	High temperature	direct contact with hot surface (e.g. motor driver boards)	80	80	160	high	no direct exposure of hot surface protected by covers. Where residual local wall temperature may be relevant dedicated warning sign will be added.	80	40	120	medium	

(ref. annex 1 guideline 2006/42/EC ed. 2.1 07/17)			ISO 14121-2 2007 A.5									
Hazard		Failure	SS1	PS1	RS1	Risk	Preventing measures	SS2	PS2	RS2	Risk	Remarks
Maintenance												
Isolation of energy sources	24VDC power supply	electrical power supply related risks (e.g. short circuit)	90	80	170	high	lower robotic platform battery pack can be disconnected by accessing the local cabinet for maintenance.Battery pack if charged must be handled following batteries prescription. Battery cells pins are in any case protected with standard insulation caps and the full battery pack is in operational condition protected with rigid removable cover	80	30	110	low	
	230VAC power supply		90	100	190	high	230VAC main switch is located at the side of the main robot cabinet. Cabinet access is protected with key	80	30	110	low	

(ref. annex 1 guideline 2006/42/EC ed. 2.1 07/17)			ISO 14121-2 2007 A.5									
Hazard		Failure	SS1	PS1	RS1	Risk	Preventing measures	SS2	PS2	RS2	Risk	Remarks
Operator intervention	Routine violation	operation of system function in hazardous conditions	90	80	170	high	maintenance mode operation can lead to dangerous operational state with exposed high energy moving parts. Maintenance must always performed on de-energized modules. For added safety each module is provided inside local cabinet with dedicated power switch. local switch must always be set to "off" before any intervention limit risks related to lost of coordination between operators working on the machine	50	50	100	low	always perform functional check at the end of the maintenance. Always perform functional tests in maintenance mode limiting process speed to the minimum needed adding where needed temporary added mechanical protections.
Mechanical isolation authorisation	interference btw operational and maintenance tasks		90	80	170	high	good coordination between control room is required to limit any potential risk. In any case set control switch to local before maintenance and disconnect batteries and local switch before intervention					

(ref. annex 1 guideline 2006/42/EC ed. 2.1 07/17)			ISO 14121-2 2007 A.5									
Hazard		Failure	SS1	PS1	RS1	Risk	Preventing measures	SS2	PS2	RS2	Risk	Remarks
Maintenance												
Cleaning	moving part	exposed actuated moving parts (e.g. rotating disc/tool) could lead to personnel injury	80	50	130	medium	cleaning of machines module require mechanical removal of protection covers. All cleaning task must be performed with device not powered/disconnected from batteries to avoid any unwanted potential risk For added safety each module is provided inside local cabinet with dedicated power switch. local switch must always be set to "off" before any intervention limit risks related to lost of coordination between operators working on the machine	30	30	60	negligible	

(ref. annex 1 guideline 2006/42/EC ed. 2.1 07/17)			ISO 14121-2 2007 A.5									
Hazard		Failure	SS1	PS1	RS1	Risk	Preventing measures	SS2	PS2	RS2	Risk	Remarks
Information												
Information and warnings on the machinery	generic hazard detailed above.											
Other	Controls and monitoring by the operator	already covered										all operator and service/maintenance technician working on the system must be trained with documented process.
Supplementary essential health and safety requirements to offset hazards due to lifting operations												
Lifting operation	falling	system modules have significant mass. correct component/modules handling can lead to high load falling	100	70	170	high	Full module lifting is currently not foreseen as standard planned operation.					NOTE. Dedicated lifting points could be added in a next release of the system to increase flexibility in deployment at operational site

(ref. annex 1 guideline 2006/42/EC ed. 2.1 07/17)			ISO 14121-2 2007 A.5									
Hazard		**Failure**	**SS1**	**PS1**	**RS1**	**Risk**	**Preventing measures**	**SS2**	**PS2**	**RS2**	**Risk**	**Remarks**
RA Team	Approvals											
Members												
1	Danieli personnel		Name	Signature	Date							
3	Danieli personnel	Company	DANIELI TELEROBOT LABS				F. Becchi	16/07/2019				
4	Danieli personnel											
5												
6												
7												
8												

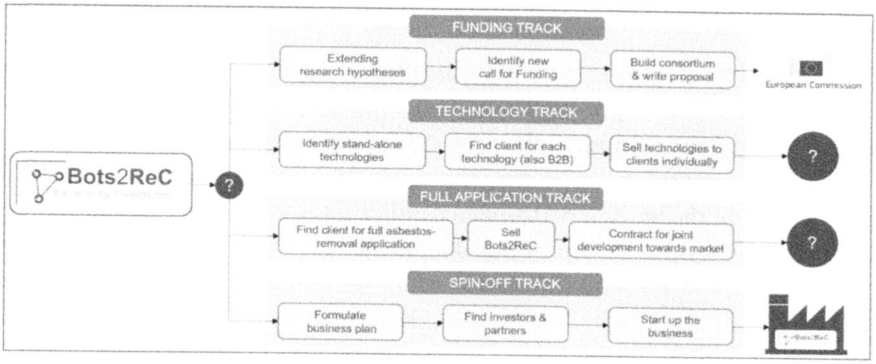

Figure 6.1. Consortium Exploitation Tracks.

Task 7.4 Project follow up strategy and business plan

As further discussed in D7.1 the follow up of Bots2ReC strategy evolved from the draft idea included in the PA. The evolution of the project and a clearer understanding of the complexity of some of the included tasks lead to two results:

- the HW level , thanks to an originally unplanned two step implementation of the robotic platform, to an industrial level prototype, suitable for a wider market exploitation;
- the complex SW infrastructure, required to structure the reference application tasks even if tested and validated at mock-up level, would require a strong and long validation process to reach a viable industrial level.

The focus of the final prototype is clearly on harsh environments but in addition to the original plan of automated asbestos removal other industries and applications are also in the interest of project continuation. A direct move to the asbestos removal market is considered unrealistic: most relevant functions are implemented and integrated but an operational validation of the fully integrated system cannot taken over by an end user at the moment. Different solutions and technologies developed within the project were already moved to a direct exploitation or used as starting point for new applications. Overall, the consortium has unanimously decided to continue the efforts and is currently reviewing the interests and funding opportunities of existing interested parties from outside the consortium.

In order to show the possibilities worked out together, the following figure shows the different exploitation tracks:

The four main tracks were called the “Funding Track”, “Technology Track”, “Full Application Track” and “Spin Off Track” and were considered to be equal in the joint discussion of the consortium.

Spin Off Track

Considering the current situation of the prototype, no consortium partner felt willing to bear the risk of a spin-off. External financing by a venture capitalist was also ruled out for the time being, as the project further underlined the need and necessity, but did not yet fully demonstrate the productivity of the system. An evaluation of the project on the basis of performance indices, as is customary, has not yet been possible for the reasons set out in this document.

Funding Track

The funding track is considered in the consortium as a clever way to preserve the different IPs of the partners and to take the desired TRL for further steps towards market-ready implementation by means of renewed funding. For this transnational consortium only EU funding was considered. A thorough research for Technology Transfer Fundings was made. An application for such a project e.g. within the RIMA Project (https://rimanetwork.eu/) will be further discussed in the consortium and to stay with the example of RIMA a call for December 2020 is a good transit phase to solve the current challenges on the prototype.

Full Application Track

The Full Application Track should mean the continuation of the basic scenario of asbestos removal, which is being implemented on behalf of an external financier. To this end, the entire expertise of the consortium would be transferred to such a service project to work towards the market launch of such a system. Based on the experience of the industrial partners, this approach was considered too risky for an external investor for the same reasons as the spin-off track, so that in view of the good opportunities offered by the technology track that now follows, the project was put on the back burner for the time being. At this point it must be explicitly stated that the continuation of asbestos removal was and is the long-term goal of this project, only that the developed system did not come close enough to the necessary TRL for another investor.

Technology Track

Even if it was not the main goal of this project, the consortium is already automatically following the Technology Track through the requests of the interest partners found. It has become clear, for example, that some players in the construction industry have similar challenges to solve, which have already been addressed in Bots2ReC on the way to automatic asbestos removal. For example, planning the fleet size based on a building plan is considered extremely interesting for the design of automated construction projects. This is already covered by the developed planning module in Bots2ReC and can be extended to other processes and robotic systems through

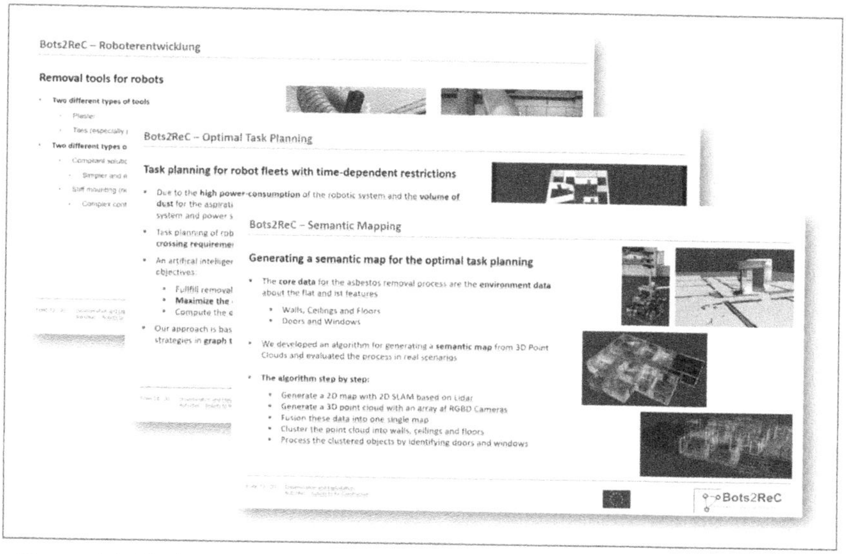

Figure 6.2. Technology Handout from Consortium to interested third parties.

the modular method of auctioning and genetic algorithm behind. In order to give our partners an idea of the technologies that can be easily decoupled from each other, the consortium has prepared a technology handout. An excerpt from this presentation is given below:

- The identified technologies that are tailored for construction sites and able to be decoupled from Bots2ReC are listed here: Construction Site robot for huge ranges and high payloads
- Removal tools for construction site robots (Tiles and Plaster)
- Semantic Mapping for simple structural buildings
- Task planning for robot fleets in construction industry
- Multi-modal control of robotized grinding
- Tandem teleoperation for two mobile robots
- Autonomous tandem following of two mobile robots
- Radar SLAM for harsh environments
- Hybrid Radar localization in 2D Maps
- Stability optimal path planning

All these technologies are currently explained and announced by the industrial partners. This includes various departments at BYCN itself as well as our partners in the industrial advisory board. The organization of the negotiations is ongoing at the time of this report and therefore cannot yet be explained with results.

In order to be able to follow this path of technology commercialization, the consortium has agreed to stay together as a community of interest. In this sense, all of the consortium's competencies can be offered and negotiated with third parties. Depending on which technology is to be pursued further, there is a permanent possibility to address all IP holding partners in form of this community of interests.

Innovations for the SMEs close or launched to market

Apart from the four different exploitation tracks mentioned, the participating companies TLABS, RA and INDURAD have already achieved success with their own products and services.

INDURAD has optimized and improved its 77GHz radar platform iLDR/iSDR to allow higher measurement rates, to reduce power requirements and to improve overall performance and usability on small robots. This includes a new ROS interface and adaptable new variants for signalprocessing, improving 2D scans. Additional, based on the experiments and experiences in the project INDURAD is about to start a new product line of very small radars operating at 122GHz. They assume that they are ready to release it to the market during 2020. INDURAD has extended its equipment test capabilities and professionalism that allows improved sensor development processes. Moreover, INDURAD has invested a lot of time in various technologies to improve the accuracy of the radio ranging technology iRTT. They have strongly evaluated new concepts and they expect by end of 2020 a new revised Radio ranging technology.

For RA the Bots2ReC project had a significant influence on the development of the new Summit Steel and in the RB2 Base. Based on the Bots2Rec V1 Prototype RA developed and launched the new RB2 Base that they are using for industrial indoor transport solutions. In addition based on the Bots2ReC V2 Protoype RA developed and launched the new Summit Steel that they sell as a general purpose platform for industrial and rough environments, and is the base of the RB-Kairos mobile manipulator. So mainly two products are already in the market as technologies throughout the developments of Bots2ReC.

Due to TLABS the developed BECKHOFF/ROS interface was used to start the JONT LAB between IIT and DANIELI TLABS to foster RD based complex control algorithm in an industrial environment. Furthermore the bevel transmission on the harmonic drive gearbox, used to reduce arm cross section, tested on the V2 arm is now used on a new set of customer driven heavy payload cleanroom robots.

DOI: 10.1561/9781680837155.ch7

Chapter 7

Discussion and Outlook

By Burkhard Corves, Tobias Haschke and Mathias Hüsing

This book examines the entire development process of two prototypes for the automatic removal of asbestos from living spaces. In addition to the mechatronic developments, the necessary algorithms and the underlying process design are illustrated. In sum, a comprehensive picture of the presented application is created with a strong focus on its technical implementation. In particular, the clear preparation of the decision-making processes across two prototypes allows the derivation of generalised basic knowledge for the continuous automation of the crafts and construction sites. In particular, the technical design of the second prototype shows that, due to the versatility of the environment on the construction site, new forms of robots must be developed which extend the state of the art both in terms of mechatronic properties (reach, manipulability, payload, movability) and through more flexible and generally valid programming. The necessity for the use of an omnidirectional mobile robot in combination with a 7-DOF kinematically redundant serial robot arm clearly shows this need for the use in multi-family houses. In addition to this large number of degrees of freedom, the required size of the work area must be emphasised in particular, in order to be able to completely process rooms from the ceiling to the floor with a single robotic system. In addition, the high volume flow of generated overburden and the continuous reduction in output represent a challenge in mobility, especially for material-intensive removal processes. With the developed tandem and the change of concept from a moving aspiration system to a central aspiration system, the presented project shows a promising approach which does not have to wait for the further development of the current battery technology.

All in all, the presented semi-autonomous asbestos removal process can be regarded as consolidated. The individual process steps are well thought-out and fulfilled by corresponding components and allow a continuous development towards an industrial standard. The developed second prototype meets the requirements of the process and also allows a continuous further development towards market introduction, which is mainly seen in the increase of robustness and the permanent process safety. In particular, continuous long-term tests must provide information about the effectiveness and economy of the overall process. In conclusion, it can be stated that the future of construction site automation requires a variety of highly flexible robotic systems and will also bring them into being in line with current robotic developments.

Index

About the Editors

Univ.-Prof. Dr.-Ing. Dr. h. c. Burkhard Corves

After completing his mechanical engineering studies at RWTH Aachen University in 1984, Prof. Corves worked as an assistant lecturer also at RWTH Aachen University, where received his doctorate in 1989. His thesis dealt with the kinematics and dynamics of industrial robots within a Collaborative Research Centre funded by the prestigious German Research Foundation.

After working at RWTH Aachen University as a Postdoc in the position of Chief-Engineer until 1991, he took a research and development position in industry in the field of special machine construction in Germany and Switzerland. In 2000, he was appointed university professor and director of the Institute for Mechanism Science, Machine Dynamics and Robotics at RWTH Aachen University, where his teaching and research is concentrating on the aforementioned institute topics. Prof. Corves has lead numerous national and European research activities such as *Bots2ReC* with more than 400 publications containing books, journal and conference papers covering different topics such as robotics, cam mechanism, glass machinery, pneumatic mechanisms, computer-aided graphical analysis and synthesis in mechanism theory, mechatronic development and machine design strategy. See also http://publications.rwth-aachen.de/search?ln=de&p=aid%3A%22P%3A%28DE-82%29IDM00228%22&jrec=1.

Since 1987 Prof. Corves is member of the German Association of Mechanical Engineers (VDI), where he holds positions as the chairman of the VDI-Scientific Board on "*Gears, Mechanisms and Machine Elements*", the chairman of the VDI-Committee "*Mechanisms for handling and manipulation*" and member of the VDI-Board of the Society of Process and Product Design. Additionally he is member of the German Society of Glass Technology (DGG), where he serves as member of the DGG Board of Directors and is engaged in the DGG technical committee FA IV "*Glass forming technology and quality assurance*". Additionally he acts as the National Chair of the German section of the International Federation for the Promotion of Mechanism and Machine Science (IFToMM) where he is also member of the Technical Committee "*Linkages & Mechanical Controls*" and the Permanent Commission for the History of Mechanism and Machine Science.

In 2013 Prof. Corves received the Honorary Doctorate from Polytechnic University of Timisoara, Romania and in 2017 the Medal of Honor of the VDI.

Prof. Dr.-Ing. Mathias Hüsing

After completing his mechanical engineering studies at Leibniz University Hannoverin 1989, Prof. Hüsing worked as an assistant lecturer at RWTH Aachen University, where he received his doctorate in 1995. His thesis dealt with tolerance analysis and syntheses of linkages.

After his doctoral thesis, he worked at RWTH Aachen University as a Postdoc in the position of Chief-Engineer, senior lecturer and senior research associate. Since 2015, he is professor at the RWTH Aachen and deputy head of the Institute of Mechanism Theory, Machine Dynamics and Robotics. Prof. Hüsing has led many national and international research projects. He had initiated and co-led the European research project *Bots2ReC*. He has published about 200 papers on kinematics, dynamics and robotics in conferences, books and journals: http://publications.rwth-aachen.de/search?ln=de&p=Mathias+H%C3%BCsing&f=&action_search=Suchen&c=RWTH+Publications&sf=&so=d&rm=&rg=10&sc=0&of=hb.

Tobias Haschke, M.Sc.

Tobias Haschke began his mechanical engineering studies at RWTH Aachen University in 2010 and completed them as Master of Science in spring 2016. During his studies, he focused on "*Development and Design*" and was already working as a student assistant at Prof. Corves' chair in the research area of robotics. His master thesis was about the generic design and analysis of multiple redundant parallel manipulators. After his studies of mechanical engineering, he started working as a Research Assistant at the Institute of Mechanism Theory, Machine Dynamics and Robotics and has been working on the EU research project *Bots2ReC* from the beginning.Therein he mainly dealt with the topics of dynamic task planning, central robot control and software architecture. In January 2019, he took over the leadership of the research project *Bots2ReC* and at the same time, he became group leader of the robotics group at the chair of Prof. Corves. In further research projects, Tobias Haschke is engaged in the agile development of cyber-physical systems and the automation of logistic scenarios using AGVs. Furthermore, he is in charge of the research area "*Construction Robotics*" at the institute of Prof. Corves and Prof. Hüsing.

Contributing Authors

Francesco Becchi
Danieli Telerobot Labs, Italy

Robin Bernau
indurad, Germany

Ramón Carrasco Jimenez
Eurecat, Spain

Sami Charaf Eddine
RWTH Aachen University, Germany

Claudia Cornely
RWTH Aachen University, Germany

Burkhard Corves
RWTH Aachen University, Germany

Tim Detert
RWTH Aachen University, Germany

Masrur Doostdar
indurad, Germany

Florian Dufour
Bouygues Construction, France

Jean-Christophe Fauroux
SIGMA Clerment, France

Diego Fogliacco
Danieli Telerobot Labs, Italy

Albert Fullana Puig
Eurecat, Spain

Néstor García
Eurecat, Spain

Martin Gritzan
indurad, Germany

Roberto Guzmán
Robotnik Automation, Spain

Tobias Haschke
RWTH Aachen University, Germany

Fabian Herb
indurad, Germany

Siddharth Maraje
SIGMA Clerment, France

Daniel Martin
Eurecat, Spain

Kamal Mohy El Dine
SIGMA Clerment, France

Raul Molina
Robotnik Automation, Spain

Miguel Moreno
Robotnik Automation, Spain

Oriol Orra
Eurecat, Spain

Matthias Rabel
indurad, Germany

Lucas Rohe
indurad, Germany